Unravelling Plant-Microbe Synergy

Developments in Applied Microbiology and Biotechnology

Unravelling Plant-Microbe Synergy

Edited by

Dinesh Chandra
Teacher in Department of School Education at Govt. Inter College Chamtola, Almora, India

Pankaj Bhatt
Visiting Scholar, Department of Agricultural & Biological Engineering, Purdue University, West Lafayette, IN, United States

ELSEVIER

ACADEMIC PRESS
An imprint of Elsevier

Academic Press is an imprint of Elsevier
125 London Wall, London EC2Y 5AS, United Kingdom
525 B Street, Suite 1650, San Diego, CA 92101, United States
50 Hampshire Street, 5th Floor, Cambridge, MA 02139, United States
The Boulevard, Langford Lane, Kidlington, Oxford OX5 1GB, United Kingdom

ISBN: 978-0-323-99896-3

For information on all Academic Press publications
visit our website at https://www.elsevier.com/books-and-journals

Publisher: Stacy Masucci
Acquisitions Editor: Linda Versteeg-Buschman
Editorial Project Manager: Tim Eslava
Production Project Manager: Niranjan Bhaskaran
Cover Designer: Victoria Pearson Esser

Typeset by STRAIVE, India

Contents

Contributors

Sajjad Ahmad Key Laboratory of Integrated Pest Management of Crop in South China, Ministry of Agriculture and Rural Affairs; Key Laboratory of Natural Pesticide and Chemical Biology, Ministry of Education, South China Agricultural University, Guangzhou, P. R. China

Jyoti Bajeli Department of Biological Sciences, CBSH, GBPUA&T, Pantnagar, Uttarakhand, India

Kalpana Bhatt Department of Botany and Microbiology, Gurukula Kangri University, Haridwar, India

Pankaj Bhatt Department of Agricultural & Biological Engineering, Purdue University, West Lafayette, IN, United States

Sandhya Bind Department of Biological Sciences, College of Basic Sciences and Humanities, Govind Ballabh Pant University of Agriculture and Technology, Pantnagar, US Nagar, Uttarakhnad, India

Sudha Bind Department of Biological Sciences, College of Basic Sciences and Humanities, Govind Ballabh Pant University of Agriculture and Technology, Pantnagar, US Nagar, Uttarakhnad, India

Dinesh Chandra Department of Biological Sciences, College of Basic Sciences and Humanities, Govind Ballabh Pant University of Agriculture and Technology, Pantnagar, US Nagar; Govt. Inter Collge Chamtola, Almora, Uttarakhand, India

Satish Chandra Department of Botany, Government Degree College Tyuni, Dehradun, India

Parul Chaudhary Department of Microbiology, Govind Ballabh Pant University of Agriculture & Technology, Pantnagar, India

Shaohua Chen State Key Laboratory for Conservation and Utilization of Subtropical Agro-bioresources, Guangdong Laboratory for Lingnan Modern Agriculture, Integrative Microbiology Research Centre, South China Agricultural University, Guangzhou, China

Hemant Dasila Department of Microbiology, G.B. Pant University of Agriculture and Technology, Pantnagar, Uttarakhand, India

Nwaga Dieudonne Soil Microbiology Laboratory, The Biotechnology Centre, Department of Microbiology, Faculty of Sciences, University of Yaoundé I, Yaoundé, Cameroon

Derya Efe Department of Medicinal and Aromatic Plants, Giresun University, Giresun, Turkey

Fekam Boyom Fabrice Antimicrobial & Biocontrol Agents Unit (AmBcAU), Laboratory for Phyto Biochemistry and Medicinal Plants Studies, Faculty of Science, University of Yaoundé I, Yaoundé, Cameroon

Sudip Kumar Ghosh Microbiology Laboratory, Department of Botany, The University of Burdwan, Purba Bardhaman, West Bengal, India

Arzu Gormez Department of Molecular Biology and Genetics, Erzurum Technical University, Erzurum, Turkey

Aparna B. Gunjal Department of Microbiology, Dr. D. Y. Patil, Arts, Commerce and Science College, Pimpri, Pune, Maharashtra, India

Neha Jeena Department of Biotechnology, Bhimtal Campus Kumaun University, Nainital, Uttarakhand, India

Amir Khan Department of Microbiology, College of Basic Sciences and Humanities, Govind Ballabh Pant University of Agriculture and Technology, Pantnagar, US Nagar, Uttarakhand, India

Bharti Kukreti Department of Microbiology, College of Basic Sciences and Humanities, Govind Ballabh Pant University of Agriculture and Technology, Pantnagar, US Nagar, Uttarakhand, India

Abhishek Kumar Forest Research Institute, Dehradun, India

Amit Kumar Forest Research Institute, Dehradun, India

Anand Kumar Department of Biological Sciences, College of Basic Sciences and Humanities, G.B. Pant University of Agriculture and Technology, Pantnagar, US Nagar, Uttarakhand, India

Kunal Kumar Amity Institute of Biotechnology, Amity University, Ranchi, Jharkhand, India

Narendra Kumar Doon (PG) College of Agriculture Science and Technology, Dehradun, India

Saurabh Kumar ICAR-Research Complex for Eastern Region, Patna, Bihar, India

Garima Kumari Forest Research Institute, Dehradun, India

Tushar Kanti Maiti Microbiology Laboratory, Department of Botany, The University of Burdwan, Purba Bardhaman, West Bengal, India

Govind Makarana ICAR-Research Complex for Eastern Region, Patna, Bihar, India

Yogita Martoliya School of Life Sciences, Jawaharlal Nehru University, New Delhi, India

Soumik Mitra Microbiology Laboratory, Department of Botany, The University of Burdwan, Purba Bardhaman, West Bengal, India

Sridev Mohapatra Department of Biological Sciences, Birla Institute of Technology and Science (Pilani), Hyderabad Campus, Secunderabad, Telangana, India

Sayanta Mondal Microbiology Laboratory, Department of Botany, The University of Burdwan, Purba Bardhaman, West Bengal, India

Tanushree Mondal Microbiology Laboratory, Department of Botany, The University of Burdwan, Purba Bardhaman, West Bengal, India

P.T. Nikhil Department of Biological Sciences, Birla Institute of Technology and Science (Pilani), Hyderabad Campus, Secunderabad, Telangana, India

Furkan Orhan Department of Molecular Biology and Genetics, Agri Ibrahim Cecen University, Agri, Turkey

Priyanka Pal Microbiology Laboratory, Department of Botany, The University of Burdwan, Purba Bardhaman, West Bengal, India

Veni Pande Cell & Molecular Biology Laboratory, Department of Zoology, Soban Singh Jeena University, Almora, Uttarakhand, India

Satish Chandra Pandey Cell & Molecular Biology Laboratory, Department of Zoology, Soban Singh Jeena University, Almora, Uttarakhand, India

Eke Pierre Department of Crop Production Technology, College of Technology, University of Bamenda, North-West Region; Antimicrobial & Biocontrol Agents Unit (AmBcAU), Laboratory for Phyto Biochemistry and Medicinal Plants Studies, Faculty of Science, University of Yaoundé I, Yaoundé, Cameroon

Krishnendu Pramanik Mycology and Plant Pathology Laboratory, Department of Botany, Siksha Bhavana, Visva-Bharati, Santiniketan, West Bengal, India

Arjita Punetha CSIR-Central Institute of Medicinal & Aromatic Plants (CIMAP), Research Centre, Pantnagar, Uttarakhand, India

N.S. Raja Gopalan Department of Biological Sciences, Birla Institute of Technology and Science (Pilani), Hyderabad Campus, Secunderabad, Telangana, India

Sunita Rawat Forest Research Institute, Dehradun, India

Fokom Raymond Department of Food Processing and Quality Control, Institute of Fisheries and Aquatic Sciences, University of Douala, Douala; Soil Microbiology Laboratory, The Biotechnology Centre, Department of Microbiology; Antimicrobial & Biocontrol Agents Unit (AmBcAU), Laboratory for Phyto Biochemistry and Medicinal Plants Studies, Faculty of Science, University of Yaoundé I, Yaoundé, Cameroon

Mukesh Samant Cell & Molecular Biology Laboratory, Department of Zoology, Soban Singh Jeena University, Almora, Uttarakhand, India

Diksha Sati Cell & Molecular Biology Laboratory, Department of Zoology, Soban Singh Jeena University, Almora, Uttarakhand, India

Ishwar Prakash Sharma Patanjali Research Institute, Haridwar, India

Raunak Sharma Department of Biological Sciences, Birla Institute of Technology and Science (Pilani), Hyderabad Campus, Secunderabad, Telangana, India

Ajay Veer Singh Department of Microbiology, College of Basic Sciences and Humanities, Govind Ballabh Pant University of Agriculture and Technology, Pantnagar, US Nagar, Uttarakhand, India

Manish Singh Forest Research Institute, Dehradun, India

Tithi Soren Microbiology Laboratory, Department of Botany, The University of Burdwan, Purba Bardhaman, West Bengal, India

Adamou Souleymanou Soil Microbiology Laboratory, The Biotechnology Centre, Department of Microbiology, Faculty of Sciences, University of Yaoundé I, Yaoundé; Faculty of Agronomy and Agricultural Sciences, Department of Agriculture, University of Dschang, Dschang, Cameroon

Deep Chandra Suyal Department of Microbiology, Akal College of Basic Sciences, Eternal University, Sirmaur, Himachal Pradesh, India

Ngo Oum Therese Soil Microbiology Laboratory, The Biotechnology Centre, Department of Microbiology, Faculty of Sciences, University of Yaoundé I, Yaoundé, Cameroon

Aakansha Verma Department of Biological Sciences, CBSH, GBPUA&T, Pantnagar, Uttarakhand, India

Shulbhi Verma Department of Biotechnology, SDAU, Dantiwada, Gujarat, India

CHAPTER 1

Multiomics strategies for alleviation of abiotic stresses in plants

Dinesh Chandra[a,b] and Pankaj Bhatt[c]

[a]*Department of Biological Sciences, College of Basic Sciences and Humanities, Govind Ballabh Pant University of Agriculture and Technology, Pantnagar, Uttarakhand, India* [b]*Govt. Inter Collge Chamtola, Almora, Uttarakhand, India* [c]*Department of Agricultural & Biological Engineering, Purdue University, West Lafayette, IN, United States*

Chapter outline

Introduction

By 2050, it is predicted that there will be 9 billion people on this planet, and, to feed this spectacular number of people, food production needs to be augmented by almost 60% of its current status (FAO, 2009). If this feels like a humongous task, then we can only imagine doing this while keeping in mind that we need to achieve this using methods that are the least

Unravelling Plant-Microbe Synergy. https://doi.org/10.1016/B978-0-323-99896-3.00002-3

hazardous to our Mother Earth. Gone are the days when mindless application of chemicals and fertilizers was the answer. Time and again, debates about a sustainable approach to combat this food deficit have taken place and making use of plant–microbe interactions has come out as the most sought out solution. The main cause of diminishing agricultural productivity is abiotic stress that is a consequence of adverse climatic conditions (Grayson, 2013). A report by the FAO (2007) shows that the areas unaffected by any environmental constraint are only 3.5% of the global land area. Abiotic stresses that hamper plant growth and productivity include droughts, salinity, flooding, anaerobiosis, nutrient starvation, light intensity, low/high temperatures, and submergence (Glick, 2012; Chandra et al., 2019a, b, 2020). About 64% of the global land area is affected by droughts (water deficit), followed by cold (57%), acidic soils (15%), floods (13%), mineral deficiency (9%), and salinity (6%) (Mittler, 2006; Cramer et al., 2011). The area under dryland agriculture in the world is 5.2 billion hectares, out of which 3.6 billion hectares of land is affected by problems such as salinity, soil degradation, and erosion (Riadh et al., 2010). These problems ultimately impact the total irrigated land and consequently diminish crop yield through loss of crops (Ruan et al., 2010: Flowers et al., 2010; Qadir et al., 2014). It is extremely difficult to precisely measure agricultural loss with respect to the loss of crop production (qualitative and quantitative) and soil health by abiotic stresses (Cramer et al., 2011).

Plants have intrinsic mechanisms to sustain environmental conditions (Simontacchi et al., 2015). Alterations in external environmental conditions cause plant metabolism out of maintaining homeostasis and necessitate that plants harbor some advanced molecular mechanisms within their cellular system to lessen the negative effects of abiotic stresses (Foyer and Noctor, 2005; Gill and Tuteja, 2010). During the course of evolution, plants have gained protective mechanisms to combat adverse environmental situations and these mechanisms cause metabolic reprogramming in the cell and enable routine biophysicochemical processes, regardless of external environmental conditions (Massad et al., 2012; Yolcu et al., 2016).

Plant responses to abiotic stress

Plants have several ways to sense the changing environmental conditions and maintain their homeostasis by tolerance, avoidance, recovery, and escape mechanisms. Their responses to environmental stimuli encompass changes at the cellular, physiological, transcriptomic, genetic, and metabolic levels (Atkinson and Urwin, 2012). Plant growth and development is severely impacted by abiotic stress, thereby leading to heavy loss of global agriculture (Verma and Deepti, 2016). Drought, salinity, frost, and heat result in decreased water content within cells, followed by the simultaneous development of phenotypic, biochemical, and molecular responses to stresses (Xu and Zhou, 2006). The most important parameter in sensing abiotic stimuli is the root architecture that is supposed to be subtler and act accordingly in soils

(Khan et al., 2016). At the same time, in the environment, a plant may face multiple stresses but its complexity of responses is different for each stress. These responses lead to the expression of several genes, followed by metabolic programming in cells. The abiotic stress-mitigating mechanism contains multiple stages of plant development (Meena et al., 2017). The chief mechanisms in the tolerance of abiotic stresses are defense, repair, acclimation, and adaptation. Plants are sensitive to water stress. Under drought stress conditions, peroxidation leads to negative consequences in antioxidant metabolism (Xu et al., 2014). The activity of enzymatic antioxidants (SOD, CAT, APX, GPX) varies from plant to plant under drought stress (Xu et al., 2015; Chandra et al., 2018a, 2019a, b).

After drought, salinity is another factor that causes distresses to modern agricultural practices. Nearly 33% of irrigated land and 20% of arable land are affected by salinity worldwide (Machado and Serralheiro, 2017). There are two ways that salinity exerts its effects: the first is through a higher concentration of salts that makes the soil harder, hence making the roots unable to extract water, and the second is through a higher concentration of salts that is toxic to plant cells. Plant health is affected by a higher concentration of salts in soils; therefore, cells in tissues respond differently to salinity stress (Voesenek and Pierik, 2008). McCue and Hanson (1990) observed that an increased level of salt decreases the osmotic potential of cells, which leads to iron toxicity and affects the vitality of plants by hampering plant growth and development and finally causing death. Salinity stress decreases aromatic amino acid levels and increases proline accumulation, polyols, and glycine betaine in cells. In addition, salinity stress increases antioxidant enzymes, modulation of hormones, and generation of nitric oxide (Gupta and Huang, 2014). A stressed condition changes the gene expression pattern of cells (Dinneny et al., 2008). SOS1, SOS2, and SOS3 proteins participate in the signaling pathway of SOS (Hasegawa et al., 2000).

Heat stress also affects crop productivity. Due to climate change, the global temperature is increasing, and this has a negative impact on the morphological, biochemical, and physiological properties of plants. A higher temperature reduces the seed germination rate, respiration, and photosynthesis and decreases membrane permeability (Xu et al., 2014). Heat stress also denatures the protein and causes inactivation of enzymes, loss of membrane integrity, and inhibition of protein synthesis (Mitra et al., 2021). Plants respond to heat stress by altering their primary and secondary metabolites and enhanced expressions of HSP and ROS (Iba, 2002). Plants sustain the impact of heat stress by the mechanisms of ROS scavenging, antioxidant metabolites, compatible solute accumulation, transcriptional modulation, and chaperone signaling (Wahid et al., 2007). Similarly, cod stress (freezing and chilling) retards the growth and productivity of crops and sometime even cause their death (Miura and Furumoto, 2013).

The occurrence of heavy metals is prevalent in agricultural soils. The main sources of contamination of agricultural lands are pesticides, untreated household and industrial wastewater, and organic and chemical fertilizers (Dhaliwal et al., 2020; Shah et al., 2020a, b; Zhang and Wang, 2020). In many countries, industrial units like textile, oil, tanneries, marbles, mining, sugar industry, paper, aluminum, and metal plating release a huge amount of unprocessed wastewater rich in heavy metals such as (arsenic), Ni (nickel), Pb (lead), Cr (chromium), and Cd (cadmium); eventually all these metals are transported to the soils through irrigation, and these exert destructive effects on crop growth and productivity, quality and safety of crops, and human and soil health (Gill et al., 2016). It has been observed that at a normal concentration, many metals have a nutritional requirement for all living organisms because of their involvement in homeostasis and protein synthesis and their roles as stimulator and enzymatic cofactors.

Plant morphological, physiological, and biochemical functions are impaired by a higher concentration of heavy metals. For example, accumulation of Pb causes phytotoxicity and reduced plant growth and inhibits seed germination, root elongation, seedling development, transpiration rate, chlorophyll synthesis, and protein and water content (Opeolu et al., 2010). Similarly, Ni is toxic to plants at a higher concentration and retards growth, metabolism, seed germination, and shoot and root growth and induces leaf spotting of plants. Cu is also essential for the normal functioning of plants, and its higher concentration is harmful to plant growth and impairs root growth and morphology (Sheldon and Menzies, 2005). In addition, Hg is lethal to plants and diminishes the transpiration rate, photosynthesis, chlorophyll synthesis, and water uptake (Singh et al., 2019). Among heavy metals, Cd is extremely lethal to plants and has overwhelming impacts on seed germination, nitrogen assimilation, plant height, leaf chlorosis, necrosis, antioxidative enzymes, and yield of crops (Ali et al., 2015; Javed et al., 2019; Wang et al., 2019). Because of its longer half-life, nonbiodegradable nature, and a higher retention rate, Cd is toxic to the metabolic process even at a low concentration (Tanwir et al., 2021).

A higher concentration of Cd in a soil–plant system diminishes Zn, Ca, Fe, Mg, Mn, and K uptake and translocation due to cationic competition at root uptake sites, and, as a result, oxidative stress is triggered, which leads to higher electrolyte leakage and production of hydrogen peroxide and malondialdehyde, whereas antioxidant activities are diminished (Rehman et al., 2017; Tao et al., 2020). Metal toxicity results in proline accumulation that provides stress tolerance to plants. Under metal stress, the role of proline in resisting ROS accumulation, osmotic adjustment, cytosolic pH buffer, peroxidation of cellular lipids to maintain cell membrane integrity and also act as a signaling molecule (Hossain et al., 2014). To a certain extent, plants are capable of coping with the negative impacts of oxidative damage caused by metal stress, but, beyond a threshold level, a higher concentration of Cd leads to a diminutive growth of plants (Bukhari et al., 2016). The removal of heavy metals from the environment requires a cost-effective and sustainable approach. At present, the techniques that are currently being used in remediation are extremely costly and toxic to the soil structure (Glick, 2010).

In addition to the abovementioned stresses, temperature and nutrient stress also hamper crop growth and productivity. A higher temperature distresses plant growth and damages cellular proteins, leading to cell death. Similarly, a low temperature lessens the metabolism due to inhibition of enzyme reactions and the interaction among macromolecules, thus modulating the membrane's properties and changes in the protein structure (Andreas et al., 2012). Compared to plants that are exposed to individual stress, those exposed to multiple stresses have more beneficial impacts because a blend of stresses decreases the detrimental effects of each other, thereby enhancing the survivability of plants. In multiple stress situations that occur concurrently with field conditions, complicated mechanisms occur in plants to help them deal with promptly changing adverse conditions.

Abiotic stress alleviation by microbes

Abiotic stress as well as biotic communities restrict plant growth and development. Plants show tolerance mechanisms to abiotic stresses by two ways: (a) avoidance of the negative impacts of stress by activation of the response system and (b) use of antistress agents (biochemical compounds) produced by microbes (Schulze, 2005; Meena et al., 2017). Abiotic stress conditions exert adverse effects on plant growth and development, and these negative impacts are alleviated by plant growth-promoting bacteria (PGPB), as shown in Fig. 1.1.

Drought stress

Droughts exert a negative effect on the productivity of crops. Enhancing food security under droughts is a crucial task for crop breeders. Therefore, utilization of microbes to combat drought-induced damage in plants is the need of the hour at present. A drought is an influential cause of constraining crop growth and yield. It is anticipated that by 2050, 50% of the world's land area will suffer from water shortage (Gupta et al., 2020). Hence, in order to ensure food security, cultivation of drought-tolerant crops is an urgent need. Several studies have demonstrated that an exogenous application of rhizobacteria and their growth-promoting traits boosts the drought tolerance of crops (Hassan et al., 2020; Huan et al., 2020).

A drought exerts its negative effects on plants by several means such as a reduced rate of photosynthesis, a low germination rate, loss of membrane integrity, and an increased production of ROS (Delshadi et al., 2017; Chandra et al., 2019a, b). PGPB improve plant growth by enhancing osmolyte production, accumulation of antioxidant photosynthetic capacity, gas exchange, relative water content, etc. under drought stress (Xiao et al., 2017; Zhang et al., 2021a, b). The previous findings of many researchers have also revealed that rhizobacteria are associated with drought tolerance (Zhang et al., 2019; Chandra et al., 2020; Goswami and Suresh, 2020). The role of PGPB in relieving drought stress and augmenting plant growth and development is summarized in Table 1.1

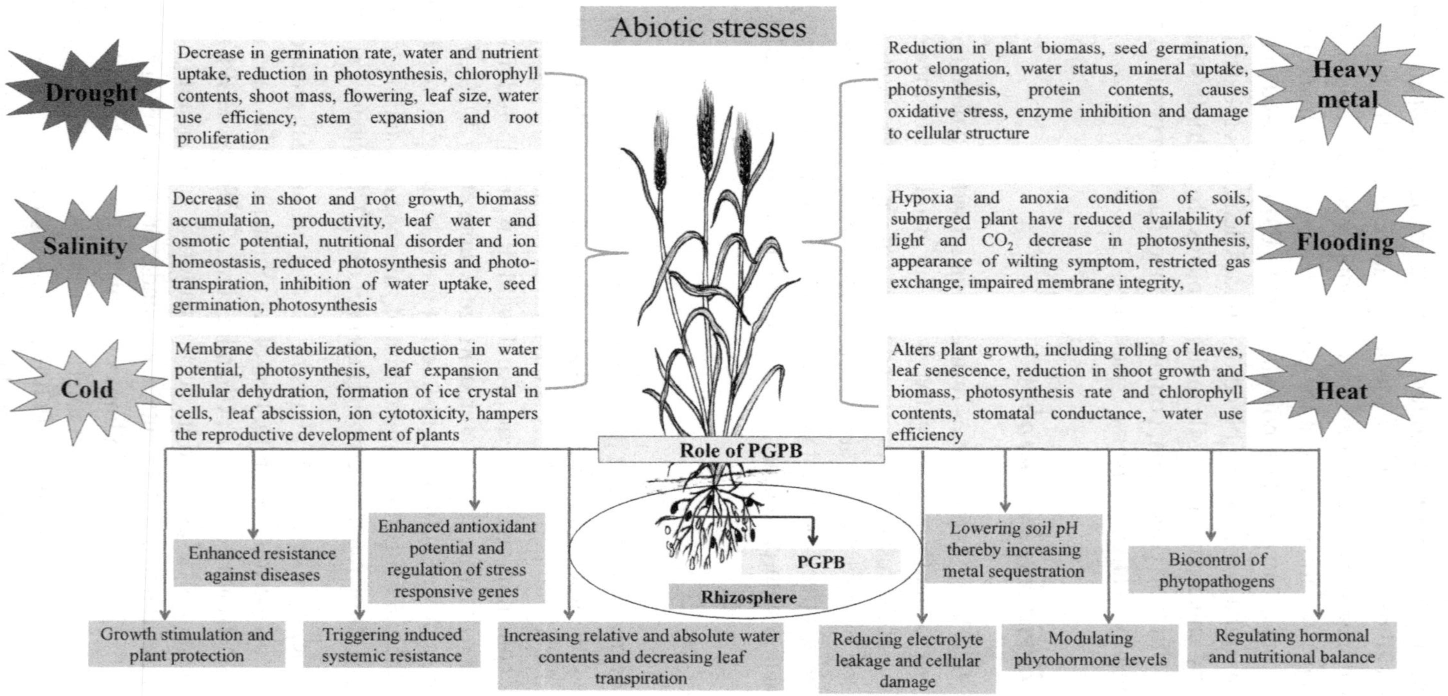

Fig. 1.1
Effects of abiotic stresses on plants and the possible role of PGPB in mitigating abiotic stresses.

Table 1.1: A summary of findings demonstrating the role of PGPB in alleviation of drought stress.

Plant	Microbes	Beneficial effect(s)	References
Maize (*Zea mays* L.)	*Cupriavidus necator* 1C2 (B1) and *Pseudomonas fluorescens* S3X (B2)	Increased aerial biomass and nitrogen and potassium use efficiency	Pereira et al. (2020)
Great millet (*Sorghum bicolor* L.)	*Streptomyces laurentii* EU-LWT3-69	Reduced lipid peroxidation and elevated accumulation of proline, glycine betaine, and chlorophyll content	Kour et al. (2020)
Rice (*Oryza sativa* L.)	*Bacillus altitudinis* FD48 and *Bacillus methylotrophicus* RABA6	Significant changes in the source-sink relationship, reduced RWC, steady increase of photosynthetic pigments and proline, and enhanced levels of ROS-quenching enzymes	Narayanasamy et al. (2020)
Mung bean (*Vigna radiata* (L.) R. Wilczek)	*Pseudomonas aeruginosa* GGRJ21	Increased dry weight, root/shoot length, and RWC and strong upregulation of *DREB2A*, *CAT1*, and *DHN*	Sarma and Saikia (2014)
Wheat (*Triticum aestivum* L.)	*Pseudomonas* spp. (DPB16, UW4) and *Variovorax paradoxus* RAA3	Increased nutrients and grain and straw yield, increased activities of antioxidants, and enhanced expression of helicases and aquaporin	Chandra et al. (2019b)
Finger millet plant (*Eleusine coracana* (L.) Gaertn.) and wheat plant (*Triticum aestivum* L.)	*Pseudomonas palleroniana* (strains DPB16 and DPB13) and *Pseudomonas fluorescens* DPB15	Improved plant growth, foliar nutrients, and antioxidant enzymes	Chandra et al. (2018a, b)
Mentha pulegium L.	*Azotobacter chroococcum* (Ac) and *Azospirillum brasilense* (Ab)	Inoculation improves the physiological and biochemical parameters	Asghari et al. (2020)
Zea mays L.	*Ochrobactrum* sp. NBRISH6	Inoculation modulates the physiological and anatomical aspects and also influences the metabolic and molecular machinery	Mishra et al. (2020)
Arabidopsis thaliana and *Brassica campestris*	*Bacillus subtilis* GOT9	Enhanced expression of drought stress- and salt stress-inducible genes	Woo et al. (2020)
Tomato	*Streptomyces* sp. IT25 and *Streptomyces* sp. C-2012	Enhanced proline, RWC, total sugar, MDA, H_2O_2, and APX and decreased activity of GPX and CAT	Abbasi et al. (2020)
Brassica juncea L.	*Bacillus cereus* NA D7 and *Bacillus* sp. MR D17	Increased expression of *DREB2* and *DREB1-2* genes	Bandeppa et al. (2019)

Continued

Table 1.1: A summary of findings demonstrating the role of PGPB in alleviation of drought stress—cont'd

Plant	Microbes	Beneficial effect(s)	References
Pea (*Pisum sativum* L.)	*Rhizobium leguminosarum* bv. *viciae* 1066S	Higher accumulation of shoot biomass, water use efficiency, and uptake of N and mineral nutrients and higher nodulation	Belimov et al. (2019)
Maize (*Zea mays* L.)	*Bacillus* (strains HYD-B17, HYTAPB18, HYDGRFB19, and RMPB44) and *Paenibacillus favisporus* BKB30	Increased plant biomass, RWC, proline, sugars, APX, GPX, and CAT and decreased leaf water and electrolyte loss	Vardharajula et al. (2011)
Foxtail millet (*Setaria italica*)	*Arthrobacter siccitolerans* 4J27, *Pseudomonas fluorescens* DR11, *Enterobacter hormaechei* DR16, and *Pseudomonas migulae* DR35	Increased rate of seed germination and seedling growth	Niu et al. (2018)
Ryegrass (*Lolium perenne* L.)	*Bacillus* sp. WM13-24 and *Pseudomonas* sp. M30-35	Improved root system architecture and drought tolerance by enhancing antioxidant enzyme activities and regulating the ABA signal	He et al. (2021)
Mentha piperita	*Pseudomonas fluorescens* WCS417 r and *Bacillus amyloliquefaciens* GB03	Higher phenolic content, enzymatic activities, and reduced lipid peroxidation	Chiappero et al. (2019)
Chickpea (*Cicer arietinum* L.)	*Pseudomonas putida* RA	Inoculation modulates the expression of miRNAs	Jatan et al. (2019)
Brassica oxyrrhina	*Pseudomonas libanensis* TR1 and *Pseudomonas reactans* Ph3R3	Bacterial treatments greatly improve organ metal concentrations, translocation, and the bioconcentration factors of Cu and Zn	Ma et al. (2016)
Arabidopsis thaliana	*Pseudomonas putida* GAP-P45	Higher accumulation of spermidine and putrescine	Sen et al. (2018)
Chickpea (*Cicer arietinum* L.)	*Pseudomonas putida* MTCC5279 (RA)	Treatment application alters the physical, physiological, and biochemical parameters and modulates the expression of stress-responsive genes	Tiwari et al. (2016)
Lactuca sativa	*Curtobacterium herbarum* strain CAH5	Inoculation reduces lipid peroxidation, Al accumulation, and oxidative stress	Silambarasan et al. (2019)

A phytohormone such as ABA, which is produced by bacteria, is helpful in the easing of drought stress (Forni et al., 2017). Increased levels of ABA in *Arabidopsis* have been observed following inoculation of *Phyllobacterium brassicacearum* STM196, thereby decreasing the rate of transpiration in leaves (Bresson et al., 2013). Treatment of wheat seedling with *Azospirillum* showed a significant increase in osmotic stress tolerance because of morphological changes in the xylem structure (Pereyra et al., 2012). The reason for enhanced tolerance is upregulation of the inodole-3-pyruvate decarboxylase gene, which results in increased synthesis of IAA in inoculant cells. Similarly, *Bacillus* sp. is also involved in imparting drought tolerance in plants. Rhizobacteria such as *Bacillus thuringiensis* are stated to increase drought resistance in French lavender plants by increasing IAA production that could enhance the metabolic activities of plants and improve the physiological and nutritional status of plants (Armada et al., 2014). Under well-watered and drought-stressed conditions, rhizobacterial strains (BN-5 and MD-23) producing EPS, IAA, and ACC deaminase increase the productivity and quality of maize.

The ACC deaminase-producing strains of bacteria play a significant role in the alleviation of drought stress impact on plants. Ethylene regulates the metabolic activities of plants, and its synthesis is regulated by both abiotic and biotic environmental conditions. The phytohormone ethylene controls the homeostasis, resulting in restricted growth of the shoot and root in a stressful environment (Glick et al., 2007). Treatment of pepper and tomato with ARV8 (*Achromobacter piechaudii*) exhibiting ACC deaminase activity enhanced the plants' fresh/dry weight (Mayak et al., 2004). Similarly, 5C-2 (*Variovorax paradoxus*)-treated plants showed enhanced plant growth and yield contributing parameters (Belimov et al., 2009). In addition, Belimov et al. (2015) also reported that in both water-deficit and well-watered conditions, ACC deaminase and auxin-producing rhizobacteria improved the growth and yield of potatoes. Hence, it is believed that execution of rhizobacteria in drought-impacted soils offers a cost-effective strategy for viable crop health as well as yield of crops.

Our study also demonstrated that treatment with *Variovorax paradoxus* RAA3 and a consortium of four strains of *Pseudomonas* spp. (DPC12, DPB13, DPB15, and DPB16) producing ACC deaminase significantly augmented wheat growth under rain-fed conditions via increased nutrient concentration and antioxidant potential (Chandra et al., 2019a). In another study, we also noticed that when wheat drought and sensitive variety treated with ACC deaminase producing strain of *Pseudomonas palleroniana* DPB16, *Pseudomonas* sp. UW4, and *Variovorax paradoxus* RAA3 enhanced the growth, yield and nutritional content under drought and rainfed conditions (Chandra et al., 2019b). Microorganisms secrete salicylic acid (SA) that is involved in the regulation of plant growth and development as well as in plant drought response. SA also acts as a signaling molecule and induces the expression of several genes that are involved in the synthesis of antioxidants, chaperones, HSPs, enzymes, and secondary metabolites under stress conditions (Kumar et al., 2019).

Salinity stress

Various PGPR genera including *Acetobacter*, *Achromobacter*, *Aeromonas*, *Azospirillum*, *Bacillus*, *Bradyrhizobium*, *Chryseobacterium*, *Flavobacterium*, *Pseudomonas*, *Sinorhizobium*, etc. have been demonstrated to increase the productivity of different crops in salt-affected soils. The chemotactic, ACC deaminase, and IAA attributes of bacteria can battle with different stresses including salinity stress (Glick, 1995). Rhizobacteria can induce induced systemic tolerance (IST) to fight against the changes in plants and to develop tolerance mechanisms in plants against salinity stress (Yang et al., 2009). Beneficial microorganisms are extremely helpful in solving the problem of salinity. Several PGPB have been described to inhabit plant roots and diminish the impact of salinity and salt stress by different mechanisms as summarized in Tables 1.2 and 1.3.

A study reported by Figueiredo et al. (2008) demonstrated that treatment of *Raphanus sativus* with *Kocuria erythromyxa* and *Staphylococcus kloosii* induced salt tolerance by producing antioxidants. Similarly, a study by Nadeem et al. (2007) stated that *Enterobacter aerogenes*-, *Pseudomonas syringae*-, and *Pseudomonas fluorescens*-inoculated maize plants show induced salt tolerance by regulation of K^+/Na^+ ratios and proline and chlorophyll levels. The IAA, phosphate-solubilizing, and ACC deaminase-producing salt-tolerant bacteria SAL-15 improve the yield contributing parameters of wheat under conditions of salinity stress (Rajput et al., 2013).

An increased level of ACC generates a higher ethylene concentration, which alters the physiological functions of plants. The mechanism that can decrease ethylene levels promotes growth of plants under salinity stress. Under salt stress, phytohormones produced by PGPR have beneficial impacts on root length, leaf area, nutrient uptake, and root tip number (Egamberdieva and Kucharova, 2009). An increased expression of the high-affinity K^+ transporter (AtHKT1) leads to increased uptake of K^+ ions under saline conditions triggered by salinity-tolerant bacteria, and, this, in turn, leads to a higher K^+/Na^+ ratio that helps in imparting tolerance to salinity (Nadeem et al., 2013). A number of mechanisms, i.e., molecular, physiological, and morphological, are employed by salt-tolerant bacteria to withstand the salinity of soil (Kumar et al., 2019; Etesami and Glick, 2020). A study by Bharti et al. (2016) revealed that *Dietzia natronolimnaea* STR1-treated wheat plants showed salinity tolerance via up/downregulation of stress-responsive genes and osmolyte production.

Heavy metal stress

PGPB are considered one of the most promising strategies for the alleviation of all heavy metal stresses due to their environmentally safe and less adverse effects (Rajkumar et al., 2012). The mechanisms utilized by PGPB to minimize the negative impact of heavy metals include efflux, volatilization, impermeability to metals, metal complexation, absorption of

Table 1.2: A summary of findings demonstrating the role of PGPB in alleviation of salinity stress.

Plant	Microbes	Beneficial effect(s)	References
Phaseolus vulgaris	PGPR consortia (*Bacillus subtilis* MTCC441 and *Pseudomonas fluorescence* MTCC 103^T)	Enhanced growth, yield, and biochemical activity	Kumar et al. (2020)
Wheat (*Triticum aestivum* L.)	*Bacillus cereus*, *Serratia marcescens*, and *Pseudomonas aeruginosa*	Treatment application increases growth and yield and also improves physiological attributes	Desoky et al. (2020)
Citrus (*Citrus macrophylla*)	*Pseudomonas putida* KT2440 and *Novosphingobium* sp. HR1a	Increased plant growth, decreased transpiration rate and stomatal conductance, and significant drop in SA and ABA	Vives-Peris et al. (2018)
Okra (*Abelmoschus esculentus* L.)	*Enterobacter* sp. UPMR18	Treated plants display a higher germination percentage, chlorophyll content, and growth parameters	Habib et al. (2016)
Wheat (*Triticum aestivum* L.)	*Bacillus siamensis* (PM13), *Bacillus* sp. (PM15), and *Bacillus methylotrophicus* (PM19)	Increased seedling germination rate, root/shoot length, and photosynthetic capacity	Amna et al. (2019)
Maize (*Zea mays* L.)	*Bacillus safensis* HL1HP11, *Bacillus pumilus* HL3RS14, *Kocuria rosea* HL1RP8, *Enterobacter aerogenes* AT1HP4, and *Aeromonas veronii* AT1RP10	Increased accumulation of glycine betaine, proline, and malondialdehyde	Mukhtar et al. (2020a)
Oat (*Avena sativa*)	*Klebsiella* sp. IG 3	Higher expression of *rbcL* and *WRKY1* genes	Sapre et al. (2018)
Common bean (*Phaseolus vulgaris*)	*Bacillus subtilis* (strains 10-4, 26D)	Inoculated strains exert positive impacts on plant growth, modulate cellular response reactions, and also regulate plant defense mechanisms	Lastochkina et al. (2021)
Wheat (*Triticum aestivum* L.)	*Bacillus subtilis* SU47 and *Arthrobacter* sp. SU18	Enhanced dry biomass, proline content, and total soluble sugars	Upadhyay et al. (2012)
Mustard (*Brassica juncea* L.)	*Pseudomonas argentinensis* HMM57 and *Pseudomonas azotoformans* JMM15	IAA and ALA production along with ACC utilization activities	Phour and Sindhu (2020)
Salicornia sp.	*Staphylococcus* sp.	Increased plant growth	Komaresofla et al. (2019)
Rice (*Oryza sativa* L.)	*Curtobacterium albidum* SRV4	Higher photosynthetic efficiency, proline content, SOD, APX content, plant growth, and K^+ uptake	Vimal et al. (2019)
Peanut (*Arachis hypogaea*)	*Ochrobactrum intermedium*	Increased dry weight and shoot/root length	Paulucci et al. (2015)

Continued

Table 1.2: A summary of findings demonstrating the role of PGPB in alleviation of salinity stress—cont'd

Plant	Microbes	Beneficial effect(s)	References
Pepper (*Capsicum annum* L.)	*Microbacterium oleivorans* KNUC7074, *Brevibacterium iodinum* KNUC7183, and *Rhizobium massiliae* KNUC7586	Elevated level of proline and chlorophyll content and increased dry/fresh biomass and shoot length	Hahm et al. (2017)
Rice (*Oryza sativa* L.)	*Bacillus pumilus* JPVS11	Bacterial strains promote photosynthetic pigment, proline, and antioxidant production	Kumar et al. (2021a)
Tomato (*Solanum lycopersicum* L.)	*Pseudomonas fluorescens* YsS6 and *P. migulae* 8R6	Increased dry/fresh weight, flower numbers, buds, and chlorophyll content	Ali et al. (2014)

Table 1.3: A summary of findings demonstrating the role of PGPB in alleviation of salt stress.

Plant	Microbes	Beneficial effect(s)	References
Zea mays L.	*Kocuria rhizophila* Y1	Increased seed germination rate, antioxidants, photosynthetic activity, relative water content, and chlorophyll content and nutrient acquisition and protection of plants by regulating IAA and ABA	Li et al. (2020)
Vigna radiata	*Kosakonia sacchari* MSK1	Decreased relative water content, Na^{+}/K^{+} ions, antioxidants, membrane injury, and stressor metabolites	Shahid et al. (2021)
Cicer arietinum	*Pseudomonas stutzeri* SGM-1	Inoculation improves plant growth by contributing macronutrients and minor essential nutrients	Mahajan et al. (2020)
Groundnut (*Arachis hypogea*)	*Pseudomonas fluorescens* TDK1	Treatment increases the yield-contributing parameters	Saravanakumar and Samiyappan (2007)
Maize (*Zea mays* L.)	*Serratia liquefaciens* KM4	Improved plant growth by maintaining the redox potential, ion homeostasis, and leaf gas exchange, enhanced expression of *APX*, *CAT*, *SOD*, *RBCS*, *RBCL*, *H^{+}-PPase*, *HKT1*, and *NHX1*, and downregulation of the ABA biosynthesis gene (*NCED*)	El-Esawi et al. (2018b)
Sunflower	*Bacillus licheniformis* AP6, *Pseudomonas plecoglossicida* PB5	Inoculant application increases the higher dry/fresh biomass and the root/shoot length and alleviates oxidative stress by CAT, SOD, and GPX	Yasmeen et al. (2020)
Soybean (*Glycine max* L.)	*Bacillus firmus* SW5	Enhanced nutrient uptake, chlorophyll synthesis, total phenolics, antioxidants, flavonoid content and growth, and biomass yield	El-Esawi et al. (2018a)

Table 1.3: A summary of findings demonstrating the role of PGPB in alleviation of salt stress—cont'd

Plant	Microbes	Beneficial effect(s)	References
Chickpea (*Cicer arietinum* L.)	*Azospirillum lipoferum* FK1	Augmented nutrient acquisition, synthesis of photosynthetic pigments, phenol and flavonoid content, antioxidant activities, osmolytes, growth, and biomass and diminished levels of H_2O_2, MDA, and electrolyte leakage	El-Esawi et al. (2019)
Potato (*Solanum tuberosum* L.)	*Bacillus* sp. (strains SR-2-1 and SR-2-1/1	Higher production of IAA, antioxidant activities, and uptake of nutrients	Tahir et al. (2019)
Tomato	*Bacillus velezensis* FMH2	Treatment application enhances the chlorophyll contents and phenols and maintains membrane integrity, reduced MDA, H_2O_2, and Na^+ accumulation, and increased uptake of K^+ and Ca^{2+}	Masmoudi et al. (2021)
Rice	*Enterobacter* sp. P23	Decreased ethylene and antioxidant enzymes	Sarkar et al. (2018a)

essential nutrient elements, and enzymatic detoxification (Kumar and Verma, 2018; Etesami, 2018). PGPB under stress condition encourage growth and development of plants via the bacterial ACC deaminase enzyme that reduces ethylene concentration (Glick, 2010). Other attributes of PGPB that help in the removal of heavy metals from soil as well as significantly enhance plant growth include siderophores, nitrogen fixation, phytohormones, phosphate solubilization, and various secondary metabolites (Rajkumar et al., 2012; Verma et al., 2013). A study by Prapagdee et al. (2013) demonstrated that Cd-resistant bacteria (*Klebsiella* sp. BAM1 and *Micrococcus* sp. MU1) boost Cd mobilization and stimulate root elongation and plant growth in contaminated soils. The bacterial strains E109 (*Bradyrhizobium japonicum*) and Az39 (*Azospirillum brasilense*) are proficiently colonized in arsenic-polluted soils and improve plant growth. PGPR-mediated heavy metal tolerance to enhance plant growth is summarized in Table 1.4.

PGPR form various organic acids that have a binding affinity to heavy metals. PGPR increase the accessibility of nutrients and organic acids and also detoxify metal ions and decrease the uptake of formation (Kavita et al., 2008). Siderophores produced by PGPR have the affinity to bind to heavy metals and form a siderophore–heavy metal complex. This complex prevents plants from absorbing heavy metals from the soil. Biosurfactants are secondary metabolites secreted extracellularly by PGPR and preferentially bind to toxic metals with a strong affinity (Pacheco et al., 2010). PGPR such as *Bacillus subtilis* produce biosurfactants that help in the elimination of heavy metals from soils, thereby reducing the effects of heavy metals (Pacwa-Płociniczak et al., 2011). Microbially produced anions and secondary metabolites

Table 1.4: A summary of findings demonstrating the role of PGPB in alleviation of heavy metal stresses.

Stress	Plant	Microbes	Beneficial effect(s)	References
Cadmium stress	Rice (*Oryza sativa* L.)	*Enterobacter aerogenes* K6	Reduced Cd uptake and oxidative stress and stress enhances Cd tolerance and rice seedling growth	Pramanik et al. (2018)
	Sesbania sesban	*Bacillus anthracis* PM21	Inoculant application maintains the level of antioxidant activities as well as enhances plant growth and biomass	Ali et al. (2021)
	Rice	*Enterobacter* sp. S2	Treatment enhances various morphological and biochemical characteristics	Mitra et al. (2018)
	Serratia sp. CP-13	Maize (*Zea mays* L.)	A noteworthy increase in photosynthetic pigments, plant biomass, antioxidant activities, content of proline and flavonoids, decreased levels of MDA, H_2O_2, and relative membrane permeability	Tanwir et al. (2021)
	Tomato (*Solanum lycopersicum* L.)	*Burkholderia* sp. N3	Treatment with N3 increases tomato seedling height, dry weight, and nutrient uptake and also promotes Fe^{3+} uptake, reduced IAA, and ZEA	Zhang et al. (2021a, b)
	Soybean	*Burkholderia contaminans* ZCC	PGP increases siderophore production, ACC deaminase activity, and IAA production and promotes soybean growth via EPS	You et al. (2021)
	Cucumis melo	*Bacillus fortis* IAGS 223	Improved plant growth as a result of decreased amount of stress markers	Shah et al. (2021a, b)
	Rice (*Oryza sativa* L.)	*Pseudomonas* sp. K32	Bacterial application reduces Cd uptake and displays biocontrol potential against pathogens	Pramanik et al. (2021)
	Alfalfa	*Bacillus subtilis*	Reduced level of MDA and enhanced activities of antioxidants and soil nutrient cycling	Li et al. (2021)
	Maize (*Zea mays* L.)	*Acinetobacter* sp. SG-5	Improved antioxidant activities and nutrients	Abbas et al. (2020)
Copper, cadmium, zinc, and lead stress	Sorghum (*Sorghum vulgare* L.)	Consortium of *Bacillus cereus* MG257494.1, *Alcaligenes faecalis* MG966440.1, and *Alcaligenes faecalis* MG257493.1	Treatment application increases microbial activities like DHA (dehydrogenase activity), decreases heavy metal bioaccumulation, and stimulates plant growth	Abou-Aly et al. (2021)

Cadmium and lead toxicity	Rice (*Oryza sativa* L.)	*Deinococcus radiodurans* Δdr2577	Reduced levels of ROS and increased antioxidant activities	Dai et al. (2021)
	Maize (*Zea mays* L.)	*Azotobacter chroococcum* CAZ3	Reduced proline content, malondialdehyde, and antioxidant enzymes, thereby increasing plant growth and yield	Rizvi and Khan (2018)
Lead stress	*Solanum melongena* L.	*Bacillus subtilis* FBL-10	Increased total soluble proteins, gas exchange, and photosynthetic rate	Shah et al. (2021b)
	Rapeseed (*Brassica napus*) and Clover (*Trifolium repens*)	*Pseudomonas fluorescens* B3, *Pseudomonas putida* B6, and *Bacillus safensis* B8	Enhanced antioxidant activities, proline and decreased MDA content, and increased plant growth and yield	Shah et al. (2020b)
	Sunflower (*Helianthus annuus*)	*Pseudomonas gessardii* BLP14, *Pseudomonas fluorescens* A506, and *Pseudomonas fluorescens* LMG 2189	Reduced MDA content and elevated contents of antioxidant activities, proline, plant yield, and growth	Saleem et al. (2018)
	Coriander (*Coriandrum sativum* L.)	*Bacillus thuringiensis* S6 and *Bacillus cereus* S19	Improved growth, photosynthesis, and antioxidant enzyme activities	Fatemi et al. (2020)
Arsenic stress	Chickpea (*Cicer arietinum* L.)	*Pseudomonas citronellolis* (PC) (KM594397)	Increased dry biomass and plant growth	Adhikary et al. (2019)
	Chickpea (*Cicer arietinum* L.)	*Acinetobacter* sp. nbri05	Reduced As uptake and increased plant growth and yield	Srivastava and Singh (2014)
	Rice seedling	*Kocuria flava* AB402 and *Bacillus vietnamensis* AB403	Reduced As uptake and increased plant growth parameters	Mallick et al. (2018)
Cooper stress	Wheat (*Triticum aestivum* L.)	*Bacillus altitudinis* WR10	Inoculant application reduces H_2O_2 levels and enhances GSH contents, ROS scavenging enzymes, and phenylpropanoid biosynthesis	Yue et al. (2021)
	Sunflower (*Helianthus annuus*)	*Pseudomonas lurida* EOO26	Increased dry weight of root and shoot and phytoremediation of Cu	Kumar et al. (2021b)

help in precipitation of heavy metals by acidification, oxidation–reduction, biosorption, bioaccumulation, and binding and accumulation of metal ions on the surface; these are also the active mechanisms employed by bacteria to decrease the availability of heavy metals to plants (Etesami, 2018). Many studies have revealed that PGPB show a potential to reduce Cd concentration in Cd-contaminated environments. Examples of PGPB are *Burkholderia*, *Cupriavidus*, and *Pseudomonas aeruginosa*, which exhibit positive effects on plant growth via biosorption and bioaccumulation of Cd (Dourado et al., 2013; Tagele et al., 2018; Shi et al., 2020).

Heat stress

Several genera such as *Azospirillum*, *Bacillus*, and *Rhizobium* have been described as heat-tolerant bacteria (Ali et al., 2009; Meena et al., 2015). Bacteria withstand the adverse effects of heat stress by producing endospores. Moreover, some of the bacteria overcome the problem of high temperature by producing exopolysaccharides (EPSs), lipopolysaccharides (LPSs), and HSPs or by alterations of saturated and unsaturated membrane lipids, thereby enhancing plant growth (Tiwari et al., 2017; Singh et al., 2019). Ali et al. (2011) demonstrated that *Pseudomonas putida* AKMP7-treated wheat plants significantly amended the resistance of plants against heat stress via membrane injury by production of various enzymatic antioxidants. The *Pseudomonas* strain produced the phytohormone IAA at a temperature of 40°C and increased the shoot/root biomass in maize (Mishra et al., 2017). Similarly, a study by Park et al. (2017) showed that plants treated with *Bacillus aryabhattai* produced greater amounts of ABA, which, in turn, provides tolerance to heat stress. Heat-tolerant PGPR can increase the availability of nutrients in the rhizosphere. At a high temperature, bacterial strains such as *Bacillus licheniformis*, *Bacillus smithii*, *Bacillus coagulans*, *Streptomyces thermonitrifica*, and *Streptococcus thermophiles* have the ability to fix atmospheric nitrogen and solubilize phosphate substances (Chang and Yang, 2009). Several nitrogen-fixing *Rhizobium* strains can synthesize HSPs that help in alleviating the impact of heat stress (Simoes-Araujo et al., 2008). Treatment of soybean with heat-tolerant *Bradyrhizobium japonicum* increased the shoot dry matter, nitrogen uptake, and yield (Rahmani et al., 2009). Srivastava et al. (2008) observed that chickpea showed tolerance to thermal stress when treated with *Pseudomonas putida* NBR10987 due to biofilm formation and overexpression of the stress sigma factor. In addition, a study by Ali et al. (2009) also revealed that when sorghum seedlings were treated with *Pseudomonas* sp. AKM-P6, they showed higher tolerance to heat stress by generation of EPSs and accumulation of HSPs. Tiwari et al. (2017) observed that under heat stress conditions when rice plants were treated with *Bacillus amyloliquefaciens* NBRI-SN12, a remarkable increase in the accumulation of osmoprotectants and an inflection in the expression of stress-responsive genes were observed. Similarly, Sarkar et al. (2018a, b) reported that *Bacillus safensis* and *Ochrobactrum pseudogrignonense* ameliorate heat stress damage by reducing ROS production and

Table 1.5: A summary of findings demonstrating the role of thermotolerant PGPB in alleviation of thermal stress.

Stress	Plant	Microbes	Beneficial effect(s)	References
Heat stress	*Solanum lycopersicum* L.	*Bacillus cereus*	Enhanced growth and heat tolerance due to ACC deaminase and plant growth regulators	Mukhtar et al. (2020b)
	Wheat (*Triticum aestivum* L.)	*Bacillus amyloliquefaciens* UCMB5113 and *Azospirillum brasilense* NO40	Reduced generation of ROS and preactivation of certain HSP transcription factors	Abd El-Daim et al. (2014)
	Glycine max	*Bacillus cereus*	Mitigation of heat stress by HSB expression, phytohormone, and amino acid production	Khan et al. (2020)
Low temperature	*Triticum* spp.	*Pseudomonas* spp.	PGP traits of treatment improve the growth of wheat	Yarzábal et al. (2018)
	Pisum sativum L.	PGPR isolates (PR-12-12 and PR-12-15)	Enhanced plant growth, PGP traits, high resistance to DNA gyrase, and low resistance to ciprofloxacin	Meena et al. (2015)
High temperature	*Triticum* spp.	*Pseudomonas putida* AKMP7	Reduced levels of microbial colonization	Ali et al. (2011)
	Cajanus cajan	*Pseudomonas*, *Bacillus*, *Serratia*, and *Rhizobia*	Maximum growth at 30°C, 40°C, and 50°C and plant growth-promoting attributes	Modi and Khanna (2018)
Chilling	*Solanum lycopersicum* cv. Mill	*Arthrobacter*, *Flavimonas*, *Flavobacterium*, *Massilia*, *Pedobacter*, and *Pseudomonas*	Higher plant growth, germination, and antioxidant activity	Subramanian et al. (2016a)

increasing the production of glycine betaine and proline in wheat plants. Bacteria-mediated thermotolerance is summarized in Table 1.5.

Mineral nutrients, water, carbon, and light are vital for plant development, optimal growth, and reproduction. PGPR provide micronutrients and macronutrients to their host. Bacteria that have the nitrogen fixation potential provide nitrogen for biosynthesis of amino acids. Abadi and Sepehri (2016) observed that *Azotobacter chroococcum*- and *Piriformospora indica*-treated wheat plants displayed a higher uptake of mineral nutrients.

Microbe-mediated alleviation of abiotic stresses in plants: The omics approaches

The consequences of climate change adversely impact plant phenotypes at the macro level and genome expressions at the molecular level. Approaches to alleviate the impacts of contrary environmental situations on crop growth entail immediate concern. In addition, conventional breeding and transgenic approaches are costly and time-consuming and require vigorous testing to ascertain their low flexibility in field conditions (Coleman-Derr and Tringe, 2014). Therefore, technology that is more efficient and adaptable to reduce the negative consequences of climate change on crop growth and productivity is needed; in this context, approaches based on the use of microbes are more reliable as these are based on the interactions between plants and microbes. The ecology of plant–microbe interactions is complex, interlinked, and dependent on ecological and environmental variables. Therefore, to understand the complexities of these interactions at molecular, biochemical, and physiological levels, multidisciplinary omics strategies such as genomics, proteomics, transcriptomics, and metabolomics are needed as these are instrumental strategies for understanding and lessening stresses in plants (Meena et al., 2017). In the environment, plants have to face multiple abiotic and biotic stresses. These multiple stress conditions generate complex defense signals in plants, and, therefore, interactions between plants and microbes can be decided by prioritization of the physiological pathways in plants (Schenk et al., 2012). Microbe and plant root interactions generate multifaceted responses in distal as well as local plant parts at molecular, physiological, and biochemical levels.

Genomics

Genomics is defined as an organism's whole genome and its interactions with the environment. On one hand, in conventional genetics, one or few genes can be studied at a time, whereas, on the other hand, genomics provides wide-ranging information about the full complement of organisms. The data generated when genomic studies are integrated with functional genomics provide comprehensive information about the functionality of genes under different sets of environmental conditions. A different approach used at the RNA level is transcriptomics, at the protein level it is proteomics, and at the metabolomic level it is metabolomics. Hence, multiomics strategies contribute to a better understanding of the mitigation strategies of different stresses. Mitigation of abiotic stress can be accomplished by manifestations of microbe–plant interactions. Kumari et al. (2015) reported that *Pseudomonas* sp. AK-1- and *Bacillus* sp. SJ-5-treated soybean increased salinity tolerance by elevating the level of proline and lipoxygenase activity. Similarly, a study by Koussevitzky et al. (2008) in *Arabidopsis thaliana* demonstrated that ascorbate peroxidase 1 (*APx1*) is explicitly necessary for heat and drought stress tolerance. An osmolyte such as ectoine is responsible for salt tolerance in *Halomonas elongata* OUT30018. For the biosynthesis of ectoine, three genes were cloned and transferred to a tobacco plant to impart tolerance to hyperosmotic shock by accumulation of ectoine and displayed normal growth (Nakayama et al., 2000).

Transcriptomics

Transcriptomics is the study of relative RNA abundance using microarray technology. In transcriptomics, a total RNA sample is extracted and converted into complementary DNA (cDNA). This cDNA is examined for upregulation and downregulation of genes. The transcriptomic technique is basically carried out in healthy versus diseased samples or in controls versus samples treated with a specific treatment. A transcriptomic analysis involves thousands of genes and generates quantitative data that help in the interpretation of mechanisms of complex biological processes that encompass many genes. This technique provides extensive information about how genes are expressed and interrelated during the course of a biological process (Vafaee et al., 2019). Under abiotic stress conditions, plant biomass production can be increased either by the increase in cell number or by the expansion of cells, as these methods require genes related to cell wall synthesis and phytohormones, which regulate cell division and cell expansion (Joshi et al., 2018). Beneficial microbes form a vibrant association with plants, as their interactions can be associated with the mitigation of abiotic stresses. Bacteria are able to produce PGP traits such as IAA production, P solubilization, and siderophores and also produce several secondary metabolites that trigger pathways and mediate induced resistance in plants by the SA and JA pathways, thereby enhancing the adaptability of plants to cope with the negative effects of environmental conditions (Chandra et al., 2019a, b; Chandran et al., 2021). For a transcriptomic analysis, another technique called metatranscriptomics is used. In this technique, RNA is directly isolated from the sample without culturing the microorganisms. This strategy leads to high-throughput analysis of the samples and an unbiased understanding of the genes (Schenk et al., 2012). The profiling of a transcriptome is extremely helpful for recognizing different sets of transcripts in different biological systems under various conditions (Bräutigam and Gowik, 2010). To obtain transcriptome-level information for the study of plant–microbe interactions, important techniques such as microarray and mRNA sequencing are used (Wang et al., 2016a, b). It has been revealed that induction of stress-responsive genes is carried out in the IAA-overproducing strain of *Sinorhizobium meliloti* through next-generation RNA sequencing. In this study, the transcript profiles of two *S. meliloti* strains (the wild-type 1021 and the IAA-overproducing derivative RD64) were compared and it was found that genes coding for sigma factor RpoH1 and other stress responses induced IAA-overproducing strains of *S. meliloti* (Defez et al., 2016). A study by Alavi et al. (2013) observed that the plant growth regulator spermidine during abiotic stress was identified by transcriptomic analysis of rapeseed and its symbiont *Stenotrophomonas rhizophila*.

A number of miRNAs have been identified in different plants such as *Medicago*, rice, *Arabidopsis*, and *Phaseolus*; they play a regulatory role (transcription factors) in abiotic stresses and in other biological processes (Trindade et al., 2010; Budak et al., 2015). miR393 has a regulatory function in imparting salinity tolerance in *Arabidopsis* (Gao et al., 2011). Similarly, miR169 plays a regulatory role by modulating the expression of NF-YA in rice and imparting tolerance to droughts and salinity (Zhao et al., 2009). Zhang et al. (2011) found that plants overexpressing miR169, which regulates the expression of stomatal activity, provide tolerance to droughts. In cucumber, Bvu-miR13 regulates WD-repeat proteins that play a significant role in stress tolerance (Li et al., 2014). Curaba et al. (2014) also reported that miRNAs target stress signaling pathways, which are liable to stress response, root development, and morphogenesis of leaves. MiR398 targets the mRNA of superoxide dismutases (SOD1 and SOD2) that have an important role in lessening ROS, consequently serving in the mitigation of abiotic stresses (Kantar et al., 2011). The stress response is mitigated by different classes of miRNAs by regulating metabolic and cellular processes (Li et al., 2010).

Metagenomics

Metagenomics is a potent tool that helps in the investigation of microbial communities based on a culture-independent approach for determination of unseen and uncultured microbial diversity in the rhizosphere (Chen and Pachter, 2005). This practice assists users in acquiring data related to the habitat-specific distribution of microbial communities with different plant growth-promoting traits including biocontrol potential and antibiotic-producing traits. The technique of high-throughput metagenomic sequencing is extremely useful for understanding rhizobacterial communities. The metagenomic sequences of endophytic microorganisms have revealed that functional features like quorum sensing and ROS scavenging play an extremely important role in enhancing tolerance to stresses (Sessitsch et al., 2012). Microbial interactions or their interactions with their host exert a direct effect on plant productivity and health. To estimate microbiome composition, its diversity, and its functions, approaches like metagenomics, metatranscriptomics, and metaproteomics are applied. Metagenomics discloses the functionality of microbial communities in terms of the richness of genes involved in individual metabolic processes associated with the stress-mitigating mechanism. The technique of metaproteomics reveals protein richness, putative proteins, and community-wide gene expressions that can be interrelated with function after bioinformatic analysis. Similarly, the technique of metatranscriptomics reveals kingdom-level changes in the rhizosphere microbiome structure (Turner et al., 2013).

Proteomics

Proteins facilitate the phenotypic expression of genes. A study of cell proteomes helps in finding out the factors that determine the gene expression of plant–microbe interactions as well as the interactions between intra- and interspecies (Meena et al., 2017). On the other hand, metabolomics reveals the overall metabolic pathways operating in cells under a different set of environmental conditions. The factors affecting plant species can be understood by datasets obtained from metabolic profiling and these datasets are linked to proteomics and transcriptomics. This information can be used for developing strategies to mitigate stress (Piasecka et al., 2019). Halotolerant bacteria are gaining importance for their ability to flourish in high-salt environments and have proven advantageous in stress-affected areas by reducing the negative impacts of salts. The applications of the metabolites of halotolerant microbes are extremely useful in crop improvement programs and also provide tolerance against stresses. On the other hand, to sort out the problem of xenobiotic compounds, in this context, the genus *Pseudomonas* is best considered because of its ability to degrade wide varieties of xenobiotic compounds. *Pseudomonas* spp. displays a number of PGP traits such as phytohormone, siderophore production, biocontrol potential, and biofilm formation, and these attributes make the *Pseudomonas* species effective inoculants for diverse environmental conditions (Wang et al., 2015; Chandra et al., 2018a, b; Chandra et al., 2019a, b). In addition, proteomic studies of methylotrophic bacteria are gaining prominence and establish a key serving of phylosphere community. The plant growth-promoting potential of these microorganisms has been successfully demonstrated by many studies under numerous stressful conditions (Tani et al., 2012; Yim et al., 2013).

Metabolomics

Microorganisms secrete a number of metabolites, and their production varies with alterations in the environmental conditions (Bundy et al., 2005). Metabolomic data are used to obtain a deep understandings of the responses to abiotic stress. The advancement in the area of molecular detection techniques has given a boost to metabolomic studies (Morrow, 2010). Several studies have demonstrated the plant metabolome comprises different bioactive molecules (Ketchum et al., 2003). Abiotic stress response leads to generation of different metabolites such as glycine betaine, trehalose, IAA, etc. in plants. The functional pathway in a microbial cell is influenced by the surrounding environment, consequently affecting the metabolome. The products of microbial metabolites are involved in both direct and indirect growth promotion of plants. Metabolites from rhizobacteria have been involved in indirect and direct growth promotion of plants. A study by Robin et al. (2006) observed that rhizobacteria have the ability to produce auxins, cytokinins, gibberelins, etc. and play a role in stimulating plant growth. Studies by Sorty et al. (2016) and Mishra et al. (2016) researched the plant growth-promoting traits of bacteria, particularly those of IAA, and the P-solubilizing potential (*Acinetobacter* sp., *Bacillus*, *Enterobacter* sp., *Pantoea* sp., *Marinobacterium* sp.) stimulates plant growth under saline conditions. In addition, microbial siderophores play a remarkable role in the availability of iron to plant roots. *Pseudomonas fluorescens* C7 produced the siderophores positively augmented the iron to *Arabidopsis thaliana* (Vansuyt et al., 2007).

Induction of abiotic stress-responsive genes for stress relief by PGPB

As sessile organisms, plants are frequently subjected to adverse environmental distresses. To overcome the responses of abiotic stress, plants have developed a complex and highly regulated defense system. During abiotic stress, thousands of genes are known to be altered that not only helps in cellular tolerance but also regulates the expression of stress-responsive genes. Furthermore, PGPB strains protect plants from abiotic stresses by producing phytohormones, organic acids, siderophores, stress-induced metabolites, and antioxidant enzymes and up- and downregulates the expression of various stress-responsive genes that provide tolerance to plants under stressful conditions. PGPB-mediated stimulation of abiotic stress-responsive genes has been described in many crops including wheat, rice, maize, pepper, tomato, okra, etc., as summarized in Table 1.6.

Table 1.6: Abiotic stress-responsive genes and their regulation in plants by PGPB.

PGPB	Plant species	Abiotic stress-responsive genes	References
Bacillus subtilis	*Arabidopsis thaliana* (L.) Heynh.	*proBA* genes	Chen et al. (2007)
Bacillus amyloliquefaciens SN13	*Oryza sativa* L.	Upregulation of *EREBP*, *SOS1*, *SERK1*, and *NADP-Me2* and downregulation of *GIG* and *SAPK4*	Nautiyal et al. (2013)
Co-inoculation with *Rhizobium tropici* and *Paenibacillus polymyxa*	*Phaseolus vulgaris*	Increased expression of the *trehalose 6-phosphate* gene	Figueiredo et al. (2008)
Bacillus subtilis GB03	*Puccinellia tenuiflora*	Upregulation of the *PtSOS1* and *PtHKT1* genes and downregulation of *PtHKT2*	Niu et al. (2016)
Bacillus amyloliquefaciens SQR9	*Zea mays* L.	Upregulation of the genes *RBCS*, *RBCL*, H(+)-*Ppase* (encoding H + pumping pyrophosphatase), *HKT1*, *NHX1*, *NHX2*, and *NHX3* and downregulation of *NCED*	Chen et al. (2016)
Enterobacter sp. EJ01	*Arabidopsis thaliana* (L.) Heynh.	Increased expression of *DREB2b*, *RD29A*, *RD29B*, *LEA* (*RAB18*), *P5CS1*, *P5CS2*, *MPK3*, and *MPK6* genes	Kim et al. (2014)
Pseudomonas putida UW4	*Solanum lycopersicum*	Upregulation of *Toc GTPase*	Yan et al. (2014)
Enterobacter sp. UPMR18	*Abelmoschus esculentus* L.	Upregulation of *CAT*, *APX*, *GR*, and *DHAR*	Habib et al. (2016)
Pseudomonas PS01	*Arabidopsis thaliana* (L.) Heynh.	Upregulation of *LOX2* and downregulation of *APX2* and *GLYI7*	Chu et al. (2019)

Table 1.6: Abiotic stress-responsive genes and their regulation in plants by PGPB—cont'd

PGPB	Plant species	Abiotic stress-responsive genes	References
Bacillus thuringiensis NEB17	*Glycine* max (L.) Merr.	Elevation of PS I and SP II proteins, *pyruvate kinase*, *glutathione-S-transferase*, *PEP carboxylase*, *isocitrate lyase*, and *RuBisCo oxygenase*	Subramanian et al. (2016b)
Dietzia natronolimnaea STR1	*Triticum aestivum* L.	Upregulation of *TaABARE*, *TaOPR1*, and *TaST*	Bharti et al. (2016)
Arthrobacter sp. EZB4 and *Bacillus* sp. EZB8	*Capsicum annuum* L.	Downregulation of *CaACCO* and *CaLTPI*	Sziderics et al. (2007)
Arthrobacter protophormiae SA3 and *B. subtilis* LDR2	*Triticum aestivum* L.	Upregulation of *TaCTR1* and *TaDRE2*	Barnawal et al. (2017)
Bacillus licheniformis K11	*Capsicum annuum* L.	Increased expression of *Cadhn*, *VA*, *sHSP*, and *CaPR-10*	Lim and Kim (2013)
Consortium of BBS (*Bacillus cereus* AR156, *Bacillus subtilis* SM21 and *Serratia* sp. XY21)	*Cucumis sativus* L.	Downregulation of *cAPX*, *rbcL*, and *rbcS*	Wang et al. (2012)
Pseudomonas putida GAP-P45	*Arabidopsis thaliana* (L.) Heynh.	Upregulation of *ornithine-Δ-aminotransferase* (*OAT*), *P5CS1*, *P5CR*, *PDH1*, and *P5CDH*	Ghosh et al. (2017)

Conclusions and future perspectives

Due to changing climatic conditions and deterioration of agricultural lands, there is an imperative need to find new means for tolerance to abiotic stresses. Abiotic stress conditions suppress the physio-biochemical properties and consequently affect the productivity of crops. To overcome the adverse consequences of abiotic stress conditions and to enhance the growth and productivity of crops in an eco-friendly manner, an effective biological agent is required. PGPB can suppress the effects of abiotic stresses and improve crop performance. Soil inoculation with stress-tolerant bacteria demonstrates their noteworthy role in enhancing yield, plant growth, nutrient uptake, induction of stress-tolerant genes, scavenging of the ROS system, and production of other stress-tolerant molecules. Omics strategies are evolving platforms for elucidating robotic insights into various stress interactions with plants. Mechanistic approaches of microbial interactions such as gene cascade, enzymes, proteins, metabolites, and up- and downregulation of stress-responsive genes are extremely useful for improving plant growth under diverse environmental stresses. More in-depth studies are required for trait characterization, methods of inoculation, identification, and compatibility assessment for mitigation of abiotic stresses in plants. In addition, we need to find the roles of microbial metabolites that are produced under environmental stress conditions.

Acknowledgments

The corresponding author of this chapter is extremely thankful to the staff members of GIC Chamtola, Almora, Uttarakhand, for providing a peaceful environment for writing this chapter and other necessary facilities.

References

Abadi, V.A.J.M., Sepehri, M., 2016. Effect of Piriformospora indica and Azotobacter chroococcum on mitigation of zinc deficiency stress in wheat (Triticum aestivum L.). Symbiosis 69 (1), 9–19.

Abbas, S., Javed, M.T., Shahid, M., Hussain, I., Haider, M.Z., Chaudhary, H.J., Tanwir, K., Maqsood, A., 2020. *Acinetobacter* sp. SG-5 inoculation alleviates cadmium toxicity in differentially cd tolerant maize cultivars as deciphered by improved physio-biochemical attributes, antioxidants and nutrient physiology. Plant Physiol. Biochem. 155, 815–827.

Abbasi, S., Sadeghi, A., Safaie, N., 2020. *Streptomyces* alleviate drought stress in tomato plants and modulate the expression of transcription factors ERF1 and WRKY70 genes. Sci. Hortic. 265, 109206.

Abd El-Daim, I.A., Bejai, S., Meijer, J., 2014. Improved heat stress tolerance of wheat seedlings by bacterial seed treatment. Plant Soil 379 (1), 337–350.

Abou-Aly, H.E., Youssef, A.M., Tewfike, T.A., El-Alkshar, E.A., El-Meihy, R.M., 2021. Reduction of heavy metals bioaccumulation in sorghum and its rhizosphere by heavy metals-tolerant bacterial consortium. Biocatal. Agric. Biotechnol. 31, 101911.

Adhikary, A., Kumar, R., Pandir, R., Bhardwaj, P., Wusirika, R., Kumar, S., 2019. *Pseudomonas citronellolis*; a multi-metal resistant and potential plant growth promoter against arsenic (V) stress in chickpea. Plant Physiol. Biochem. 142, 179–192.

Alavi, P., Starcher, M., Zachow, C., Müller, H., Berg, G., 2013. Root-microbe systems: the effect and mode of interaction of stress protecting agent (SPA) *Stenotrophomonas rhizophila* DSM14405T. Front. Plant Sci. 4, 141.

Ali, S.Z., Sandhya, V., Grover, M., Kishore, N., Rao, L.V., Venkateswarlu, B., 2009. *Pseudomonas* sp. strain AKM-P6 enhances tolerance of sorghum seedlings to elevated temperatures. Biol. Fertil. Soils 46 (1), 45–55.

Ali, S.Z., Sandhya, V., Grover, M., Linga, V.R., Bandi, V., 2011. Effect of inoculation with a thermotolerant plant growth-promoting *Pseudomonas putida* strain AKMP7 on growth of wheat (*Triticum* spp.) under heat stress. J. Plant Interact. 6 (4), 239–246.

Ali, S., Charles, T.C., Glick, B.R., 2014. Amelioration of high salinity stress damage by plant growth-promoting bacterial endophytes that contain ACC deaminase. Plant Physiol. Biochem. 80, 160–167.

Ali, B., Gill, R.A., Yang, S., Gill, M.B., Farooq, M.A., Liu, D., Daud, M.K., Ali, S., Zhou, W., 2015. Regulation of cadmium-induced proteomic and metabolic changes by 5-aminolevulinic acid in leaves of *Brassica napus* L. PLoS One 10, 1–23.

Ali, J., Ali, F., Ahmad, I., Rafique, M., Munis, M.F.H., Hassan, S.W., Sultan, T., Iftikhar, M., Chaudhary, H.J., 2021. Mechanistic elucidation of germination potential and growth of *Sesbania sesban* seedlings with *Bacillus anthracis* PM21 under heavy metals stress: an in vitro study. Ecotoxicol. Environ. Saf. 208, 111769.

Amna, Din, B.U., Sarfraz, S., Xia, Y., Kamran, M.A., Javed, M.T., Sultan, T., Munis, M.F.H., Chaudhary, H.J., 2019. Mechanistic elucidation of germination potential and growth of wheat inoculated with exopolysaccharide and ACC-deaminase producing Bacillus strains under induced salinity stress. Ecotoxicol. Environ. Saf. 183, 109466.

Andreas, T., Christophe, C., Essaïd, A.B., 2012. Physiological and molecular changes in plants at low temperatures. Planta 235, 1091–1105.

Armada, E., Roldán, A., Azcon, R., 2014. Differential activity of autochthonous bacteria in controlling drought stress in native *Lavandula* and *Salvia* plants species under drought conditions in natural arid soil. Microb. Ecol. 67 (2), 410–420.

Asghari, B., Khademian, R., Sedaghati, B., 2020. Plant growth promoting rhizobacteria (PGPR) confer drought resistance and stimulate biosynthesis of secondary metabolites in pennyroyal (*Mentha pulegium* L.) under water shortage condition. Sci. Hortic. 263, 109132.

Atkinson, N.J., Urwin, P.E., 2012. The interaction of plant biotic and abiotic stresses: from genes to the field. J. Exp. Bot. 63 (10), 3523–3543.

Bandeppa, S., Paul, S., Thakur, J.K., Chandrashekar, N., Umesh, D.K., Aggarwal, C., Asha, A.D., 2019. Antioxidant, physiological and biochemical responses of drought susceptible and drought tolerant mustard (*Brassica juncea* L) genotypes to rhizobacterial inoculation under water deficit stress. Plant Physiol. Biochem. 143, 19–28.

Barnawal, D., Bharti, N., Pandey, S.S., Pandey, A., Chanotiya, C.S., Kalra, A., 2017. Plant growth-promoting rhizobacteria enhance wheat salt and drought stress tolerance by altering endogenous phytohormone levels and TaCTR1/TaDREB2 expression. Physiol. Plant. 161 (4), 502–514.

Belimov, A.A., Dodd, I.C., Hontzeas, N., Theobald, J.C., Safronova, V.I., Davies, W.J., 2009. Rhizosphere bacteria containing 1-aminocyclopropane-1-carboxylate deaminase increase yield of plants grown in drying soil via both local and systemic hormone signalling. New Phytologist. 181 (2), 413–423.

Belimov, A.A., Dodd, I.C., Safronova, V.I., Shaposhnikov, A.I., Azarova, T.S., Makarova, N.M., Davies, W.J., Tikhonovich, I.A., 2015. Rhizobacteria that produce auxins and contain 1-amino-cyclopropane-1-carboxylic acid deaminase decrease amino acid concentrations in the rhizosphere and improve growth and yield of well-watered and water-limited potato (*Solanum tuberosum*). Ann. Appl. Biol. 167 (1), 11–25.

Belimov, A.A., Zinovkina, N.Y., Safronova, V.I., Litvinsky, V.A., Nosikov, V.V., Zavalin, A.A., Tikhonovich, I.A., 2019. Rhizobial ACC deaminase contributes to efficient symbiosis with pea (*Pisum sativum* L.) under single and combined cadmium and water deficit stress. Environ. Exp. Bot. 167, 103859.

Bharti, N., Pandey, S.S., Barnawal, D., Patel, V.K., Kalra, A., 2016. Plant growth promoting rhizobacteria *Dietzia natronolimnaea* modulates the expression of stress responsive genes providing protection of wheat from salinity stress. Sci. Rep. 6 (1), 1–16.

Bräutigam, A., Gowik, U., 2010. What can next generation sequencing do for you? Next generation sequencing as a valuable tool in plant research. Plant Biol. 12 (6), 831–841.

Bresson, J., Varoquaux, F., Bontpart, T., Touraine, B., Vile, D., 2013. The PGPR strain P hyllobacterium brassicacearum STM 196 induces a reproductive delay and physiological changes that result in improved drought tolerance in *Arabidopsis*. New Phytol. 200 (2), 558–569.

Budak, H., Kantar, M., Bulut, R., Akpinar, B.A., 2015. Stress responsive miRNAs and isomiRs in cereals. Plant Sci. 235, 1–13.

Bukhari, S.A., Zheng, W., Xie, L., Zhang, G., Shang, S., Wu, F., 2016. Cr-induced changes in leaf protein profile, ultrastructure and photosynthetic traits in the two contrasting tobacco genotypes. Plant Growth Regul. 79 (2), 147–156.

Bundy, J.G., Willey, T.L., Castell, R.S., Ellar, D.J., Brindle, K.M., 2005. Discrimination of pathogenic clinical isolates and laboratory strains of *Bacillus cereus* by NMR-based metabolomic profiling. FEMS Microbiol. Lett. 242 (1), 127–136.

Chandra, D., Srivastava, R., Glick, B.R., Sharma, A.K., 2018a. Drought-tolerant *Pseudomonas* spp. improve the growth performance of finger millet (*Eleusine coracana* (L.) Gaertn.) under non-stressed and drought-stressed conditions. Pedosphere 28, 227–240.

Chandra, D., Srivastava, R., Sharma, A., 2018b. Influence of IAA and ACC deaminase producing fluorescent pseudomonads in alleviating drought stress in wheat (*Triticum aestivum*). Agric. Res. 7, 290–299.

Chandra, D., Srivastava, R., Gupta, V.V., Franco, C.M., Sharma, A.K., 2019a. Evaluation of ACC-deaminase-producing rhizobacteria to alleviate water-stress impacts in wheat (*Triticum aestivum* L.) plants. Can. J. Microbiol. 65 (5), 387–403.

Chandra, D., Srivastava, R., Gupta, V.V., Franco, C.M., Paasricha, N., Saifi, S.K., Tuteja, N., Sharma, A.K., 2019b. Field performance of bacterial inoculants to alleviate water stress effects in wheat (*Triticum aestivum* L.). Plant Soil 441 (1), 261–281.

Chandra, D., Srivastava, R., Glick, B.R., Sharma, A.K., 2020. Rhizobacteria producing ACC deaminase mitigate water-stress response in finger millet (*Eleusine coracana* (L.) Gaertn.). 3 Biotech 10, 65.

Chandran, H., Meena, M., Swapnil, P., 2021. Plant growth-promoting rhizobacteria as a green alternative for sustainable agriculture. Sustainability 13 (19), 10986.

Chang, C.H., Yang, S.S., 2009. Thermo-tolerant phosphate-solubilizing microbes for multi-functional biofertilizer preparation. Bioresour. Technol. 100, 1648–1658.

Chen, K., Pachter, L., 2005. Bioinformatics for whole-genome shotgun sequencing of microbial communities. PLoS Comput. Biol. 1 (2), e24.

Chen, M., Wei, H., Cao, J., Liu, R., Wang, Y., Zheng, C., 2007. Expression of *Bacillus subtilis proBA* genes and reduction of feedback inhibition of proline synthesis increases proline production and confers osmotolerance in transgenic *Arabidopsis*. BMB Rep. 40 (3), 396–403.

Chen, L., Liu, Y., Wu, G., Veronican Njeri, K., Shen, Q., Zhang, N., Zhang, R., 2016. Induced maize salt tolerance by rhizosphere inoculation of *Bacillus amyloliquefaciens* SQR9. Physiol. Plant. 158 (1), 34–44.

Chiappero, J., del Rosario Cappellari, L., Alderete, L.G.S., Palermo, T.B., Banchio, E., 2019. Plant growth promoting rhizobacteria improve the antioxidant status in *Mentha piperita* grown under drought stress leading to an enhancement of plant growth and total phenolic content. Ind. Crop. Prod. 139, 111553.

Chu, T.N., Tran, B.T.H., Hoang, M.T.T., 2019. Plant growth-promoting rhizobacterium Pseudomonas PS01 induces salt tolerance in *Arabidopsis thaliana*. BMC Res. Notes 12 (1), 1–7.

Coleman-Derr, D., Tringe, S.G., 2014. Building the crops of tomorrow: advantages of symbiont-based approaches to improving abiotic stress tolerance. Front. Microbiol. 5, 283.

Cramer, G.R., Urano, K., Delrot, S., Pezzotti, M., Shinozaki, K., 2011. Effects of abiotic stress on plants: a systems biology perspective. BMC Plant Biol. 11, 163.

Curaba, J., Singh, M.B., Bhalla, P.L., 2014. miRNAs in the crosstalk between phytohormone signalling pathways. J. Exp. Bot. 65 (6), 1425–1438.

Dai, S., Chen, Q., Jiang, M., Wang, B., Xie, Z., Yu, N., Zhou, Y., Li, S., Wang, L., Hua, Y., Tian, B., 2021. Colonized extremophile *Deinococcus radiodurans* alleviates toxicity of cadmium and lead by suppressing heavy metal accumulation and improving antioxidant system in rice. Environ. Pollut. 284, 117127.

Defez, R., Esposito, R., Angelini, C., Bianco, C., 2016. Overproduction of indole-3-acetic acid in free-living rhizobia induces transcriptional changes resembling those occurring in nodule bacteroids. Mol. Plant-Microbe Interact. 29 (6), 484–495.

Delshadi, S., Ebrahimi, M., Shirmohammadi, E., 2017. Influence of plant-growth-promoting bacteria on germination, growth and nutrients' uptake of *Onobrychis sativa* L. under drought stress. J. Plant Interact. 12 (1), 200–208.

Desoky, E.S.M., Saad, A.M., El-Saadony, M.T., Merwad, A.R.M., Rady, M.M., 2020. Plant growth-promoting rhizobacteria: potential improvement in antioxidant defense system and suppression of oxidative stress for alleviating salinity stress in *Triticum aestivum* (L.) plants. Biocatal. Agric. Biotechnol. 30, 101878.

Dhaliwal, S.S., Singh, J., Taneja, P.K., Mandal, A., 2020. Remediation techniques for removal of heavy metals from the soil contaminated through different sources: a review. Environ. Sci. Pollut. Res. 27 (2), 1319–1333.

Dinneny, J.R., Long, T.A., Wang, J.Y., Jung, J.W., Mace, D., Pointer, S., Barron, C., Brady, S.M., Schiefelbein, J., Benfey, P.N., 2008. Cell identity mediates the response of *Arabidopsis* roots to abiotic stress. Science 320 (5878), 942–945.

Dourado, M.N., Martins, P.F., Quecine, M.C., Piotto, F.A., Souza, L.A., Franco, M.R., Tezotto, T., Azevedo, R.A., 2013. *Burkholderia* sp. SCMS54 reduces cadmium toxicity and promotes growth in tomato. Ann. Appl. Biol. 163 (3), 494–507.

Egamberdieva, D., Kucharova, Z., 2009. Selection for root colonising bacteria stimulating wheat growth in saline soils. Biol. Fertil. Soils 45 (6), 563–571.

El-Esawi, M.A., Alaraidh, I.A., Alsahli, A.A., Alamri, S.A., Ali, H.M., Alayafi, A.A., 2018a. *Bacillus firmus* (SW5) augments salt tolerance in soybean (*Glycine max* L.) by modulating root system architecture, antioxidant defense systems and stress-responsive genes expression. Plant Physiol. Biochem. 132, 375–384.

El-Esawi, M.A., Alaraidh, I.A., Alsahli, A.A., Alzahrani, S.M., Ali, H.M., Alayafi, A.A., Ahmad, M., 2018b. *Serratia liquefaciens* KM4 improves salt stress tolerance in maize by regulating redox potential, ion homeostasis, leaf gas exchange and stress-related gene expression. Int. J. Mol. Sci. 19 (11), 3310.

El-Esawi, M.A., Al-Ghamdi, A.A., Ali, H.M., Alayafi, A.A., 2019. *Azospirillum lipoferum* FK1 confers improved salt tolerance in chickpea (*Cicer arietinum* L.) by modulating osmolytes, antioxidant machinery and stress-related genes expression. Environ. Exp. Bot. 159, 55–65.

Etesami, H., 2018. Bacterial mediated alleviation of heavy metal stress and decreased accumulation of metals in plant tissues: mechanisms and future prospects. Ecotoxicol. Environ. Saf. 147, 175–191.

Etesami, H., Glick, B.R., 2020. Halotolerant plant growth–promoting bacteria: prospects for alleviating salinity stress in plants. Environ. Exp. Bot. 178, 104124.

FAO, 2007. http://www.fao.org/3/a1075e/a1075e00.pdf.

FAO, 2009. Food Security and Agricultural Mitigation in Developing Countries: Options for Capturing Synergies. FAO, Rome.

Fatemi, H., Pour, B.E., Rizwan, M., 2020. Isolation and characterization of lead (Pb) resistant microbes and their combined use with silicon nanoparticles improved the growth, photosynthesis and antioxidant capacity of coriander (*Coriandrum sativum* L.) under Pb stress. Environ. Pollut. 266, 114982.

Figueiredo, M.V., Burity, H.A., Martinez, C.R., Chanway, C.P., 2008. Alleviation of drought stress in the common bean (*Phaseolus vulgaris* L.) by co-inoculation with *Paenibacillus polymyxa* and *Rhizobium tropici*. Appl. Soil Ecol. 40 (1), 182–188.

Flowers, T.J., Galal, H.K., Bromham, L., 2010. Evolution of halophytes: multiple origins of salt tolerance in land plants. Funct. Plant Biol. 37 (7), 604–612.

Forni, C., Duca, D., Glick, B.R., 2017. Mechanisms of plant response to salt and drought stress and their alteration by rhizobacteria. Plant Soil 410, 335–356.

Foyer, C.H., Noctor, G., 2005. Oxidant and antioxidant signalling in plants: a re-evaluation of the concept of oxidative stress in a physiological context. Plant Cell Environ. 28 (8), 1056–1071.

Gao, P., Bai, X., Yang, L., Lv, D., Pan, X., Li, Y., Cai, H., Ji, W., Chen, Q., Zhu, Y., 2011. Osa-MIR393: a salinity- and alkaline stress-related microRNA gene. Mol. Biol. Rep. 38 (1), 237–242.

Ghosh, D., Sen, S., Mohapatra, S., 2017. Modulation of proline metabolic gene expression in *Arabidopsis thaliana* under water-stressed conditions by a drought-mitigating *Pseudomonas putida* strain. Ann. Microbiol. 67 (10), 655–668.

Gill, S.S., Tuteja, N., 2010. Reactive oxygen species and antioxidant machinery in abiotic stress tolerance in crop plants. Plant Physiol. Biochem. 48 (12), 909–930.

Gill, R.A., Zhang, N., Ali, B., Farooq, M.A., Xu, J., Gill, M.B., Mao, B., Zhou, W., 2016. Role of exogenous salicylic acid in regulating physio-morphic and molecular changes under chromium toxicity in black-and yellow-seeded *Brassica napus* L. Environ. Sci. Pollut. Res. 23 (20), 20483–20496.

Glick, B.R., 1995. The enhancement of plant growth by free-living bacteria. Can. J. Microbiol. 41 (2), 109–117.

Glick, B.R., 2010. Using soil bacteria to facilitate phytoremediation. Biotechnol. Adv. 28 (3), 367–374.

Glick, B.R., 2012. Plant growth-promoting bacteria: mechanisms and applications. Scientifica 2012.

Glick, B.R., Todorovic, B., Czarny, J., Cheng, Z., Duan, J., McConkey, B., 2007. Promotion of plant growth by bacterial ACC deaminase. Crit. Rev. Plant Sci. 26 (5–6), 227–242.

Goswami, M., Suresh, D.E.K.A., 2020. Plant growth-promoting rhizobacteria—alleviators of abiotic stresses in soil: a review. Pedosphere 30 (1), 40–61.

Grayson, M., 2013. Agriculture and drought. Nature 501, S1.

Gupta, B., Huang, B., 2014. Mechanism of salinity tolerance in plants: physiological, biochemical, and molecular characterization. Int. J. Genomics 2014, 701596.

Gupta, A., Rico-Medina, A., Caño-Delgado, A.I., 2020. The physiology of plant responses to drought. Science 368 (6488), 266–269.

Habib, S.H., Kausar, H., Saud, H.M., 2016. Plant growth-promoting rhizobacteria enhance salinity stress tolerance in okra through ROS-scavenging enzymes. Biomed. Res. Int. 2016.

Hahm, M.S., Son, J.S., Hwang, Y.J., Kwon, D.K., Ghim, S.Y., 2017. Alleviation of salt stress in pepper (*Capsicum annum* L.) plants by plant growth-promoting rhizobacteria. J. Microbiol. Biotechnol. 27 (10), 1790–1797.

Hasegawa, P.M., Bressan, R.A., Zhu, J.K., Bohnert, H.J., 2000. Plant cellular and molecular responses to high salinity. Annu. Rev. Plant Biol. 51, 463–499.

Hassan, N., Ebeed, H., Aljaarany, A., 2020. Exogenous application of spermine and putrescine mitigate adversities of drought stress in wheat by protecting membranes and chloroplast ultra-structure. Physiol. Mol. Biol. Plants 26 (2), 233–245.

He, A., Niu, S., Yang, D., Ren, W., Zhao, L., Sun, Y., Meng, L., Zhao, Q., Paré, P.W., Zhang, J., 2021. Two PGPR strains from the rhizosphere of *Haloxylon ammodendron* promoted growth and enhanced drought tolerance of ryegrass. Plant Physiol. Biochem. 161, 74–85.

Hossain, M.A., Hoque, M.A., Burritt, D.J., Fujita, M., 2014. Proline protects plants against abiotic oxidative stress: biochemical and molecular mechanisms. In: Oxidative Damage to Plants: Antioxidant Networks and Signaling. Academic Press, pp. 477–522.

Huan, L., Jin-Qiang, W., Qing, L., 2020. Photosynthesis product allocation and yield in sweet potato with spraying exogenous hormones under drought stress. J. Plant Physiol. 253, 153265.

Iba, K., 2002. Acclimative response to temperature stress in higher plants: approaches of gene engineering for temperature tolerance. Annu. Rev. Plant Biol. 53 (1), 225–245.

Jatan, R., Chauhan, P.S., Lata, C., 2019. Pseudomonas putida modulates the expression of miRNAs and their target genes in response to drought and salt stresses in chickpea (*Cicer arietinum* L.). Genomics 111 (4), 509–519.

Javed, M.T., Tanwir, K., Akram, M.S., Shahid, M., Niazi, N.K., Lindberg, S., 2019. Phytoremediation of cadmium-polluted water/sediment by aquatic macrophytes: role of plant-induced pH changes. In: Hasanuzzaman, M., Prasad, M.N.V., Fujita, M. (Eds.), Cadmium Toxicity and Tolerance in Plants. Academic Press, pp. 495–529.

Joshi, R., Singla-Pareek, S.L., Pareek, A., 2018. Engineering abiotic stress response in plants for biomass production. J. Biol. Chem. 293 (14), 5035–5043.

Kantar, M., Lucas, S.J., Budak, H., 2011. miRNA expression patterns of *Triticum dicoccoides* in response to shock drought stress. Planta 233 (3), 471–484.

Kavita, B., Shukla, S., Kumar, G.N., Archana, G., 2008. Amelioration of phytotoxic effects of Cd on mung bean seedlings by gluconic acid secreting rhizobacterium *Enterobacter asburiae* PSI3 and implication of role of organic acid. World J. Microbiol. Biotechnol. 24 (12), 2965–2972.

Ketchum, R.E., Rithner, C.D., Qiu, D., Kim, Y.S., Williams, R.M., Croteau, R.B., 2003. *Taxus* metabolomics: methyl jasmonate preferentially induces production of taxoids oxygenated at C-13 in *Taxus* x media cell cultures. Phytochemistry 62 (6), 901–909.

Khan, M.A., Gemenet, D.C., Villordon, A., 2016. Root system architecture and abiotic stress tolerance: current knowledge in root and tuber crops. Front. Plant Sci. 7, 1584.

Khan, M.A., Asaf, S., Khan, A.L., Jan, R., Kang, S.M., Kim, K.M., Lee, I.J., 2020. Thermotolerance effect of plant growth-promoting *Bacillus cereus* SA1 on soybean during heat stress. BMC Microbiol. 20 (1), 1–14.

Kim, K., Jang, Y.J., Lee, S.M., Oh, B.T., Chae, J.C., Lee, K.J., 2014. Alleviation of salt stress by *Enterobacter* sp. EJ01 in tomato and *Arabidopsis* is accompanied by up-regulation of conserved salinity responsive factors in plants. Mol. Cells 37 (2), 109.

Komaresofla, B.R., Alikhani, H.A., Etesami, H., Khoshkholgh-Sima, N.A., 2019. Improved growth and salinity tolerance of the halophyte *Salicornia* sp. by co–inoculation with endophytic and rhizosphere bacteria. Appl. Soil Ecol. 138, 160–170.

Kour, D., Rana, K.L., Kaur, T., Sheikh, I., Yadav, A.N., Kumar, V., Dhaliwal, H.S., Saxena, A.K., 2020. Microbe-mediated alleviation of drought stress and acquisition of phosphorus in great millet (*Sorghum bicolour* L.) by drought-adaptive and phosphorus-solubilizing microbes. Biocatal. Agric. Biotechnol. 23, 101501.

Koussevitzky, S., Suzuki, N., Huntington, S., Armijo, L., Sha, W., Cortes, D., Shulaev, V., Mittler, R., 2008. Ascorbate peroxidase 1 plays a key role in the response of *Arabidopsis thaliana* to stress combination. J. Biol. Chem. 283 (49), 34197–34203.

Kumar, A., Verma, J.P., 2018. Does plant—microbe interaction confer stress tolerance in plants: a review? Microbiol. Res. 207, 41–52.

Kumar, A., Patel, J.S., Meena, V.S., Srivastava, R., 2019. Recent advances of PGPR based approaches for stress tolerance in plants for sustainable agriculture. Biocatal. Agric. Biotechnol. 20, 101271.

Kumar, V., Kumar, P., Khan, A., 2020. Optimization of PGPR and silicon fertilization using response surface methodology for enhanced growth, yield and biochemical parameters of French bean (*Phaseolus vulgaris* L.) under saline stress. Biocatal. Agric. Biotechnol. 23, 101463.

Kumar, A., Singh, S., Mukherjee, A., Rastogi, R.P., Verma, J.P., 2021a. Salt-tolerant plant growth-promoting *Bacillus pumilus* strain JPVS11 to enhance plant growth attributes of rice and improve soil health under salinity stress. Microbiol. Res. 242, 126616.

Kumar, A., Voropaeva, O., Maleva, M., Panikovskaya, K., Borisova, G., Rajkumar, M., Bruno, L.B., 2021b. Bioaugmentation with copper tolerant endophyte *Pseudomonas lurida* strain EOO26 for improved plant growth and copper phytoremediation by *Helianthus annuus*. Chemosphere 266, 128983.

Kumari, S., Vaishnav, A., Jain, S., Varma, A., Choudhary, D.K., 2015. Bacterial-mediated induction of systemic tolerance to salinity with expression of stress alleviating enzymes in soybean (*Glycine max* L. Merrill). J. Plant Growth Regul. 34 (3), 558–573.

Lastochkina, O., Aliniaeifard, S., Garshina, D., Garipova, S., Pusenkova, L., Allagulova, C., Fedorova, K., Baymiev, A., Koryakov, I., Sobhani, M., 2021. Seed priming with endophytic *Bacillus subtilis* strain-specifically improves growth of *Phaseolus vulgaris* plants under normal and salinity conditions and exerts antistress effect through induced lignin deposition in roots and decreased oxidative and osmotic damages. J. Plant Physiol., 153462.

Li, Y.F., Zheng, Y., Addo-Quaye, C., Zhang, L., Saini, A., Jagadeeswaran, G., Axtell, M.J., Zhang, W., Sunkar, R., 2010. Transcriptome-wide identification of microRNA targets in rice. Plant J. 62 (5), 742–759.

Li, Q., Zhao, P., Li, J., Zhang, C., Wang, L., Ren, Z., 2014. Genome-wide analysis of the WD-repeat protein family in cucumber and *Arabidopsis*. Mol. Gen. Genomics. 289 (1), 103–124.

Li, X., Sun, P., Zhang, Y., Jin, C., Guan, C., 2020. A novel PGPR strain *Kocuria rhizophila* Y1 enhances salt stress tolerance in maize by regulating phytohormone levels, nutrient acquisition, redox potential, ion homeostasis, photosynthetic capacity and stress-responsive genes expression. Environ. Exp. Bot. 174, 104023.

Li, Q., Xing, Y., Fu, X., Ji, L., Li, T., Wang, J., Chen, G., Qi, Z., Zhang, Q., 2021. Biochemical mechanisms of rhizospheric *Bacillus subtilis*-facilitated phytoextraction by alfalfa under cadmium stress—microbial diversity and metabolomics analyses. Ecotoxicol. Environ. Saf. 212, 112016.

Lim, J.H., Kim, S.D., 2013. Induction of drought stress resistance by multi-functional PGPR *Bacillus licheniformis* K11 in pepper. Plant Pathol. J. 29 (2), 201.

Ma, Y., Rajkumar, M., Zhang, C., Freitas, H., 2016. Inoculation of *Brassica oxyrrhina* with plant growth promoting bacteria for the improvement of heavy metal phytoremediation under drought conditions. J. Hazard. Mater. 320, 36–44.

Machado, R.M.A., Serralheiro, R.P., 2017. Soil salinity: effect on vegetable crop growth. Management practices to prevent and mitigate soil salinization. Horticulturae 3 (2), 30.

Mahajan, S.G., Nandre, V.S., Salunkhe, R.C., Shouche, Y.S., Kulkarni, M.V., 2020. Chemotaxis and physiological adaptation of an indigenous abiotic stress tolerant plant growth promoting *Pseudomonas stutzeri*: amelioration of salt stress to *Cicer arietinum*. Biocatal. Agric. Biotechnol. 27, 101652.

Mallick, I., Bhattacharyya, C., Mukherji, S., Dey, D., Sarkar, S.C., Mukhopadhyay, U.K., Ghosh, A., 2018. Effective rhizoinoculation and biofilm formation by arsenic immobilizing halophilic plant growth promoting bacteria (PGPB) isolated from mangrove rhizosphere: a step towards arsenic rhizoremediation. Sci. Total Environ. 610, 1239–1250.

Masmoudi, F., Tounsi, S., Dunlap, C.A., Trigui, M., 2021. Endophytic halotolerant *Bacillus velezensis* FMH2 alleviates salt stress on tomato plants by improving plant growth and altering physiological and antioxidant responses. Plant Physiol. Biochem. 165, 217–227.

Massad, T.J., Dyer, L.A., Vega, C., G., 2012. Costs of defense and a test of the carbon-nutrient balance and growth-differentiation balance hypotheses for two co-occurring classes of plant defense. PLoS One 7 (10), e47554.

Mayak, S., Tirosh, T., Glick, B.R., 2004. Plant growth-promoting bacteria that confer resistance to water stress in tomatoes and peppers. Plant Sci. 166 (2), 525–530.

McCue, K.F., Hanson, A.D., 1990. Salt-inducible betaine aldehyde dehydrogenase from sugar beet: cDNA cloning and expression. Trends Biotechnol. 8, 358–362.

Meena, H., Ahmed, M.A., Prakash, P., 2015. Amelioration of heat stress in wheat, *Triticum aestivum* by PGPR (*Pseudomonas aeruginosa* strain 2CpS1). Biosci. Biotechnol. Res. 8 (2), 171–174.

Meena, K.K., Sorty, A.M., Bitla, U.M., Choudhary, K., Gupta, P., Pareek, A., Singh, D.P., Prabha, R., Sahu, P.K., Gupta, V.K., Singh, H.B., Krishanani, K.K., Minhas, P.S., 2017. Abiotic stress responses and microbe-mediated mitigation in plants: the omics strategies. Front. Plant Sci. 8, 172.

Mishra, B.K., Meena, K.K., Dubey, P.N., Aishwath, O.P., Kant, K., Sorty, A.M., Bitla, U., 2016. Influence on yield and quality of fennel (*Foeniculum vulgare* Mill.) grown under semi-arid saline soil, due to application of native phosphate solubilizing rhizobacterial isolates. Ecol. Eng. 97, 327–333.

Mishra, S.K., Khan, M.H., Misra, S., Dixit, V.K., Khare, P., Srivastava, S., Chauhan, P.S., 2017. Characterisation of Pseudomonas spp. and Ochrobactrum sp. isolated from volcanic soil. Antonie Van Leeuwenhoek 110 (2), 253–270.

Mishra, S.K., Khan, M.H., Misra, S., Dixit, V.K., Gupta, S., Tiwari, S., Gupta, S.C., Chauhan, P.S., 2020. Drought tolerant *Ochrobactrum* sp. inoculation performs multiple roles in maintaining the homeostasis in *Zea mays* L. subjected to deficit water stress. Plant Physiol. Biochem. 150, 1–14.

Mitra, S., Pramanik, K., Sarkar, A., Ghosh, P.K., Soren, T., Maiti, T.K., 2018. Bioaccumulation of cadmium by *Enterobacter* sp. and enhancement of rice seedling growth under cadmium stress. Ecotoxicol. Environ. Saf. 156, 183–196.

Mitra, D., Rodríguez, A.M.D., Cota, F.I.P., Khoshru, B., Panneerselvam, P., Moradi, S., Sagarika, M.S., Anđelković, S., de los Santos-Villalobos, S., Mohapatra, P.K.D., 2021. Amelioration of thermal stress in crops by plant growth-promoting rhizobacteria. Physiol. Mol. Plant Pathol., 101679.

Mittler, R., 2006. Abiotic stress, the field environment and stress combination. Trends Plant Sci. 11, 15–19.

Miura, K., Furumoto, T., 2013. Cold signaling and cold response in plants. Int. J. Mol. Sci. 14 (3), 5312–5337.

Modi, R., Khanna, V., 2018. Evaluation of thermotolerant rhizobacteria for multiple plant growth promoting traits from pigeonpea rhizosphere. J. Appl. Nat. Sci. 10 (1), 518–521.

Morrow Jr., K.J., 2010. Mass spec central to metabolomics. Genet. Eng. Biotechnol. News 30 (7), 1–3.

Mukhtar, S., Zareen, M., Khaliq, Z., Mehnaz, S., Malik, K.A., 2020a. Phylogenetic analysis of halophyte-associated rhizobacteria and effect of halotolerant and halophilic phosphate-solubilizing biofertilizers on maize growth under salinity stress conditions. J. Appl. Microbiol. 128 (2), 556–573.

Mukhtar, T., Smith, D., Sultan, T., Seleiman, M.F., Alsadon, A.A., Ali, S., Chaudhary, H.J., Solieman, T.H., Ibrahim, A.A., Saad, M.A., 2020b. Mitigation of heat stress in *Solanum lycopersicum* L. by ACC-deaminase and exopolysaccharide producing *Bacillus cereus*: effects on biochemical profiling. Sustainability 12 (6), 2159.

Nadeem, S.M., Zahir, Z.A., Naveed, M., Arshad, M., 2007. Preliminary investigations on inducing salt tolerance in maize through inoculation with rhizobacteria containing ACC deaminase activity. Can. J. Microbiol. 53 (10), 1141–1149.

Nadeem, S.M., Zahir, Z.A., Naveed, M., Nawaz, S., 2013. Mitigation of salinity-induced negative impact on the growth and yield of wheat by plant growth-promoting rhizobacteria in naturally saline conditions. Ann. Microbiol. 63 (1), 225–232.

Nakayama, H., Yoshida, K., Ono, H., Murooka, Y., Shinmyo, A., 2000. Ectoine, the compatible solute of *Halomonas elongata*, confers hyperosmotic tolerance in cultured tobacco cells. Plant Physiol. 122 (4), 1239–1248.

Narayanasamy, S., Thangappan, S., Uthandi, S., 2020. Plant growth-promoting *Bacillus* sp. cahoots moisture stress alleviation in rice genotypes by triggering antioxidant defense system. Microbiol. Res. 239, 126518.

Nautiyal, C.S., Srivastava, S., Chauhan, P.S., Seem, K., Mishra, A., Sopory, S.K., 2013. Plant growth-promoting bacteria *Bacillus amyloliquefaciens* NBRISN13 modulates gene expression profile of leaf and rhizosphere community in rice during salt stress. Plant Physiol. Biochem. 66, 1–9.

Niu, S.Q., Li, H.R., Paré, P.W., Aziz, M., Wang, S.M., Shi, H., Li, J., Han, Q.Q., Guo, S.Q., Li, J., Zhang, J.L., 2016. Induced growth promotion and higher salt tolerance in the halophyte grass *Puccinellia tenuiflora* by beneficial rhizobacteria. Plant Soil 407 (1), 217–230.

Niu, X., Song, L., Xiao, Y., Ge, W., 2018. Drought-tolerant plant growth-promoting rhizobacteria associated with foxtail millet in a semi-arid agroecosystem and their potential in alleviating drought stress. Front. Microbiol. 8, 2580.

Opeolu, B.O., Adenuga, O.O., Ndakidemi, P.A., Olujimi, O.O., 2010. Assessment of phyto-toxicity potential of lead on tomato (*Lycopersicon esculentum* L) planted on contaminated soils. Int. J. Phys. Sci. 5 (2), 68–73.

Pacheco, G.J., Ciapina, E.M.P., Gomes, E.D.B., Pereira Junior, N., 2010. Biosurfactant production by *Rhodococcus erythropolis* and its application to oil removal. Braz. J. Microbiol. 41, 685–693.

Pacwa-Płociniczak, M., Płaza, G.A., Piotrowska-Seget, Z., Cameotra, S.S., 2011. Environmental applications of biosurfactants: recent advances. Int. J. Mol. Sci. 12 (1), 633–654.

Park, Y.G., Mun, B.G., Kang, S.M., Hussain, A., Shahzad, R., Seo, C.W., Kim, A.Y., Lee, S.U., Oh, K.Y., Lee, D.Y., Lee, I.J., 2017. *Bacillus aryabhattai* SRB02 tolerates oxidative and nitrosative stress and promotes the growth of soybean by modulating the production of phytohormones. PLoS One 12, e0173203.

Paulucci, N.S., Gallarato, L.A., Reguera, Y.B., Vicario, J.C., Cesari, A.B., de Lema, M.B.G., Dardanelli, M.S., 2015. *Arachis hypogaea* PGPR isolated from argentine soil modifies its lipids components in response to temperature and salinity. Microbiol. Res. 173, 1–9.

Pereira, S.I.A., Abreu, D., Moreira, H., Vega, A., Castro, P.M.L., 2020. Plant growth-promoting rhizobacteria (PGPR) improve the growth and nutrient use efficiency in maize (*Zea mays* L.) under water deficit conditions. Heliyon 6 (10), e05106.

Pereyra, M.A., Garcia, P., Colabelli, M.N., Barassi, C.A., Creus, C.M., 2012. A better water status in wheat seedlings induced by *Azospirillum* under osmotic stress is related to morphological changes in xylem vessels of the coleoptile. Appl. Soil Ecol. 53, 94–97.

Phour, M., Sindhu, S.S., 2020. Amelioration of salinity stress and growth stimulation of mustard (*Brassica juncea* L.) by salt-tolerant *Pseudomonas* species. Appl. Soil Ecol. 149, 103518.

Piasecka, A., Kachlicki, P., Stobiecki, M., 2019. Analytical methods for detection of plant metabolomes changes in response to biotic and abiotic stresses. Int. J. Mol. Sci. 20 (2), 379.

Pramanik, K., Mitra, S., Sarkar, A., Maiti, T.K., 2018. Alleviation of phytotoxic effects of cadmium on rice seedlings by cadmium resistant PGPR strain *Enterobacter aerogenes* MCC 3092. J. Hazard. Mater. 351, 317–329.

Pramanik, K., Mandal, S., Banerjee, S., Ghosh, A., Maiti, T.K., Mandal, N.C., 2021. Unraveling the heavy metal resistance and biocontrol potential of *Pseudomonas* sp. K32 strain facilitating rice seedling growth under Cd stress. Chemosphere 274, 129819.

Prapagdee, B., Chanprasert, M., Mongkolsuk, S., 2013. Bioaugmentation with cadmium-resistant plant growth-promoting rhizobacteria to assist cadmium phytoextraction by *Helianthus annuus*. Chemosphere 92 (6), 659–666.

Qadir, M., Quillérou, E., Nangia, V., Murtaza, G., Singh, M., Thomas, R.J., Drechsel, P., 2014. Economics of salt-induced land degradation and restoration. Nat. Res. Forum 38, 282–295.

Rahmani, H.A., Saleh-Rastin, N., Khavazi, K., Asgharzadeh, A., Fewer, D., Kiani, S., Lindström, K., 2009. Selection of thermotolerant bradyrhizobial strains for nodulation of soybean (*Glycine max* L.) in semi-arid regions of Iran. World J. Microbiol. Biotechnol. 25, 591–600.

Rajkumar, M., Sandhya, S., Prasad, M.N.V., Freitas, H., 2012. Perspectives of plant-associated microbes in heavy metal phytoremediation. Biotechnol. Adv. 30 (6), 1562–1574.

Rajput, L.U.B.N.A., Imran, A., Mubeen, F.A.T.H.I.A., Hafeez, F.Y., 2013. Salt-tolerant PGPR strain *Planococcus rifietoensis* promotes the growth and yield of wheat (*Triticum aestivum* L.) cultivated in saline soil. Pak. J. Bot. 45 (6), 1955–1962.

Rehman, M.Z., Rizwan, M., Ali, S., Ok, Y.S., Ishaque, W., Nawaz, M.F., Akmal, F., Waqar, M., 2017. Remediation of heavy metal contaminated soils by using *Solanum nigrum*: a review. Ecotoxicol. Environ. Saf. 143, 236–248.

Riadh, K., Wided, M., Hans-Werner, K., Chedly, A., 2010. Responses of halophytes to environmental stresses with special emphasis to salinity. Adv. Bot. Res. 53, 117–145.

Rizvi, A., Khan, M.S., 2018. Heavy metal induced oxidative damage and root morphology alterations of maize (*Zea mays* L.) plants and stress mitigation by metal tolerant nitrogen fixing *Azotobacter chroococcum*. Ecotoxicol. Environ. Saf. 157, 9–20.

Robin, A., Mougel, C., Siblot, S., Vansuyt, G., Mazurier, S., Lemanceau, P., 2006. Effect of ferritin overexpression in tobacco on the structure of bacterial and pseudomonad communities associated with the roots. FEMS Microbiol. Ecol. 58 (3), 492–502.

Ruan, C.J., da Silva, J.A.T., Mopper, S., Qin, P., Lutts, S., 2010. Halophyte improvement for a salinized world. Crit. Rev. Plant Sci. 29, 329–359.

Saleem, M., Asghar, H.N., Zahir, Z.A., Shahid, M., 2018. Impact of lead tolerant plant growth promoting rhizobacteria on growth, physiology, antioxidant activities, yield and lead content in sunflower in lead contaminated soil. Chemosphere 195, 606–614.

Sapre, S., Gontia-Mishra, I., Tiwari, S., 2018. *Klebsiella* sp. confers enhanced tolerance to salinity and plant growth promotion in oat seedlings (*Avena sativa*). Microbiol. Res. 206, 25–32.

Saravanakumar, D., Samiyappan, R., 2007. ACC deaminase from *Pseudomonas fluorescens* mediated saline resistance in groundnut (*Arachis hypogea*) plants. J. Appl. Microbiol. 102 (5), 1283–1292.

Sarkar, A., Ghosh, P.K., Pramanik, K., Mitra, S., Soren, T., Pandey, S., Mondal, M.H., Maiti, T.K., 2018a. A halotolerant *Enterobacter* sp. displaying ACC deaminase activity promotes rice seedling growth under salt stress. Res. Microbiol. 169 (1), 20–32.

Sarkar, J., Chakraborty, B., Chakraborty, U., 2018b. Plant growth promoting rhizobacteria protect wheat plants against temperature stress through antioxidant signalling and reducing chloroplast and membrane injury. J. Plant Growth Regul. 37 (4), 1396–1412.

Sarma, R.K., Saikia, R., 2014. Alleviation of drought stress in mung bean by strain Pseudomonas aeruginosa GGRJ21. Plant Soil 377 (1), 111–126.

Schenk, P.M., Carvalhais, L.C., Kazan, K., 2012. Unraveling plant–microbe interactions: can multi-species transcriptomics help? Trends Biotechnol. 30 (3), 177–184.

Schulze, W.X., 2005. Protein analysis in dissolved organic matter: what proteins from organic debris, soil leachate and surface water can tell us—a perspective. Biogeosciences 2, 75–86.

Sen, S., Ghosh, D., Mohapatra, S., 2018. Modulation of polyamine biosynthesis in Arabidopsis *thaliana* by a drought mitigating Pseudomonas putida strain. Plant Physiol. Biochem. 129, 180–188.

Sessitsch, A.N.G.E.L.A., Hardoim, P.A.B.L.O., Döring, J., Weilharter, A., Krause, A.N.D.R.E.A.S., Woyke, T.A.N.J.A., Mitter, B., Hauberg-Lotte, L., Friedrich, F., Rahalkar, M., Reinhold-Hurek, B., 2012. Functional characteristics of an endophyte community colonizing rice roots as revealed by metagenomic analysis. Mol. Plant-Microbe Interact. 25 (1), 28–36.

Shah, A.A., Ahmed, S., Ali, A., Yasin, N.A., 2020a. 2-Hydroxymelatonin mitigates cadmium stress in *Cucumis sativus* seedlings: modulation of antioxidant enzymes and polyamines. Chemosphere 243, 125308.

Shah, T., Munsif, F., D'amato, R., Nie, L., 2020b. Lead toxicity induced phytotoxic impacts on rapeseed and clover can be lowered by biofilm forming lead tolerant bacteria. Chemosphere 246, 125766.

Shah, A.A., Aslam, S., Akbar, M., Ahmad, A., Khan, W.U., Yasin, N.A., Ali, B., Rizwan, M., Ali, S., 2021a. Combined effect of *Bacillus fortis* IAGS 223 and zinc oxide nanoparticles to alleviate cadmium phytotoxicity in *Cucumis melo*. Plant Physiol. Biochem. 158, 1–12.

Shah, A.A., Yasin, N.A., Akram, K., Ahmad, A., Khan, W.U., Akram, W., Akbar, M., 2021b. Ameliorative role of *Bacillus subtilis* FBL-10 and silicon against lead induced stress in *Solanum melongena*. Plant Physiol. Biochem. 158, 486–496.

Shahid, M., Ameen, F., Maheshwari, H.S., Ahmed, B., AlNadhari, S., Khan, M.S., 2021. Colonization of Vigna radiata by a halotolerant bacterium *Kosakonia sacchari* improves the ionic balance, stressor metabolites, antioxidant status and yield under NaCl stress. Appl. Soil Ecol. 158, 103809.

Sheldon, A.R., Menzies, N.W., 2005. The effect of copper toxicity on the growth and root morphology of Rhodes grass (*Chloris gayana* Knuth.) in resin buffered solution culture. Plant Soil 278 (1–2), 341–349.

Shi, Z., Zhang, Z., Yuan, M., Wang, S., Yang, M., Yao, Q., Ba, W., Zhao, J., Xie, B., 2020. Characterization of a high cadmium accumulating soil bacterium, *Cupriavidus* sp. WS2. Chemosphere 247, 125834.

Silambarasan, S., Logeswari, P., Valentine, A., Cornejo, P., 2019. Role of *Curtobacterium herbarum* strain CAH5 on aluminum bioaccumulation and enhancement of *Lactuca sativa* growth under aluminum and drought stresses. Ecotoxicol. Environ. Saf. 183, 109573.

Simoes-Araujo, J.L., Alves-Ferreira, M., Rumjanek, N.G., Margis-Pinheiro, M., 2008. VuNIP1 (NOD26-like) and VuHSP17.7 gene expression are regulated in response to heat stress in cowpea nodule. Environ. Exp. Bot. 63 (1–3), 256–265.

Simontacchi, M., Galatro, A., Ramos-Artuso, F., Santa-María, G.E., 2015. Plant survival in a changing environment: the role of nitric oxide in plant responses to abiotic stress. Front. Plant Sci. 6, 977.

Singh, S.K., Singh, P.P., Gupta, A., Singh, A.K., Keshri, J., 2019. Tolerance of heavy metal toxicity using PGPR strains of *Pseudomonas* species. In: Singh, A.K., Kumar, A., Singh, A.K. (Eds.), PGPR Amelioration in Sustainable Agriculture. Woodhead Publishing, pp. 239–252.

Sorty, A.M., Meena, K.K., Choudhary, K., Bitla, U.M., Minhas, P.S., Krishnani, K.K., 2016. Effect of plant growth promoting bacteria associated with halophytic weed (*Psoralea corylifolia* L) on germination and seedling growth of wheat under saline conditions. Appl. Biochem. Biotechnol. 180 (5), 872–882.

Srivastava, S., Singh, N., 2014. Mitigation approach of arsenic toxicity in chickpea grown in arsenic amended soil with arsenic tolerant plant growth promoting *Acinetobacter* sp. Ecol. Eng. 70, 146–153.

Srivastava, S., Yadav, A., Seem, K., Mishra, S., Chaudhary, V., Nautiyal, C.S., 2008. Effect of high temperature on *Pseudomonas putida* NBRI0987 biofilm formation and expression of stress sigma factor RpoS. Curr. Microbiol. 56 (5), 453–457.

Subramanian, P., Kim, K., Krishnamoorthy, R., Mageswari, A., Selvakumar, G., Sa, T., 2016a. Cold stress tolerance in psychrotolerant soil bacteria and their conferred chilling resistance in tomato (*Solanum lycopersicum* Mill.) under low temperatures. PloS One 11 (8), e0161592.

Subramanian, S., Souleimanov, A., Smith, D.L., 2016b. Proteomic studies on the effects of lipo-chitooligosaccharide and thuricin 17 under unstressed and salt stressed conditions in Arabidopsis thaliana. Front. Plant Sci. 7, 1314.

Sziderics, A.H., Rasche, F., Trognitz, F., Sessitsch, A., Wilhelm, E., 2007. Bacterial endophytes contribute to abiotic stress adaptation in pepper plants (*Capsicum annuum* L.). Can. J. Microbiol. 53 (11), 1195–1202.

Tagele, S.B., Kim, S.W., Lee, H.G., Kim, H.S., Lee, Y.S., 2018. Effectiveness of multi-trait *Burkholderia contaminans* KNU17BI1 in growth promotion and management of banded leaf and sheath blight in maize seedling. Microbiol. Res. 214, 8–18.

Tahir, M., Ahmad, I., Shahid, M., Shah, G.M., Farooq, A.B.U., Akram, M., Tabassum, S.A., Naeem, M.A., Khalid, U., Ahmad, S., Zakir, A., 2019. Regulation of antioxidant production, ion uptake and productivity in potato (*Solanum tuberosum* L.) plant inoculated with growth promoting salt tolerant Bacillus strains. Ecotoxicol. Environ. Saf. 178, 33–42.

Tani, A., Takai, Y., Suzukawa, I., Akita, M., Murase, H., Kimbara, K., 2012. Practical application of methanol-mediated mutualistic symbiosis between *Methylobacterium* species and a roof greening moss, *Racomitrium japonicum*. PLoS One 7 (3), e33800.

Tanwir, K., Javed, M.T., Abbas, S., Shahid, M., Akram, M.S., Chaudhary, H.J., Iqbal, M., 2021. *Serratia* sp. CP-13 alleviates Cd toxicity by morpho-physio-biochemical improvements, antioxidative potential and diminished cd uptake in *Zea mays* L. cultivars differing in Cd tolerance. Ecotoxicol. Environ. Saf. 208, 111584.

Tao, Q., Zhao, J., Li, J., Liu, Y., Luo, J., Yuan, S., Li, B., Li, Q., Xu, Q., Yu, X., Wang, C., 2020. Unique root exudate tartaric acid enhanced cadmium mobilization and uptake in Cd-hyperaccumulator *Sedum alfredii*. J. Hazard. Mater. 383, 121177.

Tiwari, S., Lata, C., Chauhan, P.S., Nautiyal, C.S., 2016. *Pseudomonas putida* attunes morphophysiological, biochemical and molecular responses in *Cicer arietinum* L. during drought stress and recovery. Plant Physiol. Biochem. 99, 108–117.

Tiwari, S., Prasad, V., Chauhan, P.S., Lata, C., 2017. *Bacillus amyloliquefaciens* confers tolerance to various abiotic stresses and modulates plant response to phytohormones through osmoprotection and gene expression regulation in rice. Front. Plant Sci. 8, 1510.

Trindade, I., Capitão, C., Dalmay, T., Fevereiro, M.P., Dos Santos, D.M., 2010. miR398 and miR408 are up-regulated in response to water deficit in *Medicago truncatula*. Planta 231 (3), 705–716.

Turner, T.R., Ramakrishnan, K., Walshaw, J., Heavens, D., Alston, M., Swarbreck, D., Osbourn, A., Grant, A., Poole, P.S., 2013. Comparative metatranscriptomics reveals kingdom level changes in the rhizosphere microbiome of plants. ISME J. 7 (12), 2248–2258.

Upadhyay, S.K., Singh, J.S., Saxena, A.K., Singh, D.P., 2012. Impact of PGPR inoculation on growth and antioxidant status of wheat under saline conditions. Plant Biol. 14 (4), 605–611.

Vafaee, F., Dashti, H., Alinejad-Rokny, H., 2019. Transcriptomic data normalization. In: Ranganathan, S., Gribskov, M., Nakai, K., Schönbach, C. (Eds.), Encyclopedia of Bioinformatics and Computational Biology. Academic Press, Oxford, pp. 364–371.

Vansuyt, G., Robin, A., Briat, J.F., Curie, C., Lemanceau, P., 2007. Iron acquisition from Fe-pyoverdine by *Arabidopsis thaliana*. Mol. Plant-Microbe Interact. 20 (4), 441–447.

Vardharajula, S., Zulfikar Ali, S., Grover, M., Reddy, G., Bandi, V., 2011. Drought-tolerant plant growth promoting *Bacillus* spp.: effect on growth, osmolytes, and antioxidant status of maize under drought stress. J. Plant Interact. 6 (1), 1–14.

Verma, A.K., Deepti, S., 2016. Abiotic stress and crop improvement: current scenario. Adv. Plants Agric. Res. 4 (4), 00149.

Verma, J.P., Yadav, J., Tiwari, K.N., Kumar, A., 2013. Effect of indigenous *Mesorhizobium* spp. and plant growth promoting rhizobacteria on yields and nutrients uptake of chickpea (*Cicer arietinum* L.) under sustainable agriculture. Ecol. Eng. 51, 282–286.

Vimal, S.R., Patel, V.K., Singh, J.S., 2019. Plant growth promoting *Curtobacterium albidum* strain SRV4: an agriculturally important microbe to alleviate salinity stress in paddy plants. Ecol. Indic. 105, 553–562.

Vives-Peris, V., Gómez-Cadenas, A., Pérez-Clemente, R.M., 2018. Salt stress alleviation in citrus plants by plant growth-promoting rhizobacteria *Pseudomonas putida* and *Novosphingobium* sp. Plant Cell Rep. 37 (11), 1557–1569.

Voesenek, L.A.C.J., Pierik, R., 2008. Plant stress profiles. Science 320, 880–881.

Wahid, A., Gelani, S., Ashraf, M., Foolad, M.R., 2007. Heat tolerance in plants: an overview. Environ. Exp. Bot. 61 (3), 199–223.

Wang, C.J., Yang, W., Wang, C., Gu, C., Niu, D.D., Liu, H.X., Wang, Y.P., Guo, J.H., 2012. Induction of drought tolerance in cucumber plants by a consortium of three plant growth-promoting rhizobacterium strains. PLoS One 7 (12), e52565.

Wang, L.Y., Xie, Y.S., Cui, Y.Y., Xu, J., He, W., Chen, H.G., Guo, J.H., 2015. Conjunctively screening of biocontrol agents (BCAs) against fusarium root rot and fusarium head blight caused by *Fusarium graminearum*. Microbiol. Res. 177, 34–42.

Wang, S., Wang, F., Gao, S., Wang, X., 2016a. Heavy metal accumulation in different rice cultivars as influenced by foliar application of nano-silicon. Water Air Soil Pollut. 227, 1–13.

Wang, Y., Hu, B., Du, S., Gao, S., Chen, X., Chen, D., 2016b. Proteomic analyses reveal the mechanism of *Dunaliella salina* Ds-26-16 gene enhancing salt tolerance in *Escherichia coli*. PLoS One 11 (5), e0153640.

Wang, P., Chen, H., Kopittke, P.M., Zhao, F.J., 2019. Cadmium contamination in agricultural soils of China and the impact on food safety. Environ. Pollut. 249, 1038–1048.

Woo, O.G., Kim, H., Kim, J.S., Keum, H.L., Lee, K.C., Sul, W.J., Lee, J.H., 2020. *Bacillus subtilis* strain GOT9 confers enhanced tolerance to drought and salt stresses in *Arabidopsis thaliana* and *Brassica campestris*. Plant Physiol. Biochem. 148, 359–367.

Xiao, W.A.N.G., Liu, F.L., Jiang, D., 2017. Priming: A promising strategy for crop production in response to future climate. J. Integr. Agric. 16 (12), 2709–2716.

Xu, Z.Z., Zhou, G.S., 2006. Combined effects of water stress and high temperature on photosynthesis, nitrogen metabolism and lipid peroxidation of a perennial grass *Leymus chinensis*. Planta 224, 1080–1090.

Xu, Z., Shimizu, H., Ito, S., Yagasaki, Y., Zou, C., Zhou, G., Zheng, Y., 2014. Effects of elevated CO_2, warming and precipitation change on plant growth, photosynthesis and peroxidation in dominant species from North China grassland. Planta 239 (2), 421–435.

Xu, Z., Jiang, Y., Zhou, G., 2015. Response and adaptation of photosynthesis, respiration, and antioxidant systems to elevated CO_2 with environmental stress in plants. Front. Plant Sci. 6, 701.

Yan, J., Smith, M.D., Glick, B.R., Liang, Y., 2014. Effects of ACC deaminase containing rhizobacteria on plant growth and expression of toc GTPases in tomato (*Solanum lycopersicum*) under salt stress. Botany 92 (11), 775–781.

Yang, J., Kloepper, J.W., Ryu, C.M., 2009. Rhizosphere bacteria help plants tolerate abiotic stress. Trends Plant Sci. 14 (1), 1–4.

Yarzábal, L.A., Monserrate, L., Buela, L., Chica, E., 2018. Antarctic *Pseudomonas* spp. promote wheat germination and growth at low temperatures. Polar Biol. 41 (11), 2343–2354.

Yasmeen, T., Ahmad, A., Arif, M.S., Mubin, M., Rehman, K., Shahzad, S.M., Iqbal, S., Rizwan, M., Ali, S., Alyemeni, M.N., Wijaya, L., 2020. Biofilm forming rhizobacteria enhance growth and salt tolerance in sunflower plants by stimulating antioxidant enzymes activity. Plant Physiol. Biochem. 156, 242–256.

Yim, W., Seshadri, S., Kim, K., Lee, G., Sa, T., 2013. Ethylene emission and PR protein synthesis in ACC deaminase producing *Methylobacterium* spp. inoculated tomato plants (*Lycopersicon esculentum* Mill.) challenged with *Ralstonia solanacearum* under greenhouse conditions. Plant Physiol. Biochem. 67, 95–104.

Yolcu, S., Ozdemir, F., Güler, A., Bor, M., 2016. Histone acetylation influences the transcriptional activation of POX in *Beta vulgaris* L. and *Beta maritima* L. under salt stress. Plant Physiol. Biochem. 100, 37–46.

You, L.X., Zhang, R.R., Dai, J.X., Lin, Z.T., Li, Y.P., Herzberg, M., Zhang, J.L., Al-Wathnani, H., Zhang, C.K., Feng, R.W., Liu, H., Rensing, C., 2021. Potential of cadmium resistant *Burkholderia contaminans* strain ZCC in promoting growth of soy beans in the presence of cadmium. Ecotoxicol. Environ. Saf. 211, 111914.

Yue, Z., Chen, Y., Chen, C., Ma, K., Tian, E., Wang, Y., Liu, H., Sun, Z., 2021. Endophytic *Bacillus altitudinis* WR10 alleviates Cu toxicity in wheat by augmenting reactive oxygen species scavenging and phenylpropanoid biosynthesis. J. Hazard. Mater. 405, 124272.

Zhang, Q., Wang, C., 2020. Natural and human factors affect the distribution of soil heavy metal pollution: a review. Water Air Soil Pollut. 231 (7), 1–13.

Zhang, X., Zou, Z., Gong, P., Zhang, J., Ziaf, K., Li, H., Xiao, F., Ye, Z., 2011. Over-expression of microRNA169 confers enhanced drought tolerance to tomato. Biotechnol. Lett. 33 (2), 403–409.

Zhang, M., He, S., Zhan, Y., Qin, B., Jin, X., Wang, M., Zhang, Y., Hu, G., Teng, Z., Wu, Y., 2019. Exogenous melatonin reduces the inhibitory effect of osmotic stress on photosynthesis in soybean. PLoS One 14 (12), e0226542.

Zhang, H., Sun, X., Dai, M., 2021a. Improving crop drought resistance by plant growth regulators and rhizobacteria: mechanisms, applications and perspectives. Plant Commun., 100228.

Zhang, J., Xiao, Q., Wang, P., 2021b. Phosphate-solubilizing bacterium Burkholderia sp. strain N3 facilitates the regulation of gene expression and improves tomato seedling growth under cadmium stress. Ecotoxicol. Environ. Saf. 217, 112268.

Zhao, M.G., Chen, L., Zhang, L.L., Zhang, W.H., 2009. Nitric reductase-dependent nitric oxide production is involved in cold acclimation and freezing tolerance in *Arabidopsis*. Plant Physiol. 151 (2), 755–767.

CHAPTER 2

Recent advances in the application of microbial inoculants in the phytoremediation of xenobiotic compounds

Pankaj Bhatt[a], Parul Chaudhary[b], Sajjad Ahmad[c,d], Kalpana Bhatt[e], Dinesh Chandra[f], and Shaohua Chen[g]

[a]Department of Agricultural & Biological Engineering, Purdue University, West Lafayette, IN, United States, [b]Department of Microbiology, Govind Ballabh Pant University of Agriculture & Technology, Pantnagar, India, [c]Key Laboratory of Integrated Pest Management of Crop in South China, Ministry of Agriculture and Rural Affairs, South China Agricultural University, Guangzhou, P. R. China, [d]Key Laboratory of Natural Pesticide and Chemical Biology, Ministry of Education, South China Agricultural University, Guangzhou, P. R. China, [e]Department of Botany and Microbiology, Gurukula Kangri University, Haridwar, India, [f]GIC Chamtola, Almora, Uttarakhand, India, [g]State Key Laboratory for Conservation and Utilization of Subtropical Agro-bioresources, Guangdong Laboratory for Lingnan Modern Agriculture, Integrative Microbiology Research Centre, South China Agricultural University, Guangzhou, China

Chapter outline

Introduction

Toxic chemicals such as pesticides, antibiotics, steroids, and heavy metals are increasingly being released into the environment. Recently, the release of several contaminants as a result of various applications has been found to be increased in soil and water environments. These contaminants are considered as emerging contaminants (Bhatt et al., 2019) (Fig. 2.1). Toxic chemicals affect living systems and show various cellular alterations (Bhatt et al., 2020a). Therefore, the removal of these pollutants from the environment is of utmost

Unravelling Plant-Microbe Synergy. https://doi.org/10.1016/B978-0-323-99896-3.00013-8

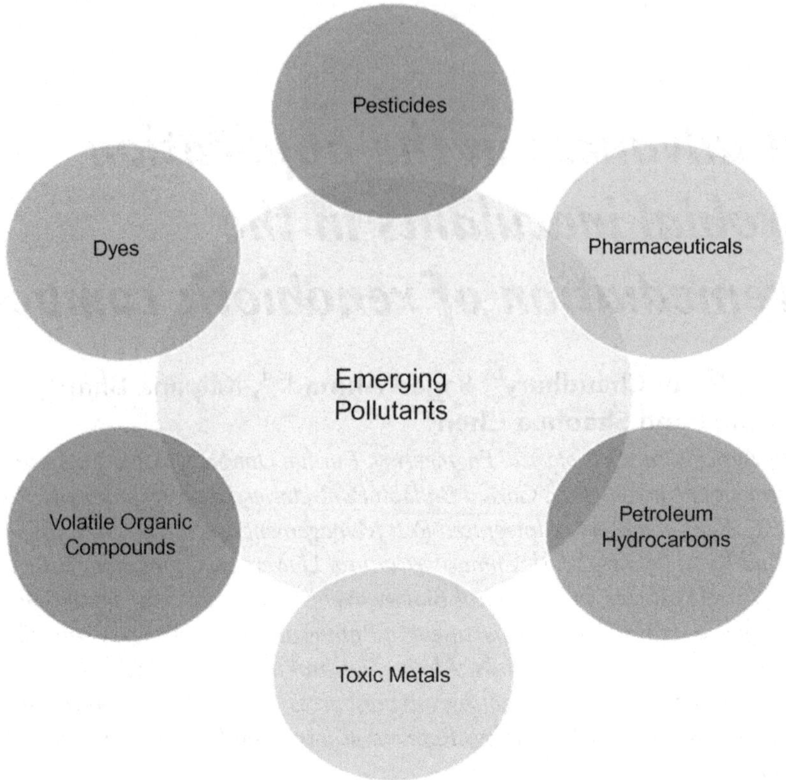

Fig. 2.1
Emerging pollutants accumulated in the environment.

importance in order to save the environment. The removal of such pollutants is possible via physicochemical, microbial, and phytoremediation-based methods. However, biological methods for the treatment of toxic pollutants are popular as they are environment-friendly (Bhatt et al., 2020b).

Phytoremediation is the method used for the removal of toxic chemicals using a combination of plants and associated microbes. The phytoremediation technology, in particular, practices the application of plants and their rhizospheric microbial floras that sequester organic and inorganic pollutants. Organic pollutants are degraded by the root zone of plants, followed by sequestration (Bhatt et al., 2022a). Organic pollutants that have been successfully phytoremediated include pesticides, organic solvents, explosives, hydrocarbons, polychlorinated biphenyls, and pharmaceuticals (Pilson-Smits, 2005). On the other hand, inorganic pollutants occur in natural environments on the earth's crust and increase due to human activities such as industrial, mining, agricultural, and military activities, thereby causing toxicity (Nriagu, 1979). Inorganic compounds such as metals do not get degraded but can be solubilized with the aid of microbes and plants (Horne, 2000).

Plants have been reported for the removal of Cr, Zn, Cu, Fe, Mo, Mn and Zn, Cd, F, Co, Hg, Se, Pb, W and V, ^{238}U, ^{137}Cs, and ^{90}Sr (Negri and Hinchman, 2000). Phytoremediation can be applied for soil, liquid, and gaseous substrates. Previously, researchers have explained the importance of phytoremediation for the removal of pollutants from soil- and water-contaminated sites. In the last two decades, phytoremediation has gained popularity in many countries for cleaning contaminated sites. The reason behind this popularity is its low cost. Developed countries focus more on removal of toxic chemicals and have a sanctioned budget for their removal from the environment (Pilson-Smits, 2005). Therefore, phytoremediation that is usually carried out in situ is cost-effective and may reduce exposure of toxic chemicals to living systems (Fig. 2.2).

The overall impact of phytoremediation on cleaning the environment is gaining significant attention. Therefore, the aim of this book chapter is to focus on the hidden areas of phytoremediation for the removal of toxic chemicals from the environment. The in-depth mechanisms of the plant–microbe-mediated removal of toxic chemicals have been well-discussed. Finally, new horizons including modern biological tools to treat the environment have also been discussed.

1. Phytoremediation of toxic pollutants
2. Phytoremediation consists of the application of plants and soil microbes for the removal of toxic chemicals from the environment. It is the most commonly accepted environmental restoration technology (Fig. 2.3).

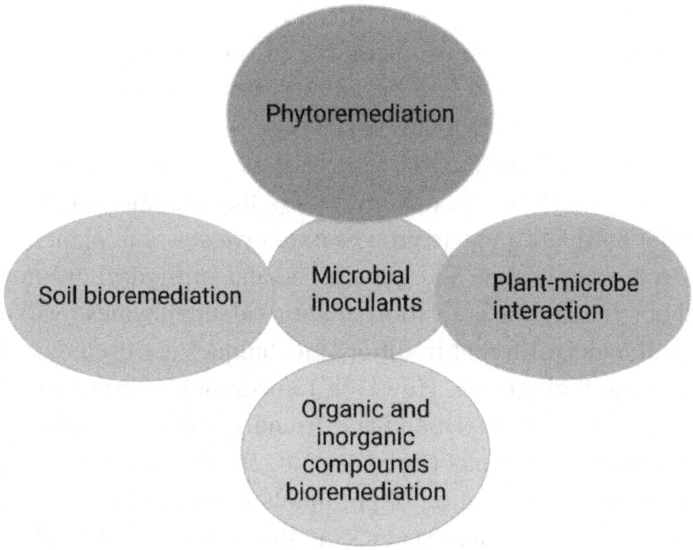

Fig. 2.2
Application of microbial inoculants in phytoremediation.

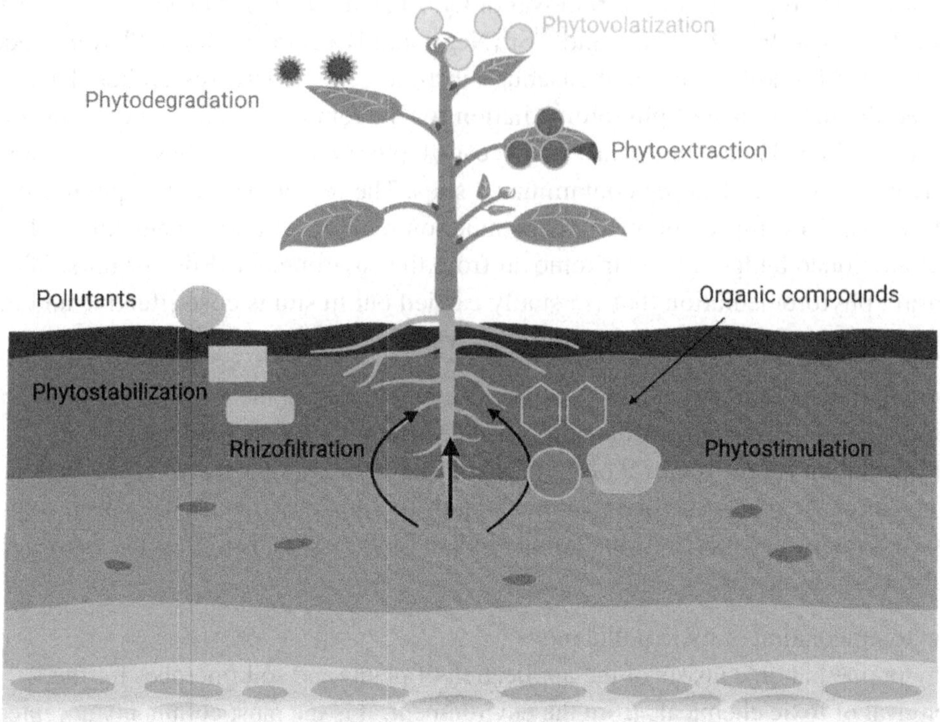

Fig. 2.3
Schematic view of the phytoremediation process.

Extensive use of organic and inorganic chemicals in various industries is the source of toxic pollutants that are released into the environment (Patra et al., 2020). The accumulation of these toxic pollutants in the agroecosystem causes hazardous diseases in aquatic and terrestrial organisms (Sarwar et al., 2010). The presence of toxic pollutants, especially heavy metals, in the environment leads to the alteration of the soil microbial species and essential nutrients and encourages oxidative stress in plants via metabolic changes such as decreasing proline and glutathione and sulfhydryl group coordination with proteins and obstructing the activation of antioxidant enzymes (Valko et al., 2005). To remove toxic pollutants, different traditional techniques are used, but these methods are time-consuming and high in cost. However, remediation of pollutants by plants is considered more effective, less costly, and eco-friendly and has the ability to restore heavy metals and other contaminants (Fulekar et al., 2009; Marques et al., 2009). Phytoremediation involves remediation of pollutants in soil and water environments through discriminating, capable plant species, as shown in Table 2.1. Therefore, after the harvesting of polluted biomass, it is converted into composites and is recycled for further use (Prasad and Freitas, 2003).

Table 2.1: Degradation of emerging pollutants by several typical plants.

Plant name	Name of the pollutant	Degradation efficiency (%)	References
Eichhornia crassipes	Tetracycline	90	Chen et al. (2012)
Water lettuce	Tetracycline	80	Chen et al. (2012)
Celery-planting aquatic system	Tetracycline	71.83	Liao et al. (2015)
Scindapsus aureus	Benzene	43-77	Treesubsuntorn and Thiravetyan (2012)
N. obliterata	Formaldehyde	90-100	Teiri et al. (2018)
Ficus elastica	Benzene	91	Chun et al. (2010)
Epipremnum aureum	Formaldehyde	96	Aydogan and Montoya (2011)
Litmus and aspen	Nitrate nitrogen and phosphorous	40-90 and 75-99	Sundaralingam and Gnanavelrajah (2014)
Zea mays	Atrazine	97	Bhatt et al. (2022b)
Medicago sativa	Polychlorinated biphenyls	43	Xu et al. (2010)
Festuca arundinacea	Lead, nickel, cadmium, and petroleum hydrocarbons	34, 23.8, 46.7, and 60.1	Steliga and Kluk (2020)
Myriophyllum aquaticum, *Spirodela oligorrhiza* L., and *Elodea Canadensis*	malathion, demeton-*S*-methyl, and crufomate	24-95	Gao et al. (2000)
Ricinus communis L.	Organochlorine pesticides	25-70	Rissato et al. (2015)
Plantago major L.	Cyanophos	94.7	Romeh (2014)
Plantago major L.	Imidacloprid	93.34	Romeh (2009)
Eichhornia crassipes	Ethion	69	Xia and Ma (2006)
Plantago major L.	Cyanophos	74.05	Romeh (2015)

The plant species that are involved in the remediation of toxic pollutants possess extensive biomass, rapidly grow in various conditions, contain highly fibrous root systems, are effortless in cultivation, and are extremely resistant to toxic pollutants (Patra et al., 2019). Generally, in phytoremediation, the widely used techniques are phytoextraction, rhizome filtration, and phytostabilization (Alkorta et al., 2004).

Phytoextraction

The severe threats posed by toxic pollutants in the environment lead to many kinds of diseases in living organisms, and thus, nowadays, their removal is gaining worldwide attention. To overcome all the issues, remediation of toxic pollutants using plants is considered a green technology, which also enhances the soil properties and are suitable socially and economically (Afonso et al., 2019). Recently, remediation of lead, which is considered a highly toxic metal as it is widely accumulated in the environment, has been carried out by adopting the phytoextraction technique. *Helianthus annuus* was selected with five different diversities, viz.,

DRSF-108, DRSF-113, LSFH171, Phule Bhaskar, and KBSH-44. The plantation was carried in pots and treated with different concentrations of lead (Pb). The results of this study revealed that after 2 months, the variety of Phule Bhaskar developed a significant ability to take up a high concentration of lead (693.69 mg/kg), whereas the lowest uptake was observed in KBSH-44 (333.16 mg/kg). Moreover, the highest accumulation of Pb residues was in the root system at a concentration of 394.32 mg/kg and then in shoots; in that duration, other findings observed that by applying a concentration of Pb, the ratios of proline and polyphenol were increased at a rate of 31.16 μmol/g and 7.15 mg/g, respectively (Chauhan et al., 2020).

In another study to remove Cd, Pb, and Zn toxic metals, a combined procedure of chemical washing and phytoextraction was carried out. First, the contaminated field was washed with $FeCl_3$ and citric acid with concentrations of 20 and 40 mmol/L, respectively. The plant species *Sedum plumbizincicola* was cultivated after applying organic fertilizers, viz., lime and sepiolite. It was reported that using $FeCl_3$ to wash the contaminated soil effectively removed the toxic metals Cd, Pb, and Zn by 35.2%, 24.3%, and 26.6%, respectively. Moreover, by the plantation of *S. plumbizincicola*, toxic metals in the polluted soil were removed at a percentage of 71, 34, and 47.7, respectively (Yu et al., 2020).

Rhizofiltration

This method involves the removal of toxic pollutants from aquatic and terrestrial environments through the adsorption process (Zhang et al., 2010). This technique is considered highly beneficial for the remediation process due to the transport of toxic pollutants to the root system and their further uptake by the aerial parts of the plants (Kvesitadze et al., 2015). Ahmad et al. (2018) revealed that the root consortium *Pseudomonas putida* is an excellent candidate for the management of pest infestation and plays a crucial role in the remediation of toxic pollutants.

In another study, two different experiments were carried to investigate the remediation of toxic pollutants by the plant root system. Two different plants, lettuces and rye grass, were cultivated in pots, and the soil was spiked with Pb with a concentration of 2000 mg/Kg for the root experiment, whereas, for the foliar experiment, both plants were kept near a smelter. The results demonstrated that Pb was significantly remediated at a concentration of 36 mg/kg in lettuce plants, whereas in the roots of rye grass, the concentration was 82 mg/kg. Moreover, the uptake of Pb by the aerial parts of the plants such as the shoot system was also recorded. The root system of lettuces and rye grass accumulated 171 and 700 mg/kg, respectively (Schreck et al., 2014). In another study, the removal of the toxic metal vanadium (V) by soybean roots was investigated. The high concentration of V enhances the bound of plant roots and shoots and their transformation dependent on the speciation of root tissues. The results showed that insoluble V compounds such as VO^{2+}, $V(OH)^{2+}$, and $V(OH)_3$ were taken up by plant roots and further translocation occurred from the roots to the shoot system (Yang et al., 2017).

Phytostabilization

Toxic pollutants are transported to the root system of plants through the adsorption or precipitation process (Lin et al., 2022). The phytostabilization method prevents the leaching of toxic substances into the soil–water system and also stops their entry into the food chains and food webs of the environment (Van Oosten and Maggio, 2015). Furthermore, this technique is also used to inhibit soil erosion, decrease the transportation of toxic substances by air, and increase soil microbial diversity (Boisson et al., 2016). To remove the toxic metal cadmium using the phytostabilization method, a halophyte plant *Acanthus ilicifolius* L. was cultivated in lab pots in a hydroponic culture environment. The results revealed that the root system of the halophyte plant significantly remediated the toxic pollutants effectively (96.4%) than did the aerial parts of the plant such as the stems with an uptake of only 1.4% and leaves with 0.6% (Shackira and Puthur, 2017).

In another study, removal of multiple toxic heavy metals such as zinc, manganese, cobalt, cadmium, lead, chromium, nickel, and cooper by the cultivation of the plant *Typha latifolia* was investigated. Root, shoot, and leaf samples were collected to examine the tolerance to heavy metals by plants. The results reported that total chlorophyll and free proline content revealed that the plant species shows great resistance against all toxic metals mainly in the root system (Varun et al., 2011). Recently, in another study, removal of lead and cadmium to prevent food contamination by phytoremediation has been investigated. An aromatic plant, *Helianthus petiolaris*, was cultivated in pots, and the soil was spiked with heavy metals with concentrations of 1000 and 50 mg/kg, respectively. The phytostabilization investigations revealed that the uptake of heavy metals greatly depended on various factors such as the microbial communities present in the rhizosphere and the adsorption of the plant's various parts. Moreover, the results showed that *H. petiolaris* exhibited great resistance against the heavy metals and more than three times the metals were accumulated in the plant root and shoot system as compared to those in the polluted soil (Saran et al., 2020). In the previous two decades, the phytoremediation technique has received extreme attention for its ability to remove toxic pollutants. Furthermore, the cultivation of plant species showed great resistance against hazardous acids, heavy metals, and other emerging toxic pollutants (Xie and van Zyl, 2020).

Rhizospheric microbes for pollutant degradation

Soil microbes play a significant role in the biogeochemical cycle, organic matter organization, mineralization, and degradation of pollutants (Chaudhary et al., 2021b,d). Pesticides, metals, petroleum, polycyclic aromatic hydrocarbons, and halogenated hydrocarbons cause stresses to human and ecosystem well-being (Bhatt et al., 2021; Zhang et al., 2022). Therefore, it is essential to remediate contaminated soils using eco-friendly technologies. Nowadays, bioremediation using bacteria and fungi is an effective method to degrade toxic compounds into simple ones (Kumar et al., 2021). Rhizoremediation is a green and cost-effective approach that involves the relationship between plants and root-associated microbes that facilitate the degradation process or removal of xenobiotic compounds through release of root exudates such as organic compounds, glucose, production of siderophores, HCN, and phytohormones (Chaudhary and Sharma, 2019; Agri et al., 2021; Singha and Pandey, 2021). These microbes also support the growth of plants and protect them from biotic and abiotic stress conditions (Chaudhary et al., 2021a,b). The microbial populations in the rhizosphere increase by two- to threefold as compared to those in the bulk soil and exhibit stronger metabolic capabilities.

Various rhizospheric microbes such as *Arthrobacter*, *Bacillus* sp., *Bacillus thuringiensis*, *Bacillus licheniformis*, *Bacillus pumilus*, *Gordonia sihwensis*, *Klebsiella pneumoniae*, *Paenibacillus*, *Pseudomonas putida*, and *Rhodococcus hoagii* degrade petroleum, xenobiotic compounds, diesel fuel, and crude oil from contaminated soils (Viesser et al., 2020). The consortium of *Achromobacter xylosoxidans*, *Ochrobactrum anthropic*, *R. ruber*, and *Stenotrophomonas maltophilia* isolated showed 85% degradation of PCBs. The association of *Arabidopsis thaliana* and *Lolium perenne* plants with *Pseudomonas* sp. significantly degrades total petroleum hydrocarbons. The *Pantoea* sp. used phenol and chromium as a carbon source and reduced Cr (VI) into the Cr (III) form. An association between the *Saccharum* sp. and *Candida* VITJzNo_4 is involved in lindane degradation (Salam et al., 2017). Lu et al. (2017) reported that glucose is significantly involved in pyrene removal by enhancing the population of *Mycobacterium*, which is dominant in pyrene degradation. *Streptomyces* sp. Z2 and *Streptomyces* sp. Z38 also degrade the lindane and chromium metal when grown with maize crops. A reduction in naphthalene, anthracene, and pyrene in oil-contaminated soils when treated with the *Amycolatopsis* sp. has been observed. Cyanide-containing compounds from contaminated soils were removed by *Sorghum bicolor* and the microorganisms present in its roots.

There are various enzymes, genes, degradative pathways, and operons, which are responsible for remediation of contaminated sites (Mishra et al., 2021a). Root exudates enhance the microbial populations of *Pseudomonas*, *Nocardioides*, and *Arthrobacter*, which contain the *nahAc* and *nidA* genes involved in polycyclic aromatic hydrocarbon degradation (Liao et al., 2021). Rhamnolipid, mannosylerythritol, cellobiose, and trehalose production by bacterial and fungal strains enhances the bioremediation of polluted sites (Mishra et al., 2021b). Laccases, lipases, cellulases, proteases, monooxygenases, dehydrogenases, ureases, FDA, and lignin peroxidase enzymes present in soil are also involved in the maintenance of soil health and biodegradation process (Kukreti et al., 2020; Kumari et al., 2021). Dehydrogenase is an intracellular enzyme used as an indicator in the dehydrogenation and transformation of organic pollutants, which are secreted by *Bacillus* sp., *Pseudomonas taiwanensis*, *Mycobacterium*, and *Pantoea agglomerans* (Khati et al., 2017;Kumari et al., 2020). Root exudates, especially carbohydrates, alter the rhizospheric bacterial population, which enhances the dehydrogenase activity and biodegradation process. Melatonin is used as a growth regulator for plants and increases the growth of *P. putida* and enhances the expression of salicylic aldehyde dehydrogenase and glutathione peroxidase enzyme activity, which is involved in PAH degradation (Rostami et al., 2021). Yoshikawa et al. (2017) reported that *Pseudomonas* sp. has the oxygenase enzyme, which converts dichloroethylene and vinyl chloride into CO_2 and chloride ions that are nontoxic compounds.

Conclusions and future perspectives

Researchers have focused on the biodegradation of toxic chemicals using indigenous microbial cultures. Less attention has been paid to plant-associated microbes for the removal of toxic chemicals from the environment. In this chapter, we have discussed how plants and their associated root bacteria are involved in the degradation of synthetic pollutants. More research is required to further investigate the degradation of toxic pollutants in the environment. The development of recent omics-based tools could be helpful in understanding the detailed molecular-level mechanism involved in the removal of toxic chemicals using the phytoremediation approach.

References

Afonso, T.F., Demarco, C.F., Pieniz, S., Camargo, F.A., Quadro, M.S., Andreazza, R., 2019. Potential of *Solanum viarum* Dunal in use for phytoremediation of heavy metals to mining areas, southern Brazil. Environ. Sci. Pollut. Res. 26 (23), 24132–24142.

Agri, U., Chaudhary, P., Sharma, A., 2021. In vitro compatibility evaluation of agriusable nanochitosan on beneficial plant growth-promoting rhizobacteria and maize plant. Natl. Acad. Sci. Lett., 1–5.

Ahmad, M., Pataczek, L., Hilger, T.H., Zahir, Z.A., Hussain, A., Rasche, F., Schafleitner, R., Solberg, S Ø, 2018. Perspectives of microbial inoculation for sustainable development and environmental management. Front. Microbiol. 9, 2992.

Alkorta, I., Hernández-Allica, J., Becerril, J.M., Amezaga, I., Albizu, I., Garbisu, C., 2004. Recent findings on the phytoremediation of soils contaminated with environmentally toxic heavy metals and metalloids such as zinc, cadmium, lead, and arsenic. Rev. Environ. Sci. Biotechnol. 3 (1), 71–90.

Aydogan, A., Montoya, L.D., 2011. Formaldehyde removal by common indoor plant species and various growing media. Atmos. Environ. 45 (16), 2675–2682.

Bhatt, P., Huang, Y., Zhan, H., Chen, S., 2019. Insight into microbial applications for the biodegradation of pyrethroid insecticides. Front. Microbiol. 10, 1778.

Bhatt, P., Bhatt, K., Huang, Y., Lin, Z., Chen, S., 2020a. Esterase is a powerful tool for the biodegradation of pyrethroid insecticides. Chemosphere 244, 125507.

Bhatt, P., Rene, E.R., Kumar, A.J., Zhang, W., Chen, S., 2020b. Binding interaction of allethrin with esterase: bioremediation potential and mechanism. Bioresour. Technol. 315, 123845.

Bhatt, P., Bhandari, G., Bhatt, K., Maithani, D., Mishra, S., Gangola, S., Bhatt, R., Huang, Y., Chen, S., 2021. Plasmid-mediated catabolism for the removal of xenobiotics from the environment. J. Hazard. Mater. 420, 126618.

Bhatt, P., Rene, E.R., Huang, Y., Wu, X., Zhou, Z., Li, J., Kumar, A.J., Sharma, A., Chen, S., 2022a. Indigenous bacterial consortium-mediated cypermethrin degradation in the presence of organic amendments and *Zea mays* plants. Environ. Res. 212, 113137.

Bhatt, P., Sethi, K., Gangola, S., Bhandari, G., Verma, A., Adnan, M., Singh, Y., Chaube, S., 2022b. Modeling and simulation of atrazine biodegradation in bacteria and its effect in other living systems. J. Biomol. Struct. Dyn. 40 (7), 3285–3295. https://doi.org/10.1080/07391102.2020.1846623.

Boisson, S., Le Stradic, S., Collignon, J., Séleck, M., Malaisse, F., Shutcha, M.N., Faucon, M.P., Mahy, G., 2016. Potential of copper-tolerant grasses to implement phytostabilisation strategies on polluted soils in South DR Congo. Environ. Sci. Pollut. Res. 23 (14), 13693–13705.

Chaudhary, P., Sharma, A., 2019. Response of nanogypsum on the performance of plant growth promotory bacteria recovered from nanocompound infested agriculture field. Environ. Ecol. 37 (1B), 363–372.

Chaudhary, P., Khati, P., Chaudhary, A., Gangola, S., Kumar, R., Sharma, A., 2021a. Bioinoculation using indigenous *Bacillus* spp. improves growth and yield of *Zea mays* under the influence of nanozeolite. 3 Biotech 11 (1), 1–12.

Chaudhary, P., Khati, P., Chaudhary, A., Maithani, D., Kumar, G., Sharma, A., 2021b. Cultivable and metagenomic approach to study the combined impact of nanogypsum and *Pseudomonas taiwanensis* on maize plant health and its rhizospheric microbiome. PLoS One 16 (4), e0250574.

Chaudhary, P., Sharma, A., Chaudhary, A., Khati, P., Gangola, S., Maithani, D., 2021d. Illumina based high throughput analysis of microbial diversity of maize rhizosphere treated with nanocompounds and *Bacillus* sp. Appl. Soil Ecol. 159, 103836.

Chauhan, P., Rajguru, A.B., Dudhe, M.Y., Mathur, J., 2020. Efficacy of lead (Pb) phytoextraction of five varieties of *Helianthus annuus* L. from contaminated soil. Environ. Technol. Innov. 18, 100718.

Chen, X.J., Li, F.Y., Hao, Y.B., 2012. The preliminary exploration of remediation the antibiotics polluted water by two hydrophytes. Subtrop. Plant Sci. 41 (4), 1–7.

Chun, S.C., Yoo, M.H., Moon, Y.S., Shin, M.H., Son, K.C., Chung, I.M., Kays, S.J., 2010. Effect of bacterial population from rhizosphere of various foliage plants on removal of indoor volatile organic compounds. Korean J. Hortic. Sci. Technol. 28 (3), 476–483.

Fulekar, M., Singh, A., Bhaduri, A.M., 2009. Genetic engineering strategies for enhancing phytoremediation of heavy metals. Afr. J. Biotechnol. 8, 529–535.

Gao, J., Garrison, A.W., Hoehamer, C., Mazur, C.S., Wolfe, N.L., 2000. Uptake and phytotransformation of organophosphorus pesticides by axenically cultivated aquatic plants. J. Agric. Food Chem. 48 (12), 6114–6120.

Horne, A.J., 2000. Phytoremediation by constructed wetlands. In: Terry, N., Bañuelos, G. (Eds.), Phytoremediation of Contaminated Soil and Water. Lewis, Boca Raton, pp. 13–40.

Khati, P., Gangola, S., Bhatt, P., Sharma, A., 2017. Nanochitosan induced growth of *Zea mays* with soil health maintenance. 3 Biotech 7 (1), 81.

Kukreti, B., Sharma, A., Chaudhary, P., Agri, U., Maithani, D., 2020. Influence of nanosilicon dioxide along with bioinoculants on. *Zea mays* and its rhizospheric soil. 3 Biotech 10, 345.

Kumar, A., Sharma, A., Chaudhary, P., Gangola, S., 2021. Chlorpyrifos degradation using binary fungal strains isolated from industrial waste soil. Biologia 76 (10), 3071–3080.

Kumari, S., Sharma, A., Chaudhary, P., Khati, P., 2020. Management of plant vigor and soil health using two agriusable nanocompounds and plant growth promotory rhizobacteria in fenugreek. 3 Biotech 10 (11), 1–11.

Kumari, H., Khati, P., Gangola, S., Chaudhary, P., Sharma, A., 2021. Performance of plant growth promotory rhizobacteria on maize and soil characteristics under the influence of TiO_2 nanoparticles. Pantnagar J. Res. 19, 28–39.

Kvesitadze, G., Khatisashvili, G., Sadunishvili, T., Kvesitadze, E., 2015. Plants for remediation: uptake, translocation and transformation of organic pollutants. In: In: Öztürk, M., Ashraf, M., Aksoy, A., Ahmad, M., Hakeem, K. (eds), Plants, Pollutants and Remediation. Springer, Dordrecht, pp. 241–308.

Liao, J., Xu, X., Liu, Y., Li, R., Liu, L., Liu, C., 2015. Removal and response of antibiotics and antibiotic resistance genes during advanced treatment of livestock wastewater by aquatic plant filter bed. Acta Sci. Circumst. 35 (8), 2464–2470.

Liao, Q., Liu, H., Lu, C., Liu, J., Waigi, M.G., Ling, W., 2021. Root exudates enhance the PAH degradation and degrading gene abundance in soils. Sci. Total Environ. 764, 144436.

Lin, Z., Pang, S., Zhou, Z., Wu, X., Li, J., Huang, Y., Zhang, W., Lei, Q., Bhatt, P., Sandhya, S., Chen, S., 2022. Novel pathway of acephate degradation by the microbial consortium ZQ01 and its potential for environmental bioremediation. J. Hazard. Mater. 426, 127841.

Lu, H., Sun, J., Zhu, L., 2017. The role of artificial root exudate components in facilitating the degradation of pyrene in soil. Sci. Rep. 7 (1), 1–10.

Marques, A.P., Rangel, A.O., Castro, P.M., 2009. Remediation of heavy metal contaminated soils: phytoremediation as a potentially promising clean-up technology. Crit. Rev. Environ. Sci. Technol. 39 (8), 622–654.

Mishra, S., Lin, Z., Pang, S., Zhang, Y., Bhatt, P., Chen, S., 2021a. Biosurfactant is a powerful tool for the bioremediation of heavy metals from contaminated soils. J. Hazard. Mater. 418, 126253.

Mishra, S., Pang, S., Zhang, W., Lin, Z., Bhatt, P., Chen, S., 2021b. Insights into the microbial degradation and biochemical mechanisms of carbamates. Chemosphere 279, 130500.

Negri, M.C., Hinchman, R.R., 2000. The Use of Plants for the Treatment of Radionuclides. John Wiley & Sons, Inc, New York, pp. 107–132.

Nriagu, J.O., 1979. Global inventory of natural and anthropogenic emissions of trace metals to the atmosphere. Nature 279 (5712), 409–411.

Patra, D.K., Pradhan, C., Patra, H.K., 2019. Chromium bioaccumulation, oxidative stress metabolism and oil content in lemon grass *Cymbopogon flexuosus* (Nees ex Steud.) W. Watson grown in chromium rich over burden soil of Sukinda chromite mine, India. Chemosphere 218, 1082–1088.

Patra, D.K., Pradhan, C., Patra, H.K., 2020. Toxic metal decontamination by phytoremediation approach: concept, challenges, opportunities and future perspectives. Environ. Technol. Innov. 18, 100672.

Pilson-Smits, E., 2005. Phytoremediation. Annu. Rev. Plant Biol. 56, 15–39.

Prasad, M.N.V., Freitas, H., 2003. Metal hyperaccumulation in plants-biodiversity prospecting for phytoremediation technology. Electron. J. Biotechnol. 6, 275–321.

Rissato, S.R., Galhiane, M.S., Fernandes, J.R., Gerenutti, M., Gomes, H.M., Ribeiro, R., de Almeida, M.V., 2015. Evaluation of *Ricinus communis* L. for the phytoremediation of polluted soil with organochlorine pesticides. Biomed. Res. Int. 2015, 1–8.

Romeh, A.A., 2009. Phytoremediation of water and soil contaminated with imidacloprid pesticide by plantago major, L. Int. J. Phytoremediation 12 (2), 188–199.

Romeh, A.A., 2014. Phytoremediation of cyanophos insecticide by plantago major L. in water. J. Environ. Health Sci. Eng. 12 (1), 1–8.

Romeh, A.A., 2015. Enhancing agents for phytoremediation of soil contaminated by cyanophos. Ecotoxicol. Environ. Saf. 117, 124–131.

Rostami, S., Azhdarpoor, A., Baghapour, M.A., Dehghani, M., Samaei, M.R., Jaskulak, M., Jafarpour, S., Samare-Najaf, M., 2021. The effects of exogenous application of melatonin on the degradation of polycyclic aromatic hydrocarbons in the rhizosphere of Festuca. Environ. Pollut. 274, 116559.

Salam, J.A., Hatha, M.A., Das, N., 2017. Microbial-enhanced lindane removal by sugarcane (*Saccharum officinarum*) in doped soil-applications in phytoremediation and bioaugmentation. J. Environ. Manag. 193, 394–399.

Saran, A., Fernandez, L., Cora, F., Savio, M., Thijs, S., Vangronsveld, J., Merini, L.J., 2020. Phytostabilization of Pb and Cd polluted soils using *Helianthus petiolaris* as pioneer aromatic plant species. Int. J. Phytoremediation 22 (5), 459–467.

Sarwar, N., Malhi, S.S., Zia, M.H., Naeem, A., Bibi, S., Farid, G., 2010. Role of mineral nutrition in minimizing cadmium accumulation by plants. J. Sci. Food Agric. 90 (6), 925–937.

Schreck, E., Dappe, V., Sarret, G., Sobanska, S., Nowak, D., Nowak, J., Stefaniak, E.A., Magnin, V., Ranieri, V., Dumat, C., 2014. Foliar or root exposures to smelter particles: consequences for lead compartmentalization and speciation in plant leaves. Sci. Total Environ. 476, 667–676.

Shackira, A.M., Puthur, J.T., 2017. Enhanced phytostabilization of cadmium by a halophyte—*Acanthus ilicifolius* L. Int. J. Phytoremediation 19 (4), 319–326.

Singha, L.P., Pandey, P., 2021. Rhizosphere assisted bioengineering approaches for the mitigation of petroleum hydrocarbons contamination in soil. Crit. Rev. Biotechnol., 1–26.

Steliga, T., Kluk, D., 2020. Application of Festuca arundinacea in phytoremediation of soils contaminated with Pb, Ni, Cd and petroleum hydrocarbons. Ecotoxicol. Environ. Saf. 194, 110409. https://doi.org/10.1016/j.ecoenv.2020.110409.

Sundaralingam, T., Gnanavelrajah, N., 2014. Phytoremediation potential of selected plants for nitrate and phosphorus from ground water. Int. J. Phytoremediation 16 (3), 275–284.

Teiri, H., Pourzamzni, H., Hajizadeh, Y., 2018. Phytoremediation of formaldehyde from indoor environment by ornamental plants: An approach to promote occupants health. Int. J. Preven. Med. 9 (70), 1–7.

Treesubsuntorn, C., Thiravetyan, P., 2012. Removal of benzene from indoor air by *Dracaena sanderiana*: effect of wax and stomata. Atmos. Environ. 57, 317–321.

Valko, M., Morris, H., Cronin, M.T.D., 2005. Metals, toxicity and oxidative stress. Curr. Med. Chem. 12 (10), 1161–1208.

Van Oosten, M.J., Maggio, A., 2015. Functional biology of halophytes in the phytoremediation of heavy metal contaminated soils. Environ. Exp. Bot. 111, 135–146.

Varun, M., D'Souza, R., Pratas, J., Paul, M.S., 2011. Evaluation of phytostabilization, a green technology to remove heavy metals from industrial sludge using *Typha latifolia* L. Biotechnol. Bioinformatics Bioeng. 1 (1), 137–145.

Viesser, J.A., Sugai-Guerios, M.H., Malucelli, L.C., Pincerati, M.R., Karp, S.G., Maranho, L.T., 2020. Petroleum-tolerant rhizospheric bacteria: isolation, characterization and bioremediation potential. Sci. Rep. 10, 2060.

Xia, H., Ma, X., 2006. Phytoremediation of ethion by water hyacinth (*Eichhornia crassipes*) from water. Bioresour. Technol. 97 (8), 1050–1054.

Xie, L., van Zyl, D., 2020. Distinguishing reclamation, revegetation and phytoremediation, and the importance of geochemical processes in the reclamation of sulfidic mine tailings: a review. Chemosphere 252, 126446.

Xu, L., Teng, Y., Li, Z.G., Norton, J.M., Luo, Y.M., 2010. Enhanced removal of polychlorinated biphenyls from alfalfa rhizosphere soil in a field study: the impact of a rhizobial inoculum. Sci. Total Environ. 408 (5), 1007–1013.

Yang, J., Wang, M., Jia, Y., Gou, M., Zeyer, J., 2017. Toxicity of vanadium in soil on soybean at different growth stages. Environ. Pollut. 231, 48–58.

Yoshikawa, M., Zhang, M., Toyota, K., 2017. Biodegradation of volatile organic compounds and their effects on biodegradability under co-existing conditions. Microbes Environ., ME16188.

Yu, G., Jiang, P., Fu, X., Liu, J., Sunahara, GI., Chen, Z., Xiao, H., Lin, F., Wang, X., 2020. Phytoextraction of cadmium-contaminated soil by Celosia argentea Linn.: A long-term field study. Environ. Pollu., 115408. https://doi.org/10.1016/j.envpol.2020.115408.

Zhang, S., Chen, M., Li, T., Xu, X., Deng, L., 2010. A newly found cadmium accumulator—*Malva sinensis* Cavan. J. Hazard. Mater. 173 (1–3), 705–709.

Zhang, W., Li, J., Zhang, Y., Wu, X., Zhou, Z., Huang, Y., Zhao, Y., Mishra, S., Bhatt, P., Chen, S., 2022. Characterization of a novel glyphosate-degrading bacterial species, *Chryseobacterium* sp. Y16C, and evaluation of its effects on microbial communities in glyphosate-contaminated soil. J. Hazard. Mater. 432, 128689.

CHAPTER 3

Multifaceted roles of root exudates in light of plant-microbe interaction

Sayanta Mondal[a], Krishnendu Pramanik[b], Priyanka Pal[a], Soumik Mitra[a], Sudip Kumar Ghosh[a], Tanushree Mondal[a], Tithi Soren[a], and Tushar Kanti Maiti[a]
[a]Microbiology Laboratory, Department of Botany, The University of Burdwan, Purba Bardhaman, West Bengal, India, [b]Mycology and Plant Pathology Laboratory, Department of Botany, Siksha Bhavana, Visva-Bharati, Santiniketan, West Bengal, India

Chapter outline

Unravelling Plant-Microbe Synergy. https://doi.org/10.1016/B978-0-323-99896-3.00003-5

Introduction

Soil, often regarded as the "black box," sustains a wide variety of macro and microorganisms forming a definitive community structure, and a part of it, i.e., the rhizosphere, is the primary point of contact between the soil and plant roots. Two Greek words, "rhiza" (meaning root) and "sphere" (meaning a field or an area of influence), are combined to form the term "rhizosphere," first coined by Hiltner in 1904 to define a plant-root consortium (Hartmann et al., 2008). The rhizosphere acts as a communication hub between roots, microorganisms, invertebrates, and even with the neighboring competitor plants. The whole rhizosphere is further subdivided into two major areas: (i) the endorhizosphere, which mainly encompasses root hairs, epidermis, and root cortical layers and (ii) the ectorhizosphere, which is the root adjacent soil and spans up to 5 mm in length (Vishwakarma et al., 2017). However, according to few other authors, the rhizospheric zone is subdivided into three areas, viz., the endorhizosphere (consisting of the endodermis and cortex), the ectorhizosphere (soil particles near the root), and the rhizoplane (consisting of the epidermis, cortex, and mucilage) (Badri and Vivanco, 2009; Bashir et al., 2016). Additionally, few other layers are also found such as the mycorrhizosphere (the mycorrhizal association point), rhizosheath (a strongly adhering dense layer comprising root hairs, soil grains, soil microbes, and mucoid layers), and bulk soil. However, the bulk soil is not considered as an integral part of the rhizosphere (Bashir et al., 2016). A diagrammatic representation of the different zones in the rhizosphere, site of root exudation, and plant-microbe interaction in the soil-root ecosystem is shown in Fig. 3.1. It is assumed that the rhizosphere may comprehend nearly 10^6–10^9 bacteria, 10^4 protozoa, 10^1–10^2 nematodes, and 10^5–10^6 fungi per gram rhizospheric soil (el Zahar Haichar et al., 2014). The overall genome size of the rhizosphere is significantly greater than those of plants. Therefore, the rhizosphere can be viewed as a repository of complex physiological, biochemical, and ecological processes that can be further pursued for sumptuous revealing.

The preliminary role of roots is to provide anchorage to the plant system, to facilitate water and nutrient absorption, and to further their conduction to the above-ground portions. Moreover, the root can also act as a storage organ as in beet and carrot or as a respiratory organ as in mangrove plants. Besides, plant roots exude a wide range of organic chemical compounds renowned as root exudates, which play multidimensional roles that include defense against pathogenic microorganisms, mobilization of nutrients, modulation of soil properties, acting as a chemoattractant and/or chemorepellent, establishing symbioses, stress tolerance, stabilization of soil aggregates, growth suppression of competitor plants, and

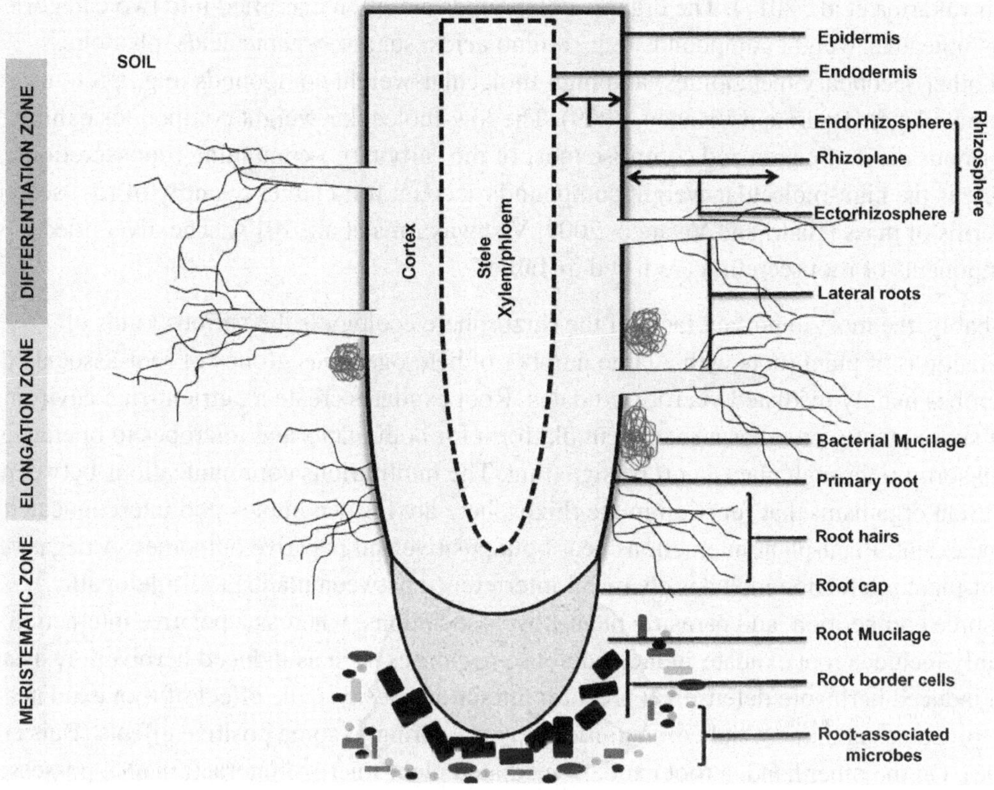

Fig. 3.1
Diagrammatic representation of the different zones in the rhizosphere, site of root exudation, and plant-microbe interaction in the soil-root ecosystem.

maintenance of soil moisture to name a few (Hayat et al., 2017; Narula et al., 2012). The oozing out of root exudates in enormous quantities into the rhizosphere may be subjected to root pressure, and exudation rates may depend upon the plant species or environmental conditions (Hayat et al., 2017). The quality and quantity of root exudates have multiple determinants like plant species, age of an individual plant, and biotic and abiotic stress factors. The presence of root-specific metabolites in root exudates plays an influential role in structuring soil microbial communities and imparts significant ecological importance (Badri and Vivanco, 2009). Approximately, 11%–27% of the net fixed C allocated to roots is secreted in the rhizosphere, and the percentage of exudation is strictly dependent upon the plant species, age of the plant species, and nutritional circumstances (Bais et al., 2006; Jones et al., 2009; Vives-Peris et al., 2020).

Plant roots incessantly ooze a wide range of chemical substances into the rhizosphere like different ions, O_2, CO_2, H_2, ethylene, water, enzymes, mucilage, border cells, secondary metabolites, and a huge number of organic compounds (el Zahar Haichar et al., 2014;

Vishwakarma et al., 2017). The organic compounds are often classified into two categories: low-molecular-weight compounds (e.g., amino acids, sugars, organic acids, phenolics, and other secondary metabolites) and high-molecular-weight compounds (e.g., proteins and mucilage) (Badri and Vivanco, 2009). The low-molecular-weight compounds exhibit enormous diversification and comprise most of the carbon (C)-containing root secretions, whereas the high-molecular-weight compounds account for a larger quantity of root secretion in terms of mass (Badri and Vivanco, 2009; Vishwakarma et al., 2017). The diversified components of root secretion are listed in Table 3.1.

Probably, the most important facet of the rhizosphere ecology is the various kinds of interactions of plant roots with a large number of heterogeneous groups of root-associated microbes mainly mediated via root exudates. Root exudates create a nutrient-rich environment that subsequently provides an excellent platform for both plants and microbes to operate as a consortium through plant-microbe signaling. The multifarious communications between different organisms that function in the rhizosphere are fairly complex and interconnected to some extent. Plant-plant interactions bear both positive and negative outcomes. A negative plant-plant interaction includes chemical interference between plants, i.e., allelopathy, resource competition, and parasitic plant-host associations, whereas a positive interaction mainly includes root exudate-induced defense responses such as induced herbivore resistance and induced herbivore defense via predator attraction. Besides, the effect of root exudates on soil processes and microbial communities may also bring in some positive effects (Bais et al., 2006). On the other hand, a root exudate-mediated plant-microbe interaction also possesses both positive and negative aspects. Root nodulation, mycorrhizal associations, endophytes, plant growth-promoting rhizobacteria (PGPR), and biotic stress tolerance through biocontrol activity fall under the positive plant-microbe interactions, whereas antimicrobial activity, biofilm inhibition, quorum-sensing mimicking, and effect on nematodes are counted as negative plant-microbe interactions (Bais et al., 2008; Vishwakarma et al., 2017). In this chapter, we will discuss the rhizodeposition concept along with the chemical nature of root exudates, the root exudation process, and various factors affecting it. The major focus of this chapter will be on understanding the multidimensional role of root exudates in plant-microbe communications. Besides, the role of root exudates in shaping microbial communities will also be discussed.

Chapter review methodology

In this chapter, a total of 246 articles with the keywords “root exudate,” “plant-microbe interaction,” “rhizosphere,” and “stress” were reviewed on the Scopus database during the period 1996–2022. VOSviewer software was used to conduct the relevant keyword analysis with a minimum keyword co-occurrence of three, and a network map of keywords was generated (Fig. 3.2). Different color variations in the network map indicate the type of cluster, and the size of the bubble presents the occurrence frequency of the individual keyword,

Table 3.1: Chemical diversity of root exudates.

Class of root exudate compounds	Examples
Carbohydrates	Arabinose, deoxyribose, glucose, galactose, fructose, maltose, mannose, mannitol, oligosaccharides, pentose, rhamnose, raffinose, ribose, sucrose, and xylose
Amino acids	α-Alanine, β-alanine, γ-aminobutyric acid, α-aminoadipic acid, arginine, asparagine, aspartic acid, citrulline, cystathionine, cysteine, cystine, deoxymugineic acid, 3-epihydroxymugineic acid, glutamine, glutamic acid, glycine, histidine, homoserine, isoleucine, leucine, lysine, L-hydroxyproline, methionine, mugineic acid, ornithine, phenylalanine, proline, serine, threonine, tryptophan, tyrosine, and valine
Organic acids	Acetic, aconitic, ascorbic, aldonic, L-aspartic, benzoic, butanoic, butyric, caffeic, chorismic, citric, *p*-coumaric, erythronic, ferulic, formic, fumaric, gallic, glutaric, glutamic, glycolic, glyoxilic, isocitric, lactic, malic, malonic, oxalacetic, oxalic, *p*-hydroxybenzoic, piscidic, propionic, protocatechuic, pyruvic, rosmarinic, salicylic, shikimic, sinapic, succinic, syringic, tartaric, tetronic, *trans*-cinnamic, valeric, and vanillic acids
Flavonols	Naringenin, kaempferol, quercetin, myricetin, naringin, rutin, genistein, strigolactone, and their substitutes with sugars
Lignins	Catechol, benzoic acid, nicotinic acid, phloroglucinol, cinnamic acid, gallic acid, ferulic acid, syringic acid, sinapoyl aldehyde, chlorogenic acid, coumaric acid, vanillin, sinapyl alcohol, quinic acid, and pyroglutamic acid
Coumarins	Umbelliferone
Aurones	Benzyl aurones synapates and sinapoyl choline
Glucosinolates	Cyclobrassinone, desuphoguconapin, desulphoprogoitrin, desulphonapoleiferin, and desulphoglucoalyssin
Anthocyanins	Cyanidin, delphinidin, pelargonidin, and their substitutes with sugar molecules
Indole compounds	Indole-3-acetic acid, brassitin, sinalexin, brassilexin, methyl indole carboxylate, and camalexin glucoside
Fatty acids	Linoleic, linolenic, oleic, palmitic, and stearic acids
Sterols	Campesterol, cholesterol, sitosterol, and stigmasterol
Allomones	Juglone, sorgoleone, 5,7,4′-trihydroxy-3′,5′-dimethoxyflavone, DIMBOA, and DIBOA
Proteins and enzymes	PR proteins, amylase, β-1,3-glucanases, hydrolases, invertase, lectins, lipase, proteases, acid/alkaline phosphatases, peroxidases, phenolase, and polygalacturonase
Purines	Adenine, guanine, and uridine/cytidine
Inorganic ions and gases	HCO_3^-, OH^-, H_2, CO_2, and H_2
Phenolics	Liquiritigenin, luteolin, daidzein (4′,7-dihydroxyflavanone), genistein (4′,5,7-trihydroxyisoflavone), coumetrol (4,4′-dihydroxy-2′-methoxychalcone), eriodictyol (4′-7-dihydroxyflavone),3,5,7,3′-tetrahydroxy-4′-methoxyflavone, naringenin, isoliquiritigenin, 7,3′-dihydroxy-4′-methoxyflavone, umbelliferone, and (+)- and (−)-catechin
Growth factors and vitamins	*p*-Amino benzoic acid, biotin, choline, inositol, *N*-methyl nicotinic acid, niacin, pathothenic acid, pantothenate, pyridoxine riboflavin, strigolactones, and thiamine
Flavonoids	Chalcone, coumarin, flavones, flavonols, flavanones, flavonones, and isoflavones

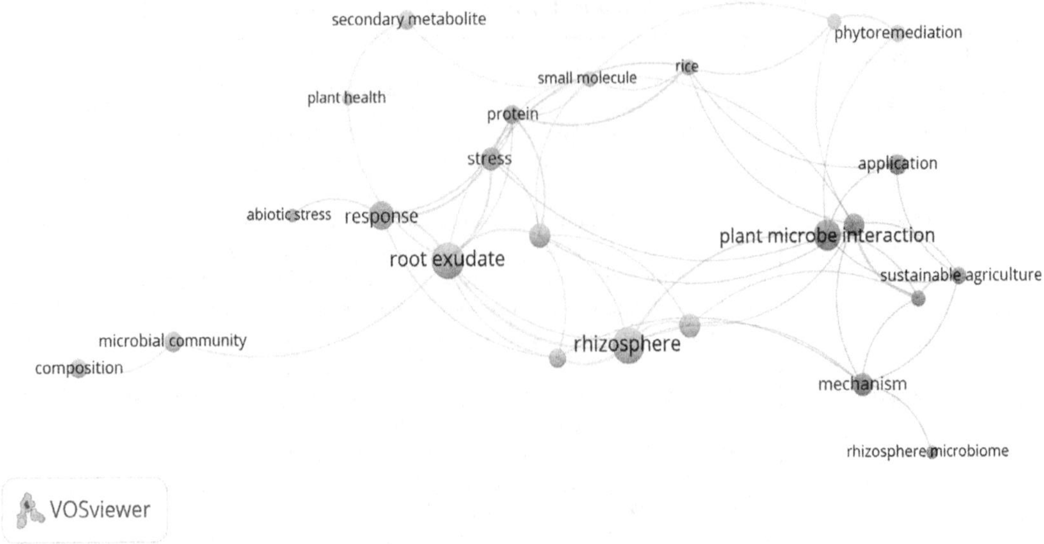

Fig. 3.2
Keyword co-occurrence analysis showed as a network map view using VOSviewer software.

whereas the line among the bubbles indicates the links. From this analysis, it can be observed that (1) root exudates are important components of a plant-microbe interaction; (2) biotic and abiotic stresses play pivotal roles in the secretion of plant root exudates; (3) root exudates have a crucial role in forming the composition of a microbial community in the rhizosphere; and (4) it is also connected with the phytoremediation strategy and sustainable agriculture. Hence, the network map analysis unquestionably pointed out the important research hotspots related to this chapter (Fig. 3.2). However, only 94 articles are screened on the "most-relevant" basis and discussed in this chapter.

Root exudates: Natural rhizodeposits of plants

Plant roots are known to modify their surrounding soil environment, a phenomenon commonly known as the rhizosphere effect. The initial stimulator of the rhizosphere effect is still not clear. However, the subsequent release of C from the roots into the surrounding soil is believed to be the major inducer of the so-called rhizosphere effect. Roots can also infuse a large amount of inorganic C into the rhizosphere, which can alter the soil biogeochemistry (Jones et al., 2009). However, the organic C released by them into the rhizosphere has the most profound role in altering the physical, chemical, and biological characteristics of the rhizospheric soil, which, in turn, helps in differentiating it from the bulk soil (el Zahar Haichar et al., 2014; Jones et al., 2009). The magnitude of the difference, however, depends on the quantity and type of C released from the roots of a particular plant species as well as on the intrinsic soil characteristics (el Zahar Haichar et al., 2014).

The term "rhizodeposition" is often used in a broad sense that primarily denotes the release of organic C into the rhizosphere (Jones et al., 2009). The process of rhizodeposition is indeed comprised of a wide array of processes whose sole purpose is to incorporate C into the soil. The processes include the loss of root cap cells and root border cells, death and lysis of the root cells (such as the cortex, root hairs, etc.), loss of C to root-associated symbionts, gaseous losses (such as CO_2, H_2, ethylene, etc.), emanation of solutes from living cells, i.e., root exudates, and insoluble polymer (mucilage) secretion from living cells. These processes are well-established and theoretically differentiable. However, at the experimental level, these loss pathways are extremely hard to distinguish from one another both in time and in space (Jones et al., 2009). Different nomenclatures for rhizodeposition have been adapted for different aspects such as the mechanism of release, biochemical nature, or functions of rhizodeposits in the rhizosphere (Jones et al., 2009).

Root exudates

A root exudate is the principal source of soil organic C released by plant roots and is a part of the rhizodeposition process. Root exudates are described as low-molecular-weight compounds, which are lost through passive diffusion and upon which plants exert little or no control at all (Bais et al., 2006). The rate of root exudate loss is somewhat higher in the root tip portion than in the mature root regions. The components of root exudates are comprised of dominant organic compounds essential for cell metabolism, which include amino acids, sugars, organic acids, and some secondary metabolites (Jones et al., 2009). However, the amount of primary metabolites such as carbohydrates, amino acids, and organic acids secreted is much higher than that of secondary metabolites like auxins, flavonoids, glucosinolates, etc. (Badri and Vivanco, 2009). Metabolites have been previously identified and subsequently quantified in many plants like soybean, rice, *Arabidopsis*, and common bean (Vives-Peris et al., 2020). A detailed component analysis of root exudates is displayed in Table 3.1. The relative concentration of root exudate components is much higher inside the roots than in the surrounding soils due to their continual removal by the microbial community and subsequent replenishment of internal pools by plants (Jones et al., 2009).

Border cells

These are metabolically active cells produced in large numbers and are programmed to detach from the root cap periphery to the external environment (Stubbs et al., 2004). They are also termed "sloughed-off cells" separated from the outer layer of the root cap and are continuously regenerated. The daily rate of border cell production is highly variable among different plant species and is dependent on external environmental conditions (Jones et al., 2009). After detachment, border cells remain active for several days (Stubbs et al., 2004). They are often surrounded by the secreted mucilage and provide protection against heavy metal toxicity. Border cells are also viewed as signaling components involved in protecting the meristem against pathogens and promoters of symbiosis. However, conflicting results have also emerged due to the inconvenience of manipulating border cell release and physiology for disease control (Jones et al., 2009).

Mucilage

The mucilage released from plants forms a gelatinous sheath around the root tips and is one of the most prominent visible signs of organic C secretion (Jones et al., 2009). The components of the mucilage include polysaccharides (10^6–10^8 Da in size), proteins, and phospholipids. The mucilage provides primary protection against heavy metal toxicity. Besides, it enhances soil aggregate stability, which, in turn, promotes soil aeration and root growth, restrains soil erosion, and maintains continuous water flow toward the rhizoplane (el Zahar Haichar et al., 2014). The in vitro synthesis of the mucilage ranges anywhere between 11 and 47 μg dm/mg root dm (Nguyen, 2003).

Gaseous components

In addition to the above-mentioned elements, plant roots also release various gaseous components such as CO_2, H_2, ethylene, etc. Due to carbohydrate respiration, plant roots release CO_2 into the soil environment, which is further stimulated by a plant and bacterial exudate molecule lumichrome (el Zahar Haichar et al., 2014). The overall CO_2 accumulation reaches up to 17.5% in the root periphery, and it can increase the overall $CaCO_3$ solubilization, which, in turn, can produce Ca^{2+} readily available for plant uptake (Dakora and Phillips, 2002). In legumes, H_2 is generated as a by-product of N_2 fixation (Dong and Layzell, 2001). Some rhizobia possess the uptake hydrogenase (*Hup*) gene, which induces H_2 oxidation to yield more energy. However, the symbioses that lack the *Hup* gene (Hup−) bear a disadvantage and therefore the H_2 produced by nitrogenase diffuses into the soil after releasing from the nodule (Dong and Layzell, 2001; Golding et al., 2012). The diffused H_2 may alter soil biogeochemistry and indirectly influence plant growth. The presence of H_2 induces H_2-oxidizing bacteria that further promote plant growth via different mechanisms such as root elongation by reducing ethylene levels in host plants (Golding et al., 2012).

In this chapter, root exudates are taken into account in terms of rhizodeposition that includes the secretion of all active and passive compounds from roots.

Factors affecting the release of root exudates

The process of root exudation is highly dependent upon several factors. These factors are broadly classified into three categories: physical factors, chemical factors, and biological factors (Vives-Peris et al., 2020).

Physical factors

Physical factors include different light intensities or different types of lights such as UV, water availability (excessive, i.e., flooding and/or scarcity, i.e., droughts), extreme temperatures (high and low), and osmotic stress to name a few. The role of light in the root exudation process is significant since a large portion of photosynthetically fixed CO_2 is allocated to root exudates. The wavelength of light, as well as the photoperiod, is influential in this regard. For example, far-IR-enriched irradiation affects the exudate composition in wild oats (*Avena fatua*) (Pomilio et al., 2000). Longer photoperiods were found to enhance the exudation of organic acids like benzoic and 4-hydroxy-benzoic acid in Japanese cucumber (*Cucumis sativus*) (Pramanik et al., 2000). The exudation of flavonoids such as catechol and catechin was also found to be affected by illumination (maximum exudation after 6h of light) in spotted knapweed (*Centaurea stoebe*) (Tharayil and Triebwasser, 2010). Alterations in root exudation, as well as reduction of beneficial microorganisms from the rhizosphere of sea grass (*Halophila ovalis*, *Halodule uninervis*, and *Cymodocea serrulata*), were reported upon low light availability (Martin et al., 2018). These studies effectively establish that photoperiods and light wavelengths affect root exudation.

Environmental temperature markedly affects processes like photosynthesis, translocation, and respiration in plants. Therefore, it subsequently manages to affect root exudates qualitatively and quantitatively. An increment in the day/night temperature in *C. sativus* from 25/20 to 30°C/25°C was found to enhance the secretion of some organic acids like benzoic, 4-hydroxy-benzoic, phthalic, and palmitic acids (Pramanik et al., 2000). Exudation of amino acids like asparagine was also found to be elevated with increasing temperature. However, this phenomenon is not universal (Shukla et al., 2011). Another important physical determinant of the root exudate composition is the water status of plants. Both excessive and scarcity of water content can increase organic C in root exudates. An increase of 71% cumulative total organic C due to drought and an increase of 45% total organic C exuded per gram dry plant due to flooding were reported in crested wheatgrass (*Agropyron cristatum*). Besides, gas chromatography-mass spectrometry (GC-MS) analysis revealed the presence of malic, fumaric, malonic, succinic, and oxalic acids in the exudates, with malic acid being the most predominant organic acid exudate (Henry et al., 2007). Osmotic stress, mainly induced by salinity, was also found to affect root exudation. Root exudates from soybean (*Glycine max*) and common bean (*Phaseolus vulgaris*) were found to excrete numerous flavonoids (naringenin, isoliquiritigenin, quercetin, umbelliferone, 7′,4-dihydroxyflavone, hesperetin, etc.) with increasing NaCl concentrations. The qualitative secretion patterns of such flavonoids were found to affected by NaCl (Dardanelli et al., 2010, 2012).

Chemical factors

Chemical factors include nutrient levels, phytohormones like salicylic acid (SA), methyl jasmonate (MeJa), and trace elements (e.g., $AgNO_3$, Cl–, Al^{3+}, etc.). The most well-explored chemical determinant of root exudates is the nutrient levels in the rhizosphere. A scarcity of N, P, K, or Fe was found to promote an increased release of sugars, amino acids, and several organic acids, e.g., glucose, glutamate, ribitol, and citrate (from Fe-deficient plants), and c-aminobutyric acid and carbohydrates (from P-deficient plants) in axenically grown maize (*Zea mays*) plants. Lower amounts of amino acids in root exudates were observed in N-deficient plants, whereas fewer carbohydrates (like sucrose, maltose, ribitol, glycerol, etc.) in root exudates were observed in K-deprived plants. Therefore, it can be hypothesized that the release of root exudates such as sugars, amino acids, and organic acids might act as an adaptive strategy to cope with nutrient-limiting conditions (Carvalhais et al., 2011). The secretion of organic compounds like citramalic acid and salicylic acid has been reported in P-deficient conditions in sugar beet. These metabolites may promote soil P solubilization (Khorassani et al., 2011). CO_2 concentration also effectively modifies root exudation (Jones et al., 2009). Different trace elements (such as heavy metals) in the form of toxic ions can also affect root exudates. In *Arabidopsis*, a higher secretion of malate and citrate was reported due to Al toxicity-induced activation of the AtMATE and AtALMT1 transporters (Liu et al., 2009). The presence of Cd in the rhizosphere was found to influence the root exudate composition of a Cd hyperaccumulator *Sedum alfredii*. The GC-MS analysis further revealed secondary metabolites like trehalose, erythritol, naphthalene, d-pinitol, and n-octacosane involved in Cd stabilization in the soil, whereas phosphoric acid, threonic acid, tetradecanoic acid, oxalic acid, and glycine were assumed to be involved in Cd mobilization in the soil (Luo et al., 2014). Besides, 15 compounds were identified upon Pb stress on the same plant species. The compounds include L-alanine, L-proline, oxalic acid, glyceric acid, 2-hydroxyacetic acid, 4-methylphenol, 2-methoxyphenol, and diethylene glycol. These compounds may stabilize Pb in the soil and can be used as potential biomarkers (Luo et al., 2017). Hydroponically grown sorghum and maize supplemented with $5\,mg\,L^{-1}$ Cd were found to secrete malate and citrate, respectively. It is also believed that organic acid secretion in the form of root exudates might have a role in reducing Cd bioavailability in the rhizosphere. The Cd-organic acid complex may enhance the stabilization of free Cd ions in the rhizosphere (Pinto et al., 2008). In a recent study, it has been revealed that rhizodeposition is likely to be affected by a decrease in total carbon, sugars, and amino acids with increasing Cd content (10, 20, and 40 μM Cd in nutrient solution) in maize plants. However, protein secretion was found to be unaltered and secretion of organic acids was found to be increased. The binding ability of many of these molecules with Cd results in Cd tolerance in maize plants (Lapie et al., 2019). Defense signaling molecules such as SA, MeJA, and nitric oxide (NO) also influence root exudate composition. Exogenous application of these compounds results in the accumulation of different secondary metabolites (Zhao et al., 2005).

Biological factors

Biological factors include different types of microorganisms (fungi, bacteria, etc.), neighboring plants, plant species, plant ecophysiology, genotype, root age, root architecture, etc. The amount, types, and quantity of compounds in root exudates vary among different plant species. Exudation rates are also variable in nature and are controlled by many factors like plant developmental stage and genotypes within the same species (Badri and Vivanco, 2009). It is speculated that plants at the seedling stage produce the lowest amount of root exudates. This amount is subsequently increased until flowering and then a decline is observed in mature plants (Aulakh et al., 2001). A positive correlation between root exudation and root growth can also be observed, in which active root growth assures more root exudation (Lucas García et al., 2001). Variations in root exudate components in two *Arabidopsis thaliana* ecotypes Ler and Col were also observed (Hoekenga et al., 2003). Root age was also found to be a decisive factor in exudate composition. More amino acids and sugars in exude composition were reported in the early days of plant growth in peas and in oats. Many other exudate components like 3-pyrazolylalalanine (in *C. sativus* L.) and tyrosine (in tomato and *Capsicum annuum* L.) also showed a similar pattern (Shukla et al., 2011).

Microorganisms have been found to influence root cell permeability, root metabolism, absorption, and secretion of certain compounds in root exudates. Several reports suggest that fungal, bacterial, and some antibiotic (such as penicillin, polymyxin) filtrates affect root exudation. However, it is improbable to conclude the actual effects or the magnitude of the effects of microorganisms on root exudates based on laboratory-based experiments. This is because in real-world scenarios, the rhizosphere supposedly sustains a wide variety of microorganisms and all of them might not be root exudate-promoting microorganisms. Besides, the microorganisms that reside in the immediate vicinity of the roots are the ones that might affect the root exudation processes (Shukla et al., 2011). In addition, root branching and root architecture may also play a determining role in root exudate composition qualitatively as well as quantitatively (Badri and Vivanco, 2009).

The mechanism of root exudation

A variety of transport pathways are utilized by plants to secrete root exudates in the rhizosphere. Generally, the process of root exudation is considered to be passive in nature and various pathways including diffusion through the root membrane, ionic channels, and vesicular transport are used. The nature of the chemical compounds to be secreted is the key factor for the selection of the exact secretion process. Low-molecular-weight compounds like sugars, carboxylic acids, amino acids, and phenolics are secreted through diffusion as a result of a differential concentration gradient between the rhizosphere and the root cell cytoplasm. This process is controlled by root membrane permeability, the integrity of root cells, and the polarity of the exudates (Badri and Vivanco, 2009). The transport of carbohydrates and specific carboxylates like malate and oxalate across the membrane is mediated by ion channels. Two anionic channels are important in this regard, namely, the SLow Anion Channels (SLACs) or the Slow-type (S-type) channels, which require several seconds to be activated, and the QUick Anion Channels (QUACs) or the Rapid-type (R-type) channels, which require only a few milliseconds (Dreyer et al., 2012). A well-studied representative of the latter type is aluminum-activated malate transporters or ALMTs. These consist of several proteins and are involved in the exudation of malate in the presence of Al^{3+} and play a significant role in Al tolerance in plants. Besides, the MATE family of active transporters (independently activated) is found to be responsible for citrate secretion during Al stress (Liu et al., 2009). High-molecular-weight metabolites are usually stored in vesicles and are secreted through vesicle transport (exocytosis) (Badri and Vivanco, 2009).

Root exudation in plants is also achieved through active transport via several root plasma membrane proteins. Two families of membrane transporters are pivotal in this context, viz., ATP-binding cassette (ABC) transporters and multidrug and toxic compound extrusion (MATE) transporters. In *A. thaliana*, ABC transporter knockout mutants *Atpdr6*, *Atpdr2*, *Atmrp2*, *Atath6*, and *Atpgp4-1* revealed the involvement of ABC transporters in the root exudation process (Badri et al., 2008). The involvement of MATE transporters in root exudation has been reported in different plants including rice (*OsFRDL4*), barley (*HvAACT1*), *Arabidopsis* (*AtMATE1*), and sorghum (*SbMATE1*). Besides, phenolic compound transporters in rice (e.g., *OsPEZ1* and *OsPEZ2*) have also been identified (Vives-Peris et al., 2020).

The role of root exudates in plant-microbe communication

Root exudates play an important role in maintaining the communication between plants and their root-associated microbes. Root exudates are comprised of a wide range of chemical compounds that act as signaling molecules that help mediate plant-microbe interactions in the rhizosphere. This interaction is primarily mediated by root exudates and has a positive influence on plant growth through a wide array of mechanisms, which include an enhanced supply of nutrients, fixation of atmospheric nitrogen, and increased stress tolerance. On the other hand, root exudates can also govern some negative interactions like secretion of antimicrobials, biofilm inhibition, and mimicking of quorum-sensing signaling. Apart from this, root exudates can modulate tri-trophic interactions between host plants, soil microbes, and nematodes. Several well-described positive and negative plant-microbe interactions that demonstrate the importance of root exudates (either directly or indirectly) along with root exudate-mediated tri-trophic interactions are discussed below.

Positive interactions: Root colonization and stress tolerance

Nitrogen-fixing symbionts

The symbiotic relationship between Gram-negative nitrogen-fixing bacteria and legumes is at the heart of plant-microbe interactions. Symbiotic bacteria include *Rhizobium*, *Bradyrhizobium*, *Sinorhizobium*, *Mesorhizobium*, and *Azorhizobium* to name a few. Plants form specialized structures commonly known as nodules within their roots to sustain these symbiotic bacteria. These structures allow symbiotic bacteria to access the carbohydrates and organic acids secreted by the plants. On the other hand, plants can access the nitrogen fixed by these symbionts within the root nodules. The secretion of several flavonoids through root exudation and its subsequent perception by bacteria initiate the establishment between plant roots and their nitrogen-fixing symbionts. Root exudates promote the expression of bacterial *nod* genes, which is essential for nodule formation. The most significant flavonoids and isoflavonoids that induce bacterial *nod* genes are flavones (luteolin, apigenin), flavanones (naringenin, hesperetin), and isoflavones (daidzein, genistein) (Narula et al., 2012). The flavonoids secreted by legumes are host-specific and help rhizobia distinguish their host plants from other legumes. For example, daidzein and genistein produced by *G. max* specifically induce *Bradyrhizobium japonicum nod* gene expression but inhibit *Sinorhizobium meliloti nod* gene expression. The *nod* genes of *S. meliloti* are induced by luteolin instead. Apart from *nod* gene expression, specific flavonoids also regulate bacterial chemotaxis. Plant flavonoids are perceived as aglycones and interact with the *nodD* gene product NodD (LysR-type regulator), which allows it to bind the *nod* box element in the promoter region of *nod* genes. The coordinated expression of *nod* genes leads to the synthesis of Nod factor molecules, which is ultimately recognized by the host plant (Bais et al., 2006). Flavonoids are viewed as the key elements that regulate nodule development through their auxin transport inhibition activity that subsequently causes local auxin accumulation at the nodulation initiation site, which ultimately leads to the formation of nodule primordia (el Zahar Haichar et al., 2014).

Several nonflavonoid molecules secreted by roots such as trigonelline, stachydrine, xanthone, vanillin, isovanillin, aldonic, erythronic, and tetronic acid are also found to elicit *nod* gene expression (el Zahar Haichar et al., 2014). On the other hand, bacterial lipo-oligosaccharides are found to elicit the expression of flavonoid biosynthesis genes in plants in a reciprocal manner (Bais et al., 2008). Bacterial infection-induced symbiotic root nodule formation comprises several steps including root hair curling, formation of infection thread followed by intussusception of bacteria, and, finally, nodule development. Another kind of symbiotic association is observed between the Gram-positive Actinobacteria genera *Frankia* and those of root tree species of eight dicot families. This association is also known as “actinorhizal symbiosis” (Vishwakarma et al., 2017). The first few steps of this process are similar to legume/*Rhizobium* symbiosis. The initial establishment is mediated via host-derived phenolic compounds such as flavanones, isoflavanones, cinnamic, benzoic, and hydroxybenzoic acids (el Zahar Haichar et al., 2014).

Mycorrhizal associations

The mutualistic relationship between mycorrhizal fungi and host plant roots is probably the best example of a symbiotic relationship in which the host roots provide organic nutrients to their mycorrhizal partner through exudation, whereas the host plant receives protection against pathogens. Just like rhizobial recognition, mycorrhizal fungi are also able to recognize their compatible host plant through root exudate detection because of their common ancestral origin (Bais et al., 2008). Root exudates are speculated to play the main communicative role in establishing mycorrhizal associations. However, the identification of definitive molecules in root exudates that influence mycorrhizal association has remained unclear. A recent development in this regard has suggested the involvement of sesquiterpene (5-deoxy-strigol, a strigolactone) in hyphal branching in dormant mycorrhizal fungi (Akiyama et al., 2005). The positive influence of root exudates in mycorrhizal growth provides plant protection against pathogens through growth suppression of the pathogenic species (Bais et al., 2008). More than 80% of land plants were found to establish an association with arbuscular mycorrhizal fungi (AMF) (Akiyama et al., 2005). AMF are obligate symbionts and are not able to complete their life cycle without a host root. After penetrating plant roots, AMF form highly specialized cross-web structures commonly known as “arbuscules.” These specialized structures serve as a primary site for nutrient exchange between the host root and AMF. The reason behind these cross-web-like branches is an inducing factor that acts as a signaling molecule that elicits hyphal morphogenesis, leading to successful root colonization. This branch-inducing factor is found in the root exudates of all mycorrhizal host plants. One of the vital steps in AMF development is the extraradical hyphae, mainly induced by a signaling molecule exuded by plant roots. This signaling molecule is called “strigolactones” (SLs) and is now considered as a phytohormone (Koltai, 2011). SLs are carotenoid-derived terpenoid lactones whose primary biosynthesis site is the root (Matusova et al., 2005). SLs play an instrumental role in recognizing and responding to low phosphate conditions (el Zahar Haichar et al., 2014).

Endophytic associations

Many plants are also found to contain nonpathogenic fungal or bacterial species internally, which are either beneficial or nondetrimental to plants and are commonly known as “endophytes.” Plant species that support endophytes impart better tolerance against several biotic and abiotic stresses (Bais et al., 2008). Besides, the presence of endophytes can affect the exudation of host plants, altering the secretion of phenolics, which further contributes to increased tolerance of mineral deficiencies (Malinowski and Belesky, 2000). A detailed chemical communication between endophytic associations and host roots is yet to be uncovered.

Plant-PGPR interactions

PGPR are a unique group of soil bacteria that generally inhabit around/on the root surface and promote plant growth directly or indirectly. The mechanism of growth promotion through PGPR is fascinating, and it involves a complex signaling cascade between the host plant and its root-associated PGPR. The characteristic features of PGPR include plant growth-promoting ability, root colonization efficiency, survival, multiplication, and competition with other microbiota while expressing some or all of their plant growth-promoting activities. The mechanism of action of PGPR is broadly classified into two categories, viz., direct mechanisms, which include nutrient acquisition (nitrogen, phosphorus, and other essential minerals) or modulating phytohormone levels, and indirect mechanisms, which include biocontrol activities (Ahemad and Kibret, 2014). Some examples of PGPR include *Agrobacterium*, *Arthrobacter*, *Azospirillum*, *Bacillus*, *Burkholderia*, *Chromobacterium*, *Enterobacter*, *Erwinia*, *Flavobacterium*, *Klebsiella*, *Micrococcus*, *Pseudomonas*, and *Serratia* to name a few. PGPR hold the key to plant growth promotion in normal conditions as well as in stressed (biotic and abiotic) conditions and therefore are considered as the key players in reducing global dependency on hazardous agrochemicals. Exudation of organic compounds from roots helps assemble microbes in the rhizosphere. Moreover, plant-microbe interactions mediated by root exudates have been found to be critical in root colonization and biocontrol activities. Root exudates are also capable of modifying the biotic and abiotic factors of the root zone. Therefore, it can regulate the activity of different PGPR genera, which might respond differentially upon sensing the root exudate of a specific plant species (Hassan et al., 2019).

The chemotactic response of bacteria to root exudates may serve as the initial communication between plants and microbes. The chemotactic responses of many bacterial genera including *Pseudomonas*, *Rhizobium*, and *Agrobacterium* have been found to increase their root colonization efficiency. Several bacteria were found to have a positive chemotactic movement toward different chemicals exuded by plants, which include sugars, amino acids, succinate, malate, fumarate, shikimate, quinate, protocatechuate, vanillate, acetosyringone, gallate, catechol, and luteolin (Brencic and Winans, 2005). Secondary metabolites like benzoxazinoids present in the root exudates of maize were found to attract *Pseudomonas putida* (Neal et al., 2012). Moreover, studies with *S. meliloti* revealed the role of certain flavonoids as chemoattractants and their role in promoting bacterial movement toward roots (el Zahar Haichar et al., 2014). The flagellar motility of some rhizobacteria is also found to be regulated by root exudates. The *cheA* mutants of four strains of the well-known PGPR *Pseudomonas fluorescens* have shown reduced root colonization efficiency than have the wild-type. The bacterial major outer membrane protein (MOMP) plays a crucial role in the early detection of host plants. The MOMP of *Azospirillum brasilense* Cd has been experimentally found to bind to the membrane-immobilized root extracts of different plants with different affinities. Besides, the MOMP from *A. brasilense* Cd showed strong adhesion to extracts of cereals when compared to extracts of legumes and tomatoes. It is assumed that the bacterial MOMP may function as an adhesin and may be involved in root adsorption and cell aggregation (Burdman et al., 2001).

Biotic stress tolerance: Biocontrol

It has been already discussed that many indigenous beneficial soil microorganisms can protect plants against different kinds of pathogens. Mycorrhizal symbiosis is one of the prime examples of the biocontrol activity of rhizospheric microorganisms. Many bacterial genera like *Agrobacterium*, *Azotobacter*, *Bacillus*, *Enterobacter*, *Erwinia*, *Pseudomonas* (fluorescent strains), *Serratia*, *Streptomyces*, etc. are also capable of protecting plants against pathogens. However, the exact mechanism of the biocontrol activity of these microorganisms is extremely hard to determine. They do so by either killing or suppressing pathogens or by competitive inhibition. Plants themselves secrete a wide range of chemical compounds that act as antimicrobials. These antimicrobial compounds are mostly broad-spectrum and thus are considered among root exudate-mediated negative interactions.

Abiotic stress tolerance: Bioremediation

Rhizospheric bacteria are well-known for their bioremediation effects as they have an immense potential for eliminating and/or restricting environmental contaminants to host plants. In return, the host plants provide them with nutrients in the form of root exudates. The utilization of root exudates for efficient growth of root-colonizing bacteria and subsequent degradation of contaminants has been established in many works. For example, in a study, the bacterium *P. putida* PCL1444 was isolated from the grass cv. Barmultra root rhizosphere. The screening was based on its root-colonizing efficiency and ability to grow on the pollutant naphthalene. It was found that the strain effectively utilizes root exudates for its growth and the high transcription of the naphthalene catabolic genes. A sixfold higher expression of naphthalene degradation pathway genes was observed in the early bacterial log phase with two different carbon sources, glucose and succinic acid (two major root exudate components found in grass roots) (Kuiper et al., 2002).

In a separate study, the Cu-binding capacity of the root exudates of cultivated plants (*Triticum aestivum* and *Brassica napus*) and wheat-associated weeds (*Matricaria inodora* and *Centaurea cyanus*) were estimated in vitro hydroponic and sterile conditions. The results suggest that the nature and amount of organic root exudates determine the extent of Cu complexation (Dousset et al., 2001). A glasshouse experiment using the root-bag technique with root exudates and rhizospheric Zn fractions, at different soil Zn concentrations to elucidate Zn accumulation in two ryegrass cultivars (*Lolium perenne* L. cvs. Airs and Tede), revealed that different root exudate compositions and rhizospheric Zn fractions were the primary reason for differential Zn accumulation in two genotypes (Wei-Hong et al., 2007). A chemotactic study of polycyclic aromatic hydrocarbon (PAH)-degrading rhizospheric bacteria (*Pseudomonas alcaligenes* 8A, *P. putida* 10D, and *P. stutzeri* 9A) demonstrated the relevance of chemotaxis in PAH-degrading bacteria. Therefore, it can be said that root exudate-mediated chemoattraction might play a role in improving bioavailability and enhancing degradation of PAHs in the rhizosphere (Ortega-Calvo et al., 2003). A decrease in the phenanthrene (a PAH)-degrading activity of *P. putida* after exposure to root exudates and root extractions was reported. However, the ability of *P. putida* to grow on root extracts and root exudates as the sole carbon and energy source may compensate for the partial repression because larger microbial populations patronize faster degradation rates (Rentz et al., 2004). On the other hand, high-molecular-weight PAHs do not serve as carbon and energy sources for rhizospheric microbes during their degradation. Rather, a microbial population primarily depends upon root products for carbon and energy during high-molecular-weight PAH degradation as found in the case of benzo[*a*]pyrene removal by *Sphingomonas yanoikuyae* (Rentz et al., 2005). A study conducted on sorghum root exudates revealed oxidase, peroxidase, and tyrosinase activities in phenanthrene-contaminated soil conditions. Interestingly, the activities were found to be enhanced with increased soil phenanthrene levels. The results are indicative of the involvement of root-secreted oxidoreductases in rhizospheric degradation of PAHs and/or their derivatives (Muratova et al., 2009).

The role of root exudates in heavy metal tolerance in plants has also been explored by many researchers. The modulation and proliferation of root exudation may assist plants in tolerating a high level of metal toxicity by altering soil pH, chelating toxic metal ions, and assimilating essential phytonutrients. The exudation of different phytochemicals can stop heavy metal mobilization and can reduce the risk of leaching. Besides, root exudates favor the colonization of beneficial microbiota along with an increase in plant biomass by increasing soil organic matter. In this manner, root exudates can contribute to phytoremediation and/or microbe-assisted bioremediation of heavy metal-contaminated soils. The secretion of root exudates like low-molecular-weight organic acids (e.g., malic acid, citric acid, oxalic acid, tartaric acid, acetic acid, glycolic acid, formic acid, lactic acid, succinic acid, fumaric acid, oleic acid, maleic acid, threonic acid, tetradecanoic acid, phosphoric acid, 3-hydroxybutanoic acid, etc.), phytosiderophores, phenols (cinnamic acid derivatives, m-coumaric acid derivatives, p-coumaric acid derivatives, vanillic acid, apigenin, ferulic acid, cinnamic acid, catechin, quercetin, gallic acid, luteolin, benzoic acid, chlorogenic acid, caffeic acid, sinapic acid, salicylic acid, p-hydroxybenzoic acid, syringic acid, ellagic acid, etc.), amino acids (tyrosine, methionine, cysteine, isoleucine, arginine, asparagine, lysine, proline, hydroxylysine, glutamic acid, histidine, L-valine, L-alanine, L-serine, L-glycine, etc.), proteins, and mucilage can modulate heavy metal mobilization/immobilization in soils and/or plants. Low-molecular-weight organic acids/siderophores (phytosiderophores and/or bacterial siderophores) may form complexation with heavy metals. The heavy metal-chelant stable complex can immobilize heavy metal ions in the rhizosphere or can enhance mobilization of less toxic forms of heavy metals into plant tissues accordingly.

Negative interactions: Root exudate-mediated antagonistic activities

Secretion of antimicrobials

Plants are capable of synthesizing a wide range of chemical compounds known as secondary metabolites. The number of secondary metabolites synthesized is estimated to be approximately 12,000, and, of these, somewhere around 2% are found in root exudates (Bais et al., 2008). In most cases, these exudated metabolites serve as defense molecules against pathogenic microorganisms, insects, and herbivores. It has already been discussed that root exudates also serve as a nutrient source for rhizospheric microorganisms, and a positive correlation between root exudates and microbial activity in the rhizosphere can be drawn. Research studies on *Arabidopsis*, rice, corn, soybean, and *Medicago truncatula* reveal that plants are rich sources of a wide array of antimicrobial compounds like indoles, terpenoids, benzoxazine, and flavonoids/isoflavonoids. These defense chemicals may be specific for a group of taxa (isoflavonoids in Fabaceae, sesquiterpenes in Solanaceae) or universal across taxa (e.g., phenylpropanoids). However, most antimicrobials secreted by plants are broad-spectrum and specificity is often determined by the detoxification capacity of the pathogen of that particular antimicrobial compound. Many research studies in the last few decades have conferred the antimicrobial properties of root exudates. Rosmarinic acid (a caffeic acid ester) was identified in the root secretion of *Ocimum basilicum*, which was found to be elicited by the cell wall extracts of *Phytophthora cinnamomi* (Bais et al., 2002).

Besides, *O. basilicum* roots were also induced to exude rosmarinic acid by in situ fungal infection by *Pythium ultimum*. Rosmarinic acid also showed antimicrobial activity against several soil-borne microbes including the opportunistic plant pathogen *Pseudomonas aeruginosa* (Bais et al., 2002). Barley (*Hordeum vulgare*) was found to secrete defense root exudates upon infection by the soil-borne pathogen *Fusarium graminearum*. Liquid chromatography with photodiode array detection (LC-DAD) confirmed the presence of t-cinnamic, p-coumaric, ferulic, syringic, and vanillic acids among these defense root exudates (Lanoue et al., 2010). Similarly, in *Lithospermum erythrorhizon* hairy roots, cell-specific production of the pigmented naphthoquinone derivatives of shikonin was reported upon elicitation. These derivatives showed antimicrobial activity against *Rhizoctonia solani*, *Pythium aphanidermatum*, and *Nectria haematococca* (Brigham et al., 1999). Constitutive production of antimicrobial compounds in the root secretions of gladiolus showed resistance against *Fusarium oxysporum f.* sp. *gladioli* (Taddei et al., 2002). These studies suggest that the battle against pathogens begins in the rhizosphere even before the pathogens come into contact with the host roots. Plants can produce both constitutive and inducible defense metabolites. For instance, rhizathalene-A (diterpene) was found to be constitutively produced and released by noninfected *Arabidopsis* (Vaughan et al., 2013). The inducible metabolites are often localized inside cells upon a pathogen attack. Interestingly, intracellular antimicrobial secretions in root cells are quite different from those of antimicrobials in root exudates (Walker et al., 2003). The role of root exudates is important in defining the pathogenicity of plant-microbe interactions (Bais et al., 2006). Plant roots also secret defense proteins upon pathogenic infection. For example, the *A. thaliana-Pseudomonas syringae* (DC3000) interaction elicits several plant defense-related proteins (De-la-Peña et al., 2010). The hairy roots of pokeweed (*Phytolacca americana*) were found to secrete various defense-related proteins including PAP-H (a ribosome-inactivating protein or RIP). The in vitro *N*-glycosidase activity of PAP-H can be lethal against pathogenic fungal ribosomes. The PAP-H activity may be facilitated by PR proteins such as chitinase, protease, and β-1,3-glucanase (Park et al., 2002). The enzymes chitinase and β-1,3-glucanase are widely distributed in higher plants and are well-known for their antifungal activity.

Biofilm inhibition

A biofilm is a complex structure that is comprised of different bacterial colonies or a single type of cell on its surface and is composed of extracellular polymeric substances (EPSs), nucleic acids, and proteins (Mondal et al., 2021). Several soil bacteria tend to form these microaggregates while colonizing the root surface. Examples of such bacterial genera include *Acinetobacter*, *Azospirillum*, *Bacillus*, *Enterobacter*, and *Pseudomonas* to name a few (Bais et al., 2008; Mondal et al., 2021). Many of these bacterial genera act as potent PGPR. Studies on bacterial root colonization patterns explain that bacteria form such microaggregates primarily at the site of root exudation (Rovira et al., 1974). It has already been discussed that the secretion of antimicrobials from roots may hamper the bacterial QS signaling and thereby the biofilm-forming potential. The synthesis of exopolysaccharides or other exopolymers may increase the chances of survival of the bacteria by enhancing biofilm formation. This might help in active root colonization by bacteria in adverse conditions. The antimicrobials secreted from roots could easily be quenched in these types of rhizospheric biofilms, leading to loss of activity of antimicrobials. On the contrary, if these antimicrobials can diffuse in a controlled manner to avoid the barrier of the biofilms or develop the competency to destroy the biofilm matrix, then antimicrobial activity could be highly intensive and way more effective. A study on *P. aeruginosa*-plant root interaction revealed that rosmarinic acid (a multifunctional caffeic acid ester) from *O. basilicum* inhibited the planktonic form of the opportunistic pathogen *P. aeruginosa*. However, it failed to penetrate the biofilm formed by *P. aeruginosa* on the root surfaces of *O. basilicum* (Walker et al., 2004). However, other researchers subsequently discovered the biofilm inhibitors of *P. aeruginosa* from human mucous secretions (lactoferrin) as well as from garlic extracts (Bjarnsholt et al., 2005; Singh et al., 2002). The root exudates of *O. basilicum* can be further screened to isolate and characterize potential biofilm inhibitors during the *P. aeruginosa*-*O. basilicum* interaction. The effectiveness of usual antimicrobials used against *P. aeruginosa* can be further increased using these biofilm inhibitory compounds. More research should be carried out to explore biofilm inhibitors against plant pathogenic bacteria and also the regulatory mechanisms of biofilm formation in plant pathogens.

Quorum-sensing mimics

Quorum sensing (QS) is the ability of bacteria to detect and respond in a cell density-dependent manner through gene expression. The regulatory system is particularly important for virulence in bacterial pathogens. The QS regulatory system has been reported in both Gram-positive and Gram-negative bacteria. Both groups include some well-known plant pathogenic genera such as *Agrobacterium*, *Erwinia*, and *Pseudomonas*. In Gram-negative bacteria, the QS system is comprised of an autoinducer (produces a free diffusible molecule) and a receptor/transcriptional activator protein, which checks the concentration of the autoinducer. With an increase in bacterial population, the level of the autoinducer is also increased in the environment (Newton and Fray, 2004). The QS signals released by these bacteria can be mimicked by plants and red algae through the secretion of structurally similar molecules. Halogenated furanones are the most studied QS mimicking compounds and are secreted by the marine red alga *Delisea pulchra*. The furanones produced by this red alga show structural similarity to α-homoserine lactones (AHLs). AHLs function as autoinducers of Gram-negative bacteria unlike peptide autoinducers of Gram-positive bacteria and are perceived to be the most common QS signals of Gram-negative bacteria. *Delisea*-derived furanones have been found to inhibit AHL-regulated behaviors in bacteria (Givskov et al., 1996). In angiosperms like rice, pea, tomato, and *M. truncatula*, root exudates were found to affect QS signaling in bacteria (Teplitski et al., 2000). The re-establishment of pathogenicity by AHL-producing transgenic plants was found to be an avirulent AHL-deficient *Erwinia carotovora* (Fray, 2002). In a screening of 33 *Penicillium* sp., several QS inhibitory (QSI) compounds were reported, among which patulin and penicillic acid were identified as biologically active QSI compounds. Besides, their effect on QS-regulated gene expression in *P. aeruginosa* was also identified through DNA microarray transcriptomics (Rasmussen et al., 2005). Soil bacteria may use plant hormonal signals as a cue for triggering QS signaling to infect a host. The natural infection between a host plant and a bacterium might involve disruption of QS signaling of other bacteria as well as inhibition of QS mimic signals from the host plant itself. The rhizosphere is a region of overlapping, communicating populations of bacteria and can be defined by mutual QS signal recognition and can be affected specifically by particular QS mimics and other allelochemicals secreted by plants. There is no direct evidence of secretion of bacterial AHL-degrading enzymes by plants, and such enzymes might also be harmful to beneficial microbes. However, it is obvious that plants may have developed defense strategies to inhibit bacterial QS signals, particularly of the pathogenic ones, by secreting chemical compounds that may act as QS signal mimics, blockers, and/or signal-degrading enzymes (Bais et al., 2008). Further studies are required to isolate and characterize those compounds from root exudates of different plant species.

Tripartite interactions between plants, microbes, and nematodes

Root exudates contain a large amount of organic carbon that provides nutrients to the soil microbial population, leading to an abundance of soil microbes in the rhizospheric vicinity. Microbe-feeding nematodes tend to flourish in such environments, and a subsequent increase in microbial turnover and nutrient supply to plants can be observed. Environmental cues and specific plant species affect the quality and quantity of the organic carbon and nutrient supply to the rhizosphere, which, in turn, alter microbial community structure. However, the effects of these factors on microbe-nematode interactions are unclear (Bais et al., 2006). Root-dwelling nematodes are believed to be engaged in complex interactions with roots and soil microbes. A study revealed that soil bacteria occupy $<10\%$ of the root surface, whereas the fungal hyphae cover only 12–14 mm m^{-2} root, leaving a large space for nematodes to move freely (Rovira et al., 1974). Thus, they can easily avoid microbial pathogens and root-secreted nematocidal chemicals and can take possession of the uncolonized territory of the rhizosphere. A study on *Trifolium repens* revealed that upon infection with the root-dwelling nematode *Heterodera trifolii*, an increase of photosynthetically fixed carbon in the rhizosphere (partially via root exudation) can be observed. Therefore, it can be said that plant roots infected with parasitic nematodes may supply extra organic carbon to the rhizosphere and may increase the microbial turnover in the rhizospheric soil (Yeates, 1999). Rhizospheric nematodes were also found to affect the quality and quantity of root exudates, which, in turn, affects beneficial soil microorganisms as well as pathogenic ones. Plant roots infected with *Meloidogyne incognita* function as a metabolic sink, and symplastic nutrient transport is diverted toward nematodes, causing increased nutrient leakage into the rhizosphere compared to healthy roots (Bais et al., 2006). The microbe-nematode interactions derived from research studies on rhizobia, mycorrhiza, fungi, and plant pathogens showed complex tri-trophic webs, whereby microorganisms and nematodes act in a competitive, additive, or synergistic manner to affect the plant (Bais et al., 2006). The soil-dwelling nematode *Caenorhabditis elegans* is believed to regulate the nodulation, thereby positively modulating the interaction between rhizobia and plant. Moreover, *C. elegans* was found to shift *S. meliloti* toward the roots of *M. truncatula* in response to plant root-released volatiles. Therefore, it can be said that root-secreted volatiles play a key role in tri-trophic interactions (Horiuchi et al., 2005). A similar kind of study reported the attraction of entomopathogenic nematodes to insect-damaged maize roots mediated by a plant volatile, (*E*)-β-caryophyllene (Rasmann et al., 2005). Further research on tripartite interactions between plants, nematodes, and other soil microorganisms is required for a deeper understanding of the underlying signaling cascade mediated by root exudates.

The effects of root exudates on shaping rhizospheric microbial communities

It is somewhat well known that plant roots house a vast array of soil microbes, both beneficial and pathogenic, at rhizospheric interfaces as a result of the microbial drive toward metabolite exudation of roots. Root exudates are the hub of carbon sources for the soil microflora that causes the sinking of more microbial population to the rhizospheric interfaces as evident by the higher number of microorganisms in the rhizosphere than in bulk soil (Nannipieri et al., 2007). Plants do selectively entice a particular clutch of microbes and, consequently, influence the spatial distribution of microbiomes in their respective rhizospheric niche, somewhat in a plant root exudate-specific manner (Broeckling et al., 2008). For example, a diverse range of rhizospheric microbial populations are found in different ecotypes of *Arabidopsis* sp., which could be the result of the unique exudation profiling of each ecotype (Micallef et al., 2009). Similarly, the carboxylate-rich rhizosphere and low pH produced significant changes to create a specialized citrate- and oxalate-metabolizing rhizobiome with a predominance of the *Burkholderia* species (Weisskopf et al., 2011). An alteration of the whole set of bacterial communities with a change in plant secretions has also been observed in the rhizosphere of the *Arabidopsis* ABC transporter mutant (Badri et al., 2009). Many studies have demonstrated that different types of plant secretions have a profound role in shaping the rhizospheric microbial communities.

The addition of p-coumaric acid to the rhizospheric soil of cucumber seedlings caused changes in the abundance and composition of the residing rhizospheric microbial and fungal communities. It was observed that the addition of p-coumaric acid resulted in the selective abundance of microflora like *Firmicutes*, β-*Proteobacteria*, γ-*Proteobacteria*, *Sordariomycete*, and *Zygomycota* that can thrive and grow on p-coumaric acid as the sole carbon source (Zhou and Wu, 2012). Similarly, the appearance of distinct atrazine-degrading rhizospheric microbial communities or symbiotic bacteria can be witnessed with the addition of distinct mixtures of root derivatives of the *Arabidopsis* sp. (Badri et al., 2013a). Another study showed that jasmonic acid-derived shifting of the rhizospheric microbial community, with the common enrichment of bacterial species like *Bacillus*, *Bacillales*, *Paenibacillus*, and *Lysinibacillus*, is primarily involved in conferring systemic resistance on the host plant species (Carvalhais et al., 2013). Application of a mixture of phytochemicals alone (a natural constituent of root exudates) can solely bring about a compositional change in the residing soil microbiome, which is so evident at various levels (Badri et al., 2013b). In addition, several factors such as plant species type, root architecture, root age, development, stage, edaphic perturbation, soil type, etc. can influence the impact of plants on the composition and dynamicity of the inhabiting rhizobiome, driving to completely transformed microbial communities (Huang et al., 2014). A descriptive study on the compositional diversity of microbial communities in the rhizosphere of wheat, pea, and oat makes it evident that the rhizospheric diversity of each plant species is completely distinct from the other one, showing the complexity of the plant-microflora interaction at the rhizospheric level (Turner et al., 2013). Moreover, it has been observed that distinct root portions harbor different sets of the microbial population as a cause of distinctive exudation profiling of root parts. Generally, the root tips are associated with the active bacteria differing from elongation and mature root zones, which are commonly inhabited by *Bacillus subtilis* and decomposer communities (Sasse et al., 2018). Besides, the genetic variation in a plant's genotype alone had a significant impact on rhizomicrobial colonization, abundance, richness, and dynamicity as witnessed in the rhizosphere of 27 maize inbred varieties (Peiffer et al., 2013).

Conclusions

From this detailed discussion, we can conclude that root exudate-mediated mechanisms play an instrumental role in governing plant-microbe interactions in the rhizosphere. Root exudates act as a buffering agent to tone down the effects of hostile environments in the rhizosphere. However, there is a significant knowledge gap in exudate molecules, their identification and characterization, exudation patterns (spatial and temporal) and processes, exudation chemistry, and their mode of action that leaves a lacuna on our current understanding of plant-microbe interactions in the rhizosphere. Moreover, most of the research studies on which our current understanding is based are in vitro. It is extremely difficult to carry out experiments on root exudates in a natural setup due to their trace amounts and labile nature. Therefore, future research studies should be aimed at these points to decipher the detailed role of root exudates. Besides, in-depth analyses of complex molecular signaling between plants and microbes elicited by root exudates, the correlation between root exudate secretion and plant growth cycle in different soil conditions, fate and turnover of root exudates, and the role of root exudates in microbial physiology are required. An interdisciplinary approach by plant physiologists, microbiologists, and ecologists is required to analyze all of these. This will lead to a better understanding of root exudate chemistry, its role in promoting beneficial plant-microbe interactions, and biotic and abiotic stress tolerance. In a nutshell, the full potential of root exudates is yet to be uncovered and it will keep the researchers busy in the coming years.

Acknowledgments

The authors are grateful to the Council of Scientific and Industrial Research (CSIR), India, for necessary financial assistance (sanction letter no. 38(1469)/18/EMR-II dated 04/04/2018). Krishnendu Pramanik gratefully acknowledges the University Grants Commission (UGC), New Delhi, for UGC—Dr. D. S. Kothari Post-Doctoral Fellowship (No.F.4-2/2006 (BSR)/BL/19-20/0072 dated October 21, 2019). Priyanka Pal expresses gratitude to the Council of Scientific and Industrial Research (CSIR), India, for awarding the CSIR fellowship (File No. 09/025(0264)/2018-EMR-I).

References

Ahemad, M., Kibret, M., 2014. Mechanisms and applications of plant growth promoting rhizobacteria: current perspective. J. King Saud Univ. Sci. 26 (1), 1–20. https://doi.org/10.1016/j.jksus.2013.05.001.

Akiyama, K., Matsuzaki, K.I., Hayashi, H., 2005. Plant sesquiterpenes induce hyphal branching in arbuscular mycorrhizal fungi. Nature 435 (7043), 824–827. https://doi.org/10.1038/nature03608.

Aulakh, M., Wassmann, R., Bueno, C., Kreuzwieser, J., Rennenberg, H., 2001. Characterization of root exudates at different growth stages of ten rice (*Oryza sativa* L.) cultivars. Plant Biol. 3 (02), 139–148. https://doi.org/10.1055/s-2001-12905.

Badri, D.V., Chaparro, J.M., Zhang, R., Shen, Q., Vivanco, J.M., 2013a. Application of natural blends of phytochemicals derived from the root exudates of *Arabidopsis* to the soil reveal that phenolic-related compounds predominantly modulate the soil microbiome. J. Biol. Chem. 288 (7), 4502–4512. https://doi.org/10.1074/jbc.M112.433300.

Badri, D.V., Loyola-Vargas, V.M., Broeckling, C.D., De-la-Peña, C., Jasinski, M., Santelia, D., Martinoia, E., Sumner, L.W., Banta, L.M., Stermitz, F., Stermitz, F., 2008. Altered profile of secondary metabolites in the root exudates of *Arabidopsis* ATP-binding cassette transporter mutants. Plant Physiol. 146 (2), 323–324. https://doi.org/10.1104/pp.107.109587.

Badri, D.V., Quintana, N., El Kassis, E.G., Kim, H.K., Choi, Y.H., Sugiyama, A., Verpoorte, R., Martinoia, E., Manter, D.K., Vivanco, J.M., 2009. An ABC transporter mutation alters root exudation of phytochemicals that provoke an overhaul of natural soil microbiota. Plant Physiol. 151 (4), 2006–2017. https://doi.org/10.1104/pp.109.147462.

Badri, D.V., Vivanco, J.M., 2009. Regulation and function of root exudates. Plant Cell Environ. 32 (6), 666–681. https://doi.org/10.1111/j.1365-3040.2009.01926.x.

Badri, D.V., Zolla, G., Bakker, M.G., Manter, D.K., Vivanco, J.M., 2013b. Potential impact of soil microbiomes on the leaf metabolome and on herbivore feeding behavior. New Phytol. 198 (1), 264–273. https://doi.org/10.1111/nph.12124.

Bais, H.P., Broeckling, C.D., Vivanco, J.M., 2008. Root exudates modulate plant—microbe interactions in the rhizosphere. In: Karlovsky, P. (Ed.), Secondary Metabolites in Soil Ecology. Soil Biology, vol. 14. Springer, Berlin, Heidelberg, pp. 241–252.

Bais, H.P., Walker, T.S., Schweizer, H.P., Vivanco, J.M., 2002. Root specific elicitation and antimicrobial activity of rosmarinic acid in hairy root cultures of *Ocimum basilicum*. Plant Physiol. Biochem. 40 (11), 983–995. https://doi.org/10.1016/S0981-9428(02)01460-2.

Bais, H.P., Weir, T.L., Perry, L.G., Gilroy, S., Vivanco, J.M., 2006. The role of root exudates in rhizosphere interactions with plants and other organisms. Annu. Rev. Plant Biol. 57, 233–266. https://doi.org/10.1146/annurev.arplant.57.032905.105159.

Bashir, O., Khan, K., Hakeem, K.R., Mir, N.A., Rather, G.H., Mohiuddin, R., 2016. Soil microbe diversity and root exudates as important aspects of rhizosphere ecosystem. In: Akhtar, M., Hakeem, K. (Eds.), Plant, Soil and Microbes. vol. 2. Springer, pp. 337–357.

Bjarnsholt, T., Jensen, P.Ø., Rasmussen, T.B., Christophersen, L., Calum, H., Hentzer, M., Hougen, H.P., Rygaard, J., Moser, C., Eberl, L., Høiby, N., 2005. Garlic blocks quorum sensing and promotes rapid clearing of pulmonary *Pseudomonas aeruginosa* infections. Microbiology 151 (12), 3873–3880. https://doi.org/10.1099/mic.0.27955-0.

Brencic, A., Winans, S.C., 2005. Detection of and response to signals involved in host-microbe interactions by plant-associated bacteria. Microbiol. Mol. Biol. Rev. 69 (1), 155–194. https://doi.org/10.1128/MMBR.69.1.155-194.2005.

Brigham, L.A., Michaels, P.J., Flores, H.E., 1999. Cell-specific production and antimicrobial activity of naphthoquinones in roots of *Lithospermum erythrorhizon*. Plant Physiol. 119 (2), 417–428. https://doi.org/10.1104/pp.119.2.417.

Broeckling, C.D., Broz, A.K., Bergelson, J., Manter, D.K., Vivanco, J.M., 2008. Root exudates regulate soil fungal community composition and diversity. Appl. Environ. Microbiol. 74 (3), 738–744. https://doi.org/10.1128/AEM.02188-07.

Burdman, S., Dulguerova, G., Okon, Y., Jurkevitch, E., 2001. Purification of the major outer membrane protein of *Azospirillum brasilense*, its affinity to plant roots, and its involvement in cell aggregation. Mol. Plant-Microbe Interact. 14 (4), 555–561. https://doi.org/10.1094/MPMI.2001.14.4.555.

Carvalhais, L.C., Dennis, P.G., Badri, D.V., Tyson, G.W., Vivanco, J.M., Schenk, P.M., 2013. Activation of the jasmonic acid plant defence pathway alters the composition of rhizosphere bacterial communities. PLoS One 8 (2). https://doi.org/10.1371/journal.pone.0056457, e56457.

Carvalhais, L.C., Dennis, P.G., Fedoseyenko, D., Hajirezaei, M.R., Borriss, R., von Wirén, N., 2011. Root exudation of sugars, amino acids, and organic acids by maize as affected by nitrogen, phosphorus, potassium, and iron deficiency. J. Plant Nutr. Soil Sci. 174 (1), 3–11. https://doi.org/10.1002/jpln.201000085.

Dakora, F.D., Phillips, D.A., 2002. Root exudates as mediators of mineral acquisition in low-nutrient environments. In: Adu-Gyamfi, J.J. (Ed.), Food Security in Nutrient-Stressed Environments: Exploiting Plants' Genetic Capabilities. Springer, Dordrecht, pp. 201–213.

Dardanelli, M.S., de Córdoba, F.J.F., Estévez, J., Contreras, R., Cubo, M.T., Rodríguez-Carvajal, M.Á., Gil-Serrano, A.M., López-Baena, F.J., Bellogín, R., Manyani, H., Ollero, F.J., Manyani, H., 2012. Changes in flavonoids secreted by *Phaseolus vulgaris* roots in the presence of salt and the plant growth-promoting rhizobacterium *Chryseobacterium balustinum*. Appl. Soil Ecol. 57, 31–38. https://doi.org/10.1016/j.apsoil.2012.01.005.

Dardanelli, M.S., Manyani, H., González-Barroso, S., Rodríguez-Carvajal, M.A., Gil-Serrano, A.M., Espuny, M.R., López-Baena, F.J., Bellogín, R.A., Megías, M., Ollero, F.J., 2010. Effect of the presence of the plant growth promoting rhizobacterium (PGPR) *Chryseobacterium balustinum* Aur9 and salt stress in the pattern of flavonoids exuded by soybean roots. Plant Soil 328 (1), 483–493. https://doi.org/10.1007/s11104-009-0127-6.

De-la-Peña, C., Badri, D.V., Lei, Z., Watson, B.S., Brandão, M.M., Silva-Filho, M.C., Sumner, L.W., Vivanco, J.M., 2010. Root secretion of defense-related proteins is development-dependent and correlated with flowering time. J. Biol. Chem. 285 (40), 30654–30665. https://doi.org/10.1074/jbc.M110.119040.

Dong, Z., Layzell, D., 2001. H_2 oxidation, O_2 uptake and CO_2 fixation in hydrogen treated soils. Plant Soil 229 (1), 1–12. https://doi.org/10.1023/A:1004810017490.

Dousset, S., Morel, J., Jacobson, A., Bitton, G., 2001. Copper binding capacity of root exudates of cultivated plants and associated weeds. Biol. Fertil. Soils 34 (4), 230–234. https://doi.org/10.1007/s003740100404.

Dreyer, I., Gomez-Porras, J.L., Riaño Pachón, D.M., Hedrich, R., Geiger, D., 2012. Molecular evolution of slow and quick anion channels (SLACs and QUACs/ALMTs). Front. Plant Sci. 3, 263. https://doi.org/10.3389/fpls.2012.00263.

el Zahar Haichar, F., Santaella, C., Heulin, T., Achouak, W., 2014. Root exudates mediated interactions belowground. Soil Biol. Biochem. 77, 69–80. https://doi.org/10.1016/j.soilbio.2014.06.017.

Fray, R.G., 2002. Altering plant–microbe interaction through artificially manipulating bacterial quorum sensing. Ann. Bot. 89 (3), 245–253. https://doi.org/10.1093/aob/mcf039.

Givskov, M., de Nys, R., Manefield, M., Gram, L., Maximilien, R., Eberl, L.E.O., Molin, S., Steinberg, P.D., Kjelleberg, S., 1996. Eukaryotic interference with homoserine lactone-mediated prokaryotic signalling. J. Bacteriol. 178 (22), 6618–6622. https://doi.org/10.1128/jb.178.22.6618-6622.1996.

Golding, A.L., Zou, Y., Yang, X., Flynn, B., Dong, Z., 2012. Plant growth promoting H_2-oxidizing bacteria as seed inoculants for cereal crops. Agric. Sci. https://doi.org/10.4236/as.2012.34060.

Hartmann, A., Rothballer, M., Schmid, M., 2008. Lorenz Hiltner, a pioneer in rhizosphere microbial ecology and soil bacteriology research. Plant Soil 312 (1), 7–14. https://doi.org/10.1007/s11104-007-9514-z.

Hassan, M.K., McInroy, J.A., Kloepper, J.W., 2019. The interactions of rhizodeposits with plant growth-promoting rhizobacteria in the rhizosphere: a review. Agriculture 9 (7), 142. https://doi.org/10.3390/agriculture9070142.

Hayat, S., Faraz, A., Faizan, M., 2017. Root exudates: composition and impact on plant-microbe interaction. In: Iqbal Ahmad, F.M.H. (Ed.), Biofilms in Plant Soil Health. John Wiley Sons Ltd, pp. 179–193.

Henry, A., Doucette, W., Norton, J., Bugbee, B., 2007. Changes in crested wheatgrass root exudation caused by flood, drought, and nutrient stress. J. Environ. Qual. 36 (3), 904–912. https://doi.org/10.2134/jeq2006.0425sc.

Hoekenga, O.A., Vision, T.J., Shaff, J.E., Monforte, A.J., Lee, G.P., Howell, S.H., Kochian, L.V., 2003. Identification and characterization of aluminum tolerance loci in *Arabidopsis* (*Landsberg erecta* × *Columbia*) by quantitative trait locus mapping. A physiologically simple but genetically complex trait. Plant Physiol. 132 (2), 936–948. https://doi.org/10.1104/pp.103.023085.

Horiuchi, J.-I., Prithiviraj, B., Bais, H.P., Kimball, B.A., Vivanco, J.M., 2005. Soil nematodes mediate positive interactions between legume plants and rhizobium bacteria. Planta 222 (5), 848–857. https://doi.org/10.1007/s00425-005-0025-y.

Huang, X.-F., Chaparro, J.M., Reardon, K.F., Zhang, R., Shen, Q., Vivanco, J.M., 2014. Rhizosphere interactions: root exudates, microbes, and microbial communities. Botany 92 (4), 267–275. https://doi.org/10.1139/cjb-2013-0225.

Jones, D.L., Nguyen, C., Finlay, R.D., 2009. Carbon flow in the rhizosphere: carbon trading at the soil-root interface. Plant Soil 321 (1), 5–33. https://doi.org/10.1007/s11104-009-9925-0.

Khorassani, R., Hettwer, U., Ratzinger, A., Steingrobe, B., Karlovsky, P., Claassen, N., 2011. Citramalic acid and salicylic acid in sugar beet root exudates solubilize soil phosphorus. BMC Plant Biol. 11 (1), 1–8. https://doi.org/10.1186/1471-2229-11-121.

Koltai, H., 2011. Strigolactones' ability to regulate root development may be executed by induction of the ethylene pathway. Plant Signal. Behav. 6 (7), 1004–1005. https://doi.org/10.4161/psb.6.7.15501.

Kuiper, I., Kravchenko, L.V., Bloemberg, G.V., Lugtenberg, B.J., 2002. *Pseudomonas putida* strain PCL1444, selected for efficient root colonization and naphthalene degradation, effectively utilizes root exudate components. Mol. Plant-Microbe Interact. 15 (7), 734–741. https://doi.org/10.1094/MPMI.2002.15.7.734.

Lanoue, A., Burlat, V., Henkes, G.J., Koch, I., Schurr, U., Röse, U.S., 2010. *De novo* biosynthesis of defense root exudates in response to *Fusarium* attack in barley. New Phytol. 185 (2), 577–588. https://doi.org/10.1111/j.1469-8137.2009.03066.x.

Lapie, C., Leglize, P., Paris, C., Buisson, T., Sterckeman, T., 2019. Profiling of main metabolites in root exudates and mucilage collected from maize submitted to cadmium stress. Environ. Sci. Pollut. Res. 26 (17), 17520–17534. https://doi.org/10.1007/s11356-019-05168-0.

Liu, J., Magalhaes, J.V., Shaff, J., Kochian, L.V., 2009. Aluminum-activated citrate and malate transporters from the MATE and ALMT families function independently to confer *Arabidopsis* aluminum tolerance. Plant J. 57 (3), 389–399. https://doi.org/10.1111/j.1365-313X.2008.03696.x.

Lucas García, J., Barbas, C., Probanza, A., Barrientos, M., Gutierrez Mañero, F., 2001. Low molecular weight organic acids and fatty acids in root exudates of two *Lupinus* cultivars at flowering and fruiting stages. Phytochem. Anal. 12 (5), 305–311. https://doi.org/10.1002/pca.596.

Luo, Q., Sun, L., Hu, X., Zhou, R., 2014. The variation of root exudates from the hyperaccumulator *Sedum alfredii* under cadmium stress: metabonomics analysis. PLoS One 9 (12). https://doi.org/10.1371/journal.pone.0115581, e115581.

Luo, Q., Wang, S., Sun, L.-N., Wang, H., 2017. Metabolic profiling of root exudates from two ecotypes of *Sedum alfredii* treated with Pb based on GC-MS. Sci. Rep. 7 (1), 1–9. https://doi.org/10.1038/srep39878.

Malinowski, D.P., Belesky, D.P., 2000. Adaptations of endophyte-infected cool-season grasses to environmental stresses: mechanisms of drought and mineral stress tolerance. Crop Sci. https://doi.org/10.2135/cropsci2000.404923x.

Martin, B.C., Gleeson, D., Statton, J., Siebers, A.R., Grierson, P., Ryan, M.H., Kendrick, G.A., 2018. Low light availability alters root exudation and reduces putative beneficial microorganisms in seagrass roots. Front. Microbiol. 8, 2667. https://doi.org/10.3389/fmicb.2017.02667.

Matusova, R., Rani, K., Verstappen, F.W., Franssen, M.C., Beale, M.H., Bouwmeester, H.J., 2005. The strigolactone germination stimulants of the plant-parasitic *Striga* and *Orobanche* spp. are derived from the carotenoid pathway. Plant Physiol. 139 (2), 920–934. https://doi.org/10.1104/pp.105.061382.

Micallef, S.A., Shiaris, M.P., Colón-Carmona, A., 2009. Influence of *Arabidopsis thaliana* accessions on rhizobacterial communities and natural variation in root exudates. J. Exp. Bot. 60 (6), 1729–1742. https://doi.org/10.1093/jxb/erp053.

Mondal, S., Pramanik, K., Ghosh, S.K., Pal, P., Mondal, T., Soren, T., Maiti, T.K., 2021. Unraveling the role of plant growth-promoting rhizobacteria in the alleviation of arsenic phytotoxicity: a review. Microbiol. Res. https://doi.org/10.1016/j.micres.2021.126809, 126809.

Muratova, A., Pozdnyakova, N., Golubev, S., Wittenmayer, L., Makarov, O., Merbach, W., Turkovskaya, O., 2009. Oxidoreductase activity of sorghum root exudates in a phenanthrene-contaminated environment. Chemosphere 74 (8), 1031–1036. https://doi.org/10.1016/j.chemosphere.2008.11.011.

Nannipieri, P., Ascher, J., Ceccherini, M.T., Landi, L., Pietramellara, G., Renella, G., Valori, F., 2007. Microbial diversity and microbial activity in the rhizosphere. Cienc. Suelo 25 (1), 89–97.

Narula, N., Kothe, E., Behl, R.K., 2012. Role of root exudates in plant-microbe interactions. J. Appl. Bot. Food Qual. 82 (2), 122–130.

Neal, A.L., Ahmad, S., Gordon-Weeks, R., Ton, J., 2012. Benzoxazinoids in root exudates of maize attract *Pseudomonas putida* to the rhizosphere. PLoS One 7 (4). https://doi.org/10.1371/journal.pone.0035498, e35498.

Newton, J., Fray, R., 2004. Integration of environmental and host-derived signals with quorum sensing during plant-microbe interactions. Cell. Microbiol. 6 (3), 213–224. https://doi.org/10.1111/j.1462-5822.2004.00362.x.

Nguyen, C., 2003. Rhizodeposition of organic C by plants: mechanisms and controls. Agronomie 23 (5–6), 375–396. https://doi.org/10.1051/agro:2003011.

Ortega-Calvo, J., Marchenko, A., Vorobyov, A., Borovick, R., 2003. Chemotaxis in polycyclic aromatic hydrocarbon-degrading bacteria isolated from coal-tar-and oil-polluted rhizospheres. FEMS Microbiol. Ecol. 44 (3), 373–381. https://doi.org/10.1016/S0168-6496(03)00092-8.

Park, S.W., Lawrence, C.B., Linden, J.C., Vivanco, J.M., 2002. Isolation and characterization of a novel ribosome-inactivating protein from root cultures of pokeweed and its mechanism of secretion from roots. Plant Physiol. 130 (1), 164–178. https://doi.org/10.1104/pp.000794.

Peiffer, J.A., Spor, A., Koren, O., Jin, Z., Tringe, S.G., Dangl, J.L., Buckler, E.S., Ley, R.E., 2013. Diversity and heritability of the maize rhizosphere microbiome under field conditions. Proc. Natl. Acad. Sci. 110 (16), 6548–6553. https://doi.org/10.1073/pnas.1302837110.

Pinto, A., Sim Es, I., Mota, A., 2008. Cadmium impact on root exudates of sorghum and maize plants: a speciation study. J. Plant Nutr. 31 (10), 1746–1755. https://doi.org/10.1080/01904160802324829.

Pomilio, A.B., Leicach, S.R., Grass, M.Y., Ghersa, C.M., Santoro, M., Vitale, A.A., 2000. Constituents of the root exudate of *Avena fatua* grown under far-infrared-enriched light. Phytochem. Anal. 11 (5), 304–308. https://doi.org/10.1002/10991565(200009/10)11:5%3C304::AIDPCA531%3E3.0.CO;2-G.

Pramanik, M., Nagai, M., Asao, T., Matsui, Y., 2000. Effects of temperature and photoperiod on phytotoxic root exudates of cucumber (*Cucumis sativus*) in hydroponic culture. J. Chem. Ecol. 26 (8), 1953–1967. https://doi.org/10.1023/A:1005509110317.

Rasmann, S., Köllner, T.G., Degenhardt, J., Hiltpold, I., Toepfer, S., Kuhlmann, U., Gershenzon, J., Turlings, T.C., 2005. Recruitment of entomopathogenic nematodes by insect-damaged maize roots. Nature 434 (7034), 732–737. https://doi.org/10.1038/nature03451.

Rasmussen, T.B., Skindersoe, M.E., Bjarnsholt, T., Phipps, R.K., Christensen, K.B., Jensen, P.O., et al., 2005. Identity and effects of quorum-sensing inhibitors produced by *Penicillium* species. Microbiology 151 (5), 1325–1340. https://doi.org/10.1099/mic.0.27715-0.

Rentz, J.A., Alvarez, P.J., Schnoor, J.L., 2004. Repression of *Pseudomonas putida* phenanthrene-degrading activity by plant root extracts and exudates. Environ. Microbiol. 6 (6), 574–583. https://doi.org/10.1111/j.1462-2920.2004.00589.x.

Rentz, J.A., Alvarez, P.J., Schnoor, J.L., 2005. Benzo [a] pyrene co-metabolism in the presence of plant root extracts and exudates: implications for phytoremediation. Environ. Pollut. 136 (3), 477–484. https://doi.org/10.1016/j.envpol.2004.12.034.

Rovira, A., Newman, E., Bowen, H., Campbell, R., 1974. Quantitative assessment of the rhizoplane microflora by direct microscopy. Soil Biol. Biochem. 6 (4), 211–216. https://doi.org/10.1016/0038-0717(74)90053-4.

Sasse, J., Martinoia, E., Northen, T., 2018. Feed your friends: do plant exudates shape the root microbiome? Trends Plant Sci. 23 (1), 25–41. https://doi.org/10.1016/j.tplants.2017.09.003.

Shukla, K.P., Sharma, S., Singh, N.K., Singh, V., Tiwari, K., Singh, S., 2011. Nature and role of root exudates: efficacy in bioremediation. Afr. J. Biotechnol. 10 (48), 9717–9724. https://doi.org/10.5897/AJB10.2552.

Singh, P.K., Parsek, M.R., Greenberg, E.P., Welsh, M.J., 2002. A component of innate immunity prevents bacterial biofilm development. Nature 417 (6888), 552–555. https://doi.org/10.1038/417552a.

Stubbs, V., Standing, D., Knox, O., Killham, K., Bengough, A., Griffiths, B., 2004. Root border cells take up and release glucose-C. Ann. Bot. 93 (2), 221–224. https://doi.org/10.1093/aob/mch019.

Taddei, P., Tugnoli, V., Bottura, G., Dallavalle, E., Zechini D'Aulerio, A., 2002. Vibrational, ^{1}H-NMR spectroscopic, and thermal characterization of gladiolus root exudates in relation to *Fusarium oxysporum f.* sp. *gladioli* resistance. Biopolymers 67 (6), 428–439. https://doi.org/10.1002/bip.10170.

Teplitski, M., Robinson, J.B., Bauer, W.D., 2000. Plants secrete substances that mimic bacterial N-acyl homoserine lactone signal activities and affect population density-dependent behaviors in associated bacteria. Mol. Plant-Microbe Interact. 13 (6), 637–648. https://doi.org/10.1094/MPMI.2000.13.6.637.

Tharayil, N., Triebwasser, D.J., 2010. Elucidation of a diurnal pattern of catechin exudation by *Centaurea stoebe*. J. Chem. Ecol. 36 (2), 200–204. https://doi.org/10.1007/s10886-010-9749-7.

Turner, T.R., Ramakrishnan, K., Walshaw, J., Heavens, D., Alston, M., Swarbreck, D., Osbourn, A., Grant, A., Poole, P.S., 2013. Comparative metatranscriptomics reveals kingdom level changes in the rhizosphere microbiome of plants. ISME J. 7 (12), 2248–2258. https://doi.org/10.1038/ismej.2013.119.

Vaughan, M.M., Wang, Q., Webster, F.X., Kiemle, D., Hong, Y.J., Tantillo, D.J., Coates, R.M., Wray, A.T., Askew, W., O'Donnell, C., Tokuhisa, J.G., Tholl, D., 2013. Formation of the unusual semivolatile diterpene rhizathalene by the *Arabidopsis* class I terpene synthase TPS08 in the root stele is involved in defense against belowground herbivory. Plant Cell 25 (3), 1108–1125. https://doi.org/10.1105/tpc.112.100057.

Vishwakarma, K., Sharma, S., Kumar, V., Upadhyay, N., Kumar, N., Mishra, R., Yadav, G., Verma, R.K., Tripathi, D.K., 2017. Current scenario of root exudate-mediated plant-microbe interaction and promotion of plant growth. In: Kumar, M., Kumar, V., Sharma, S., Prasad, R. (Eds.), Probiotics in Agroecosystem. Springer, pp. 349–369.

Vives-Peris, V., de Ollas, C., Gómez-Cadenas, A., Pérez-Clemente, R.M., 2020. Root exudates: from plant to rhizosphere and beyond. Plant Cell Rep. 39 (1), 3–17. https://doi.org/10.1007/s00299-019-02447-5.

Walker, T.S., Bais, H.P., Déziel, E., Schweizer, H.P., Rahme, L.G., Fall, R., Vivanco, J.M., 2004. *Pseudomonas aeruginosa*-plant root interactions. Pathogenicity, biofilm formation, and root exudation. Plant Physiol. 134 (1), 320–331. https://doi.org/10.1104/pp.103.027888.

Walker, T.S., Bais, H.P., Grotewold, E., Vivanco, J.M., 2003. Root exudation and rhizosphere biology. Plant Physiol. 132 (1), 44–51. https://doi.org/10.1104/pp.102.019661.

Wei-Hong, X., Huai, L., Qi-Fu, M., Xiong, Z.-T., 2007. Root exudates, rhizosphere Zn fractions, and Zn accumulation of ryegrass at different soil Zn levels. Pedosphere 17 (3), 389–396. https://doi.org/10.1016/S1002-0160(07)60047-2.

Weisskopf, L., Heller, S., Eberl, L., 2011. *Burkholderia* species are major inhabitants of white lupin cluster roots. Appl. Environ. Microbiol. 77 (21), 7715–7720. https://doi.org/10.1128/AEM.05845-11.

Yeates, G., 1999. Effects of plants on nematode community structure. Annu. Rev. Phytopathol. 37 (1), 127–149. https://doi.org/10.1146/annurev.phyto.37.1.127.

Zhao, J., Davis, L.C., Verpoorte, R., 2005. Elicitor signal transduction leading to production of plant secondary metabolites. Biotechnol. Adv. 23 (4), 283–333. https://doi.org/10.1016/j.biotechadv.2005.01.003.

Zhou, X., Wu, F., 2012. p-Coumaric acid influenced cucumber rhizosphere soil microbial communities and the growth of *Fusarium oxysporum* f. sp. *cucumerinum* Owen. PLoS One 7 (10). https://doi.org/10.1371/journal.pone.0048288, e48288.

CHAPTER 4

Elicitins as microbe-associated molecular patterns and their role in plant defense

Satish Chandra[a] and Ishwar Prakash Sharma[b]

[a]*Department of Botany, Government Degree College Tyuni, Dehradun, India,* [b]*Patanjali Research Institute, Haridwar, India*

Chapter outline

Introduction

Plants play a central role in the different ecosystems of the earth. The ability of plants to convert solar energy into chemical energy makes them important autotrophic organisms. Different life forms on Earth depend on plants for their energy requirements, and these life forms interact with plants in various ways in order to meet their energy demands. Some of these interactions are herbivorous, i.e., an animal eating a plant; pathogenic, i.e., an organism acting as a parasite on a plant; and symbiotic, i.e., mutually beneficial to both the plant and the organism with the latter providing nutrients and getting carbohydrates as an energy source from the former. Pathogen–host and symbiotic–host organisms share an evolutionary history. The interactions between different pathogens and the host plant and between symbiotic microbes and the host plant are crucial in the establishment of a relationship. Each pathogen uses a highly specialized mechanism to use the resources of the host. These mechanisms are broadly categorized into three groups: biotrophs, necrotrophs, and hemibiotrophs. Biotrophs secrete less cell wall-degrading enzymes (CWDEs) and produce fewer toxins. These, with the help of specialized structures known as haustoria, obtain nutrients from the living host

Unravelling Plant-Microbe Synergy. https://doi.org/10.1016/B978-0-323-99896-3.00001-1

without destroying the host cell. Necrotrophs produce various phytotoxic compounds and cell wall-degrading enzymes (CWDEs) at the site of infection and, thereby, destroy the host cell. Hemibiotrophs initially absorb nutrients through the biotrophic mode of infection that is subsequently replaced by the necrotrophic mode of nutrition in the host (Pieterse et al., 2009, 2014; Mengiste, 2012).

Plants have also evolved various mechanisms to counter different pathogens. Thick cuticles and various other structures on the plant surface protect against different pathogens and serve as the first line of defense. However, pathogens manage to invade these structures and cause infections in plants. To counter such pathogens, plants have active immune responses, a more sophisticated mechanism of defense. Plants recognize the common features of pathogens termed 'pathogen-associated molecular patterns (PAMPs)', also called microbe-associated molecular patterns (MAMPs), to initiate diverse downstream signaling events that lead to the activation of innate resistance termed 'PAMP-triggered immunity (PTI)' also known as MAMP-triggered immunity (MTI). Pathogens have also devised counterstrategies to overcome PTI. Pathogens synthesize a group of molecules called effectors to counter the PTI and promote virulence called effector-triggered susceptibility (ETS). Furthermore, plants identify the effectors and launch their second immunity action called effector-triggered immunity (ETI). The final result of this host–pathogen conflict depends on the ability of pathogens to destroy the plant's immune system and the ability of the plant to identify the pathogens and activate an effective immune response against them (Pieterse et al., 2009). In addition to the pathogen attack, damages to host cells and tissues caused by pathogens also turn on immune response in the host. Inducers of such responses are called host-derived damage-associated molecular patterns (DAMPs) (Albert et al., 2020; Choi and Klessig, 2016). Here, we provide an overview of PAMPs, PTI, ETI, development of systemic acquired resistance (SAR) and induced systemic resistance (ISR), and the role of elicitins in plant defense.

Pathogen-associated molecular patterns (PAMPs)

Pathogen-associated molecular patterns (PAMPs) are molecular structures, which are conserved in nature and are produced by microorganisms. These structures are recognized to be foreign by the cell surface receptors of the host cell (Silva-Gomes et al., 2014). Most of the microbes during infection exhibit such patterns and are hence also termed 'microbe-associated molecular patterns (MAMPs)'. In *Arabidopsis* and other plant species, numerous bacterial MAMPs such as flagellin, peptidoglycan (PGN), bacterial cold-shock proteins (RNP1 motif), elongation factor TU (EF-TU), lipopolysaccharides (LPSs), and bacterial superoxide dismutase (SOD) have been identified (Newman et al., 2013). Depending on the growth stage, morphotype, and species, different species of fungi express different PAMPs, which are recognized by the cell surface receptors of the host cell to trigger an immune response (Jannuzzi, et al., 2020). Elicitins, β-glucan, mannan, chitin, and glycopeptides are some of the MAMPs expressed by fungi (Dodds and Rathjen, 2010; Oßwald et al., 2014; Ye and Murata, 2016).

MAMPs are identified on the plant surface by unique receptor proteins called pattern recognition receptors (PRRs). These are localized on the plasma membrane and receptor-like proteins (RLPs) and receptor-like kinases (RLKs) with modular functional domains. Receptor-like kinases comprise an intracellular kinase domain, a transmembrane domain, and an extracellular domain. RLPs possess a transmembrane domain and an extracellular domain, whereas an intracellular kinase domain is absent. Epidermal growth factor (EGF), leucine-rich repeats (LRRs), lysin motifs (LysMs), and lectin-like domains are among the ligand-binding ectodomains. Perception of MAMPs by cognate PRRs leads to the dissociation of the PRRs' receptor-like cytoplasmic kinase (RLCK) complex and transphosphorylation of PRRs, co-receptors, and RLCKs. This initiates PAMP-triggered immunity (PTI) with downstream signaling in the host cell (Saijo et al., 2018). Perception of PAMPs induces an instant immune response at the host cell plasma membrane, and different phosphorylation reactions (auto- and transphosphorylation) occur (Bigeard et al., 2015).

PAMP-triggered immunity (PTI)

Signals derived from PRRs are further transmitted through phosphorylation cascades such as calcium-dependent protein kinases (CDPKs) and mitogen-activated protein kinases (MAPKs) to downstream targets such as NADPH/respiratory burst oxidase homolog D (RBOHD), transcriptional factors (TFs), and the plasma membrane (PM) resident H + ATPases. This leads to an increase in reactive oxygen species (ROS) production at the site of infection and influx of extracellular Ca_2^+ in the cytosol (Ca_2^+ burst). The Ca_2^+ influx triggers opening of influx of H^+ and efflux of K^+, Cl^-, and NO_3^- like plasma membrane transporters. This causes rapid alkalinization (~ 1 min) of extracellular medium and host plasma membrane a depolarized (1–3 min) (Bigeard et al., 2015; Saijo et al., 2018).

ROS production occurs rapidly at ~2–3 min and reaches a peak at around 10–14 min after perception of MAPKs. The NADPH oxidase localized on the plasma membrane called respiratory burst oxidase homolog D (RBOHD) in *Arabidopsis* is responsible for this. The membrane-impermeable superoxide O_2^- is produced by RBOHD in the apoplast. Superoxide O_2^-, with the help of superoxide dismutases, is converted into hydrogen peroxide (H_2O_2) and can enter the cytosol and different organelles. Cytosolic Ca_2^+ can be elevated by H_2O_2. The ROS burst triggers, through MAMPs, positively regulate cytosolic Ca_2^+ levels by inducing an additive cytosolic Ca_2^+ elevation. In signal transduction of PTI, nitric oxide (NO) and its derivatives are involved through various steps. Upon pathogen infection, it is also postulated that phosphatidic acid (PA) and ceramides also function as signaling molecules (Bigeard et al., 2015). The mitogen-activated protein kinase (MAPK) cascade includes a MAPK kinase kinase (MAPKKK or MEKK) signal transduction pathway for the activation of different transcriptional factors (TFs) (Cui, et al., 2018). Finally, all this leads to upregulation of ethylene production and stomatal closure and increases callose deposition and metabolic reprogramming (Saijo et al., 2018).

Effector-triggered immunity (ETI)

To win the battle against the host plant, pathogens have evolved a mechanism to inhibit PTI. They release virulence effector proteins into host cells, which suppress PTI and spread the disease. The plant immune system has also evolved some receptor proteins (called resistance (R) proteins) to counteract this strategy of pathogens. By doing so, plant receptors detect the effector proteins and trigger disease resistance called effector-triggered immunity (ETI). Avirulence (Avr) proteins are the effectors that are recognized by R proteins. An Avr protein is carried by an avirulent pathogen that induces plant immunity. A hypersensitive response (HR) at the local tissue level by rapid programmed cell death (PCD) is caused by an avirulent pathogen. An immune signal plant hormone, salicylic acid (SA), is generated after the interaction of Avr proteins with the host plant, which further activates a highly specific systemic acquired resistance (SAR) (Wu et al., 2014; Zhang and Zhou, 2010). The R proteins are intracellular nucleotide-binding leucine-rich repeat (NB-LRR) proteins. On the basis of different patterns, coils, and kinase activities, these proteins are further classified into 13 different classes. The variability in the R genes or proteins is due to gene clustering, tandem duplication, and insertions/deletions caused by transposable elements (Silva and Micheli, 2020).

In its inactive state, an R protein binds to ADP. R proteins undergo conformational change when they detect an effector protein and release the hidden nucleotide-binding site from intramolecular interactions, leading to ADP/ATP nucleotide exchange and hydrolysis in the central nucleotide-binding site (NBS) domain. R proteins induce production and accumulation of SA, expression of pathogenesis-related (PR) genes, oxidative burst, nitric oxide (NO), and PCD. ETI is a prominent plant immune process that leads to extensive transcriptional reprogramming. Some of the R proteins also have nuclear localization signals (NLSs) and regulate the ETI-related signal transduction. ETI causes localized PCD or programmed cell death at the site of infection and is formulized to kill biotrophic pathogens by depriving nutrients from living cells. The fusion of the central vacuole membrane with the plasma membrane causes the release of antibacterial proteins into the apoplast, and this fusion is specific to R protein-triggered PCD. R protein signaling also triggers SA synthesis. A low level of SA at the infection site suppresses cell death; cell death can be increased by overaccumulation of SA (Wu et al., 2014).

Systemic acquired resistance (SAR)

Systemic acquired resistance (SAR) provides broad-spectrum resistance throughout the plant to counter related or unrelated pathogens. Effector-triggered immunity (ETI) plays a pivotal role in this process, and signals from the infected leaves are transported to the uninfected distal parts. The notable molecules are the phytohormone salicylic acid (SA), azelaic acid (AzA), and the phosphorylated sugar derivative glycerol-3-phosphate (G3P) (Singh et al., 2017).

Salicylic acid plays a pivotal role in SAR. The biosynthesis of SA occurs through chorismic acid in the shikimic acid pathway. After infection, SA concentration is increased in both infected sites and distal uninfected tissues. However, in the distal uninfected tissues, the level of SA is far less. Most of the SA is synthesized after infection, but some derivatives of SA are also present in plants that are believed to be crucial in SAR. Some of these are methyl SA (MeSA) and methylated ester. MeSA is a volatile, biologically inactive substance and functions only after conversion to SA. Therefore, MeSA is considered to be an airborne defense signal that strengthens SAR. At the sites of infections, SA is converted to MeSA with the help of the SA/benzoic acid (BA) methyltransferases (SA/BA-MT), and MeSA accumulated in the phloem in an SAMT-dependent manner. When translocated to the distal parts, MeSA is further converted to SA by the MeSA esterase activity of the SA-binding protein 2 (SABP2). Moreover, MeSA induction occurs at 48 and 72 h postprimary infection, and, thus, it is believed that MeSA is one of the key components of SAR but is not the primary signal of SAR. A cytosolic protein NPR1 (nonexpressor of PR1) is considered to be a regulator of SA-derived SAR. NPR1 exists as an oligomer due to intermolecular disulfide bonding in the cytoplasm in the absence of SA. While, increased concentration of the SA triggers dissociation of the NPR1 monomers form the oligomeric state. Monomers of NPR1 are transported to the nucleus to activate various proteins and transcription factors required for the establishment of SAR (Singh et al., 2017).

After the infection, SA, reactive oxygen species (ROS), and nitrous oxide (NO) accumulate at the site of infection. NO further increases the concentration of the ROS in the loop. Azelaic acid (AzA) is derived from the C18 oleic acid (18:1), linoleic acid (18:2) and linolenic acid (18:3), which has double bond between carbons 9 and 10 that after cleavage form a C9 compound monocarboxylic acid called 9-oxononanoic acid (ONA) that is converted into AzA. ROS is considered important for the generation of AzA that induces SAR by triggering biosynthesis of glycerol-3-phosphate (G3P). G3P-induced SAR is reported in the hemibiotrophic fungus *Colletotrichum higginsianum*. SAR is compromised in defective G3P synthesis mutants; exogenous application of G3P can restore the defects (Singh et al., 2017). Transport of the signal molecules for SAR from the site of infection to the distal sites occurs via the phloem through two main routes. G3P and AzA are transported through the symplastic routes to the distal tissues; contrary to this, SA moves via the apoplastic routes (Singh et al., 2017). SAR, epigenetically, via DNA methylation and chromatin remodeling, regulates the gene expression and immunity of plants (Luna et al., 2012).

Induced systemic resistance (ISR)

Plants are the ultimate source of food for terrestrial heterotrophic organisms. Considering their prime importance, nature has provided them with various ways to strengthen their immunity. One such way is SAR in which some parts of plants can activate an immune response without getting infected. Thus, a disease can be suppressed locally, without infecting the whole plant. Another such a mechanism is induced systemic resistance (ISR). ISR is triggered in plants by beneficial microbes living in the rhizosphere. These microbes can be bacteria or fungi known as plant growth-promoting rhizobacteria (PGPR) and plant growth-promoting fungi (PGPF). Some of these microbes are *Paenibacillus polymyxa*, *Trichoderma* spp., *Serratia* spp., *Bacillus* spp., *Pseudomonas simiae* (syn. *Pseudomonas fluorescens*), etc. These microbes live in the rhizosphere in mutualistic relationships. The diversity and concentration of microbes near the plant roots is determined by the availability of carbon resources. About 40% of phothosynthetically fixed carbon is deposited by plants in the rhizosphere via roots. This increases microbial density and composition in the rhizopshere, which have a significant effect on plant health. In the vicinity of their roots, plants select certain microbes that are beneficial to plant growth or compounds secreted by these microbes that suppress harmful microbial growth. Plants and PGPR share a co-evolutionary relationship (Pieterse et al., 2014; Romera et al., 2019).

The initial interaction between plant roots and beneficial microbes is that in which the microbes suppress the defense system of the host for successful establishment inside the roots. These also produce some elicitors that trigger systemic immunity in the host. Some of these effectors that induce systemic immunity are 2,4-diacetylphloroglucinol (DAPG), *N*-acyl homoserine lactones, pyocyanin, flagella, iron-regulated siderophores, biosurfactants, 2R,3R-butanediol, xylanases, and cellulases. MYB72 is a root-specific transcription factor that plays an important role in beneficial microbe-induced systemic resistance, and it is also important for plant growth under low Fe conditions. It is also observed that Fe deficiency responses in their host roots can be induced by ISR-eliciting microbes. This causes acidification of the rhizosphere, increased ferric reductase activity, release of phenolics and flavins, and development of root hairs in the Fe deficiency responses by the host plant. For establishment of ISR, jasmonic acid (JA) and ethylene (ET) play an important role. They also act on the cytosolic NPR1 for the induction of immunity. It is salicylic acid (SA)-independent SAR. However, some of the beneficial microbes also exhibit salicylic acid-dependent SAR (Pieterse et al., 2014; Romera et al., 2019).

Elicitins

Elicitin proteins are members of multigene families in oomycetes, e.g., *Phytophthora* and *Pythium*, and were first discovered in the late 1980s. Some oomycete members are not able to synthesize their own sterols and need external sterol sources to complete their life cycle. Elicitins function as sterol transporters to these members and fulfill their requirement. These are broadly categorized into subclasses, namely, elicitin (ELI) and elicitin-like (ELL) genes. Elicitin proteins vary from plant proteins, in that they are structurally more conserved with characteristic six cysteines that form three disulfide bonds at the C-terminal domains of different sizes that may have high serine, threonine, or proline content and lack tryptophan, histidine, and arginine residues (Fig. 4.1). Some of the elicitins of the *Phytophthora* species are enlisted in Table 4.1 (Oßwald et al., 2014). Elicitins also cause induction of SAR and hypersensitive response (HR) in some plant species. These properties of elicitins resemble those of MAPs (Oßwald et al., 2014; Derevnina et al., 2016; Noman et al., 2020).

Elicitins, i.e., parasiticein (PAR-A), cryptogein (CRY-B elicitin), INF1, and capsaicin (CAP-A), isolated from *Phytophthora capsici*, *Phytophthora cryptogea*, *Phytophthora infestans*, and *Phytophthora parasitica*, induce potent defense responses in several plant species of *Brassicaceae* and *Solanaceae*. Sterol is necessary for the growth and sporulation of the *Pythium* and *Phytophthora* species. These genera are unable to synthesize sterol. Elicitins act as carriers for the sterol that scavenges them from the plasma membrane of the host plant species. Elicitins also induce cell death of the host due to disruption of the plasma membrane. However, the elicitin mutant also induces cell death, and, hence, this aspect of elicitins needs further investigation (Derevnina et al., 2016).

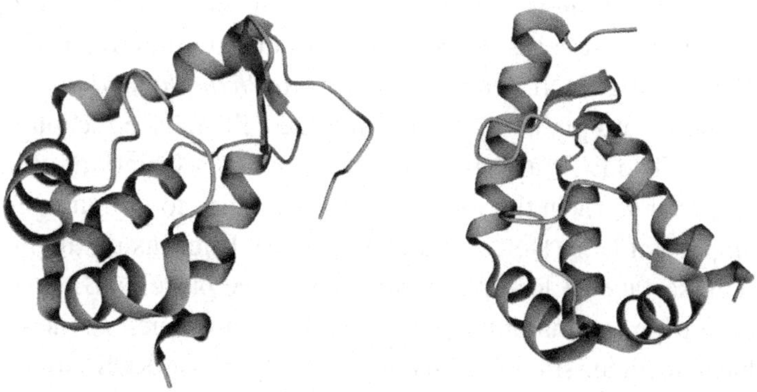

Sylvaticin (Uniprot id: D0VWX7) extracted from *Pythium sylvaticum*

Beta-cinnamomin (beta-CIN) (Uniprot id: P15569) extracted from *Phytophthora cryptogea*

Fig. 4.1
Structure of elicitins.

Table 4.1: Elicitins extracted from different *Phytophthora* species.

Phytophthora species	Elicitins
P. alni	AE1.1, AE1.2, AE2, BE1, BE2, HAE1
P. cambivora	AE1.1, AE2
P. cactorum	CAC-A (cacto)
P. capsici	Capsaicin (α), CAP-Pa28
P. cinnamomi	Cinnamomin, α-CIN, β-CIN, HAE1-cin, HAE2-cin
P. citrophthora	Citro
P. cryptogea	Cryptogein, Cry-α, Cry-β, HAE1-cry, HAE2-cry
P. drechsleri	Dre-α, Dre-β
P. fragariae	AE1.1, AE2
P. infestans	INF1, INF2a, INF2b, INF4, INF5, INF6
P. hibernalis	hib1 = syringicin, hib2, hib3
P. megasperma	α-megaspermin, β-megaspermin, ϒ-megaspermin
P. nicotianae	172 = parasiticein
P. palmivora	Palmivorein
P. parasitica	PARA1 (Parasiticein), 310 = Parasiticein
P. plurivora (formerly, *P. citricola*)	Citricolin = plurivorin
P. quercina	Quercinin
P. ramorum	(ram-alpha 1 and ram-alpha 2) + 3 elicitins
P. syringae	Syringicin, SOJA, SOJB = Sojein1, Cojein2

Elicitins exhibit many characteristics of MAMPs, as they all have some conserved regions among them; however, there is no sequence similarity of elicitins with plant proteins and they are recognized as nonself-molecules by the host. They are important for the biological activities of the *Phytophthora* and *Pythium* species as sterol transporters. Elicitins are expressed during interaction with the host species. In the host cell, these proteins are recognized by host cell surface PRRs and initiate immune responses. Four elicitins, namely, INF1, INF2a, INF5, and INF6P, from *Phytophthora infestans* trigger cell death in transgenic potato plants by expressing the cell surface receptor-like protein elicitin response (ELR). The ELR also induces responses to a variety of elicitins in wild potato (*Solanum microdontum*). In the induction of resistance to various oomycetes, the ELR plays an important role (Derevnina et al., 2016). Elicitins on the host cell are recognized by the cell surface ELR and the transmembrane proteins BAK1/SERK3 and SOBIR1. Recognition triggers immunity processes in the host such as the production of ROS in the biphasic mechanism. Elicitins induce early-phase ROS burst via calcium influx, NADPH oxidases, and respiratory burst oxidase homolog protein A (RBOHA) and respiratory burst oxidase homolog protein B (RBOHB) activation. The late-phase ROS burst includes MAPK cascade activation, which leads to phosphorylation of the transcription factor WRKY7/8/9/11 and, consequently, RBOHB gene upregulation. Production of ROS leads to the hypersensitive response (HR)-induced cell death of the host (Derevnina et al., 2016).

Elicitins trigger immunity in the host species not only through HR-induced cell death but also through other mechanisms such as ethylene (ET)- and jasmonic acid (JA)-mediated defense pathways. INF1 isolated from *P. infestans* activates defense pathways in tomato for the bacterial pathogen *Ralstonia solanacearum* through ethylene- and jasmonic acid-mediated defense response without triggering an HR. However, plants remain susceptible to late blight by *P. infestans* (Kawamura et al., 2009; Derevnina et al., 2016). The proper mechanism of elicitin-mediated JA and ET activation still needs to be elucidated (Derevnina et al., 2016). Elicitin-responsive plants once inoculated with elicitins exhibit more resistance to elicitin-producing pathogens. Elicitin pretreatment of tomato, radish, and tobacco plants results in enhanced resistance to *R. solanacearum*, *Xanthomonas campestris* pv. *Armoraciae*, and *P. parasitica*, respectively. When the ELR is transferred from wild potato to cultivated potato plants, it enhances resistance of the cultivated tomato to different *P. infestans* strains (Du et al., 2015). Elicitins induce SAR of the host plant and reduce the chances of a subsequent pathogen attack. Oligandrin (an ELL) activates induced systemic resistance to *Fusarium* crown and root rot disease in tomato plants. Oligandrin is isolated or secreted by *Pythium oligandrum* (Benhamou et al. 2001). When *Quercus suber* roots are pretreated with α- and β-cinnamomin, they reduce pathogen biomass in root cells against *P. cinnamomi* (Oßwald et al., 2014). *Phytophthora palmivora* is a pathogen to *Hevea brasiliensis*. When an elicitin is extracted from *P. palmivora*, it induces defense responses in cell culture tissues and resistance in *H. brasiliensis* (Dutsadee and Nunta, 2008).

Conclusions

Elicitins are highly conserved small proteins secreted by oomycetes (*Phytophthora*, *Pythium* sp.) and have extracellular sterol-binding activity. These molecules also act as PAMPs or MAMPs and induce immunity in the host plant species. Elicitin-pretreated plants show immunity against some pathogens, whereas the host species remains prone to elicitin-secreting pathogens. However, elicitins trigger some immunity against pathogen reinfection. A proper mechanism of elicitin-induced immunity and HR activities remains to be elucidated. Various proteins involved in this process and other molecular cascades are a matter of further investigation.

References

Albert, I., Hua, C., Nürnberger, T., Pruitt, R.N., Zhang, L., 2020. Surface sensor systems in plant immunity. Plant Physiol. 182 (4), 1582–1596.

Benhamou, N., Belanger, R.R., Rey, P., Tirilly, Y., 2001. Oligandrin, the elicitin-like protein produced by the mycoparasite *Pythium oligandrum*, induces systemic resistance to *Fusarium* crown and root rot in tomato plants. Plant Physiol. Biochem. 39, 681–696.

Bigeard, J., Colcombet, J., Hirt, H., 2015. Signaling mechanisms in pattern-triggered immunity (PTI). Mol. Plant 8 (4), 521–539.

Choi, H.W., Klessig, D.F., 2016. DAMPs, MAMPs, and NAMPs in plant innate immunity. BMC Plant Biol. 16 (1), 1–10.

Cui, F., Sun, W., Kong, X., 2018. Rlcks bridge plant immune receptors and mapk cascades. Trends Plant Sci. 23 (12), 1039–1041.

Derevnina, L., Dagdas, Y.F., De la Concepcion, J.C., Bialas, A., Kellner, R., Petre, B., Domazakis, E., Du, J., Wu, C.H., Lin, X., Kamoun, S., 2016. Nine things to know about elicitins. New Phytol. 212 (4), 888–895.

Dodds, P.N., Rathjen, J.P., 2010. Plant immunity: towards an integrated view of plant–pathogen interactions. Nat. Rev. Genet. 11 (8), 539–548.

Du, J., Verzaux, E., Chaparro-Garcia, A., Bijsterbosch, G., Keizer, L.P., Zhou, J.I., Liebrand, T.W., Xie, C., Govers, F., Robatzek, S., Van Der Vossen, E.A., Vleeshouwers, V.G., 2015. Elicitin recognition confers enhanced resistance to Phytophthora infestans in potato. Nat. Plants 1 (4), 1–5.

Dutsadee, C., Nunta, C., 2008. Induction of peroxidase, scopoletin, phenolic compounds and resistance in *Hevea brasiliensis* by elicitin and a novel protein elicitor purified from *Phytophthora palmivora*. Physiol. Mol. Plant Pathol. 72, 179–187.

Jannuzzi, G.P., De Almeida, J.R.F., Paulo, L.N.M., De Almeida, S.R., Ferreira, K.S., 2020. Intracellular PRRs activation in targeting the immune response against fungal infections. Front. Cell. Infect. Microbiol. 10, 562.

Kawamura, Y., Hase, S., Takenaka, S., Kanayama, Y., Yoshioka, H., Kamoun, S., Takahashi, H., 2009. INF1 elicitin activates jasmonic acid- and ethylene-mediated signalling pathways and induces resistance to bacterial wilt disease in tomato. J. Phytopathol. 157, 287–297.

Luna, E., Bruce, T.J., Roberts, M.R., Flors, V., Ton, J., 2012. Next-generation systemic acquired resistance. Plant Physiol. 158 (2), 844–853.

Mengiste, T., 2012. Plant immunity to necrotrophs. Annu. Rev. Phytopathol. 50, 267–294.

Newman, M.A., Sundelin, T., Nielsen, J.T., Erbs, G., 2013. MAMP (microbe-associated molecular pattern) triggered immunity in plants. Front. Plant Sci. 4, 139.

Noman, A., Aqeel, M., Irshad, M.K., Qari, S.H., Hashem, M., Alamri, S., AbdulMajeed, A.M., Al-Sadi, A.M., 2020. Elicitins as molecular weapons against pathogens: consolidated biotechnological strategy for enhancing plant growth. Crit. Rev. Biotechnol. 40 (6), 821–832.

Oßwald, W., Fleischmann, F., Rigling, D., Coelho, A.C., Cravador, A., Diez, J., Dalio, R.J., Horta Jung, M., Pfanz, H., Robin, C., Werres, S., 2014. Strategies of attack and defence in woody plant–*Phytophthora* interactions. For. Pathol. 44 (3), 169–190.

Pieterse, C.M., Leon-Reyes, A., Van der Ent, S., Van Wees, S.C., 2009. Networking by small-molecule hormones in plant immunity. Nat. Chem. Biol. 5 (5), 308–316.

Pieterse, C.M., Zamioudis, C., Berendsen, R.L., Weller, D.M., Van Wees, S.C., Bakker, P.A., 2014. Induced systemic resistance by beneficial microbes. Annu. Rev. Phytopathol. 52, 347–375.

Romera, F.J., García, M.J., Lucena, C., Martínez-Medina, A., Aparicio, M.A., Ramos, J., Alcántara, E., Angulo, M., Pérez-Vicente, R., 2019. Induced systemic resistance (ISR) and Fe deficiency responses in dicot plants. Front. Plant Sci. 10, 287.

Saijo, Y., Loo, E.P.I., Yasuda, S., 2018. Pattern recognition receptors and signaling in plant–microbe interactions. Plant J. 93 (4), 592–613.

Silva, R.J.S., Micheli, F., 2020. RRGPredictor, a set-theory-based tool for predicting pathogen-associated molecular pattern receptors (PRRs) and resistance (R) proteins from plants. Genomics 112 (3), 2666–2676.

Silva-Gomes, S., Decout, A., Nigou, J., 2014. Pathogen-associated molecular patterns (PAMPs). In: Parnham, M. (Ed.), Encyclopedia of Inflammatory Diseases. Birkhäuser, Basel, https://doi.org/10.1007/978-3-0348-0620-6_35-1.

Singh, A., Lim, G.H., Kachroo, P., 2017. Transport of chemical signals in systemic acquired resistance. J. Integr. Plant Biol. 59 (5), 336–344.

Wu, L., Chen, H., Curtis, C., Fu, Z.Q., 2014. Go in for the kill: how plants deploy effector-triggered immunity to combat pathogens. Virulence 5 (7), 710–721.

Ye, W., Murata, Y., 2016. Microbe associated molecular pattern signaling in guard cells. Front. Plant Sci. 7, 583.

Zhang, J., Zhou, J.M., 2010. Plant immunity triggered by microbial molecular signatures. Mol. Plant 3 (5), 783–793.

CHAPTER 5

Molecular insights into stress-responsive genes in the mitigation of environmental stresses

Narendra Kumar[a], Shulbhi Verma[b], Amit Kumar[c], Hemant Dasila[d], Deep Chandra Suyal[e], Garima Kumari[c], Sunita Rawat[c], Neha Jeena[f], Manish Singh[c], and Abhishek Kumar[c]

[a]Doon (PG) College of Agriculture Science and Technology, Dehradun, India, [b]Department of Biotechnology, SDAU, Dantiwada, Gujarat, India, [c]Forest Research Institute, Dehradun, India, [d]Department of Microbiology, G.B. Pant University of Agriculture and Technology, Pantnagar, Uttarakhand, India, [e]Department of Microbiology, Akal College of Basic Sciences, Eternal University, Sirmaur, Himachal Pradesh, India, [f]Department of Biotechnology, Bhimtal Campus Kumaun University, Nainital, Uttarakhand, India

Chapter outline

Unravelling Plant-Microbe Synergy. https://doi.org/10.1016/B978-0-323-99896-3.00005-9

Introduction

Abiotic stresses caused by adverse climatic conditions are a significant constraint for agricultural output loss. Crop plants must employ their own biological processes to deal with the external constraints imposed by edaphic and environmental factors (such as soil acidity, salinity, and alkalinity) or else their growth, development, and, ultimately, productivity would deteriorate. Abiotic stresses include low/high temperatures, droughts, submergence, salt, light intensity, nutritional insufficiency, and anaerobiosis (Bailey-Serres and Voesenek, 2008; Chandra et al., 2019a, b). Light, water, carbon, and mineral nutrients are required by plants for maximum growth, development, and reproduction. Plant growth, development, and productivity are hampered by extreme circumstances. An adverse environment characterized by dramatic temperature extremes, salinity, and drought creates a complicated collection of stress factors. Plants can detect and respond to stress in a variety of ways that benefit their survival. Unfavorable conditions have the greatest evident influence at the cellular level, followed by physiological symptoms (Grayson, 2013; Jiang et al., 2016).

Abiotic stressors such as heat and drought reduce agricultural yields and jeopardize global food security. Plants have different dynamic morphological, physiological, and biochemical responses that allow them to adapt to a variety of environmental conditions. Heat and drought stresses, even at their mildest levels, have a negative impact on agricultural yields. A few of the components that influence stress responses include signaling, hormones, transcription factors, and secondary metabolites. In various crops, the reproductive phase of development, which leads to grain production, has been demonstrated to be more sensitive to heat stress. Biotechnological advances, such as improvements in genetics and information technology, may reduce the negative effects of heat and drought through the use of agronomic management strategies and the development of crop varieties with enhanced yield under stress (Lamaoui et al., 2018).

Similarly, cold temperature is a major abiotic stressor that has a significant impact on plant physiology and productivity. Cold stress is classified according to the ambient temperature as cold stress (< 20°C) or freezing stress (< 0°C). Cold stress has a considerable impact on the geographic distribution of plants, in addition to influencing their growth and development. Plants deal with two circumstances in harsh environmental settings: stress avoidance and stress tolerance. Earlier cold stress and cold acclimatization studies using *Arabidopsis* as a model species have greatly enhanced our understanding of the molecular mechanisms of cold acclimation. In addition, various investigations in different plant species have been conducted to uncover the molecular mechanisms and gene regulatory networks (Mickelbart et al., 2015). Furthermore, light stress alters the antioxidant status through regulating the flavonoid, glutathione, phenylpropanoid, cinnamate, and pyridoxine production pathways, and, in severe cases, it causes programmed cell death. Exposure to toxic heavy metals is the major cause of adverse health effects in both animals and plants. They also alter the expression of genes because of their high reactivity; they can have a direct impact on growth, aging, and energy synthesis. Enhanced levels of heavy metals, on the other hand, are related to an increase in the formation of reactive oxygen species (ROS) such as superoxide, hydroxyl, and nonfree radical species (singlet oxygen and hydrogen peroxide), which can induce oxidative stress by interrupting the balance between prooxidant and antioxidant equilibrium within plants (Dalvi and Bhalerao, 2013). Notwithstanding this, the continual and rapid evolution of biotic agents like viruses, bacteria, fungi, insects, and nematodes causes major problems in plant growth, development, and productivity. Because of biotic interactions, plants attenuate stress responses by producing or regulating stress-sensitive genes and transcription factors (Fraire-Velázquez et al., 2011).

Stress: Abiotic and biotic

Abiotic stress

Drought, heat, salt, cold, and pathogen infections are all examples of environmental stress conditions that may wreak havoc on plant development and output in the field (Fig. 5.1). Nonetheless, the impacts of these stressors on plants are generally investigated in the laboratory under controlled growth circumstances. Due to their sessile nature, plants are constantly exposed to a wide range of environmental stresses. Abiotic stressors that influence plants and crops in the field are being investigated in depth. Drought, salt, heat, cold, chilling, freezing, nutrition, high light intensity, and agricultural locations across the world are just a few of the factors that might affect crop yield. Environmental stress is a significant challenge in our quest for agricultural sustainability, as it decreases agricultural plant yield by up to 70%. Water stress, which is produced by droughts and temperature extremes, is the most frequent abiotic stress that limits plant productivity. Plants react to these conditions by going through a series of biochemical and physiological changes (Mittler and Blumwald, 2010; Chandra et al., 2018a, b).

Many signaling pathways regulate plant stress responses, and the patterns of gene expression created in response to various stress stimuli overlap. Understanding how plants detect environmental cues and transmit them to the cellular machinery to activate adaptive responses is critical for developing transgenic strategies and rational breeding that improve crop stress tolerance. Plants are challenged by diseases (including viruses, bacteria, nematodes, and fungi) and herbivore pests in natural settings, in addition to abiotic stressors (Atkinson and Urwin, 2012). Climate change has the potential to affect the habitat range of pests and diseases. For example, disease transmission is known to be aided by rising temperatures (Nicol et al., 2011).

Furthermore, several abiotic stress situations have been proven to impair plant defense systems, making them more susceptible to pathogen invasion (Atkinson and Urwin, 2012). As a result, the main crops in future are exposed to a larger range of abiotic and

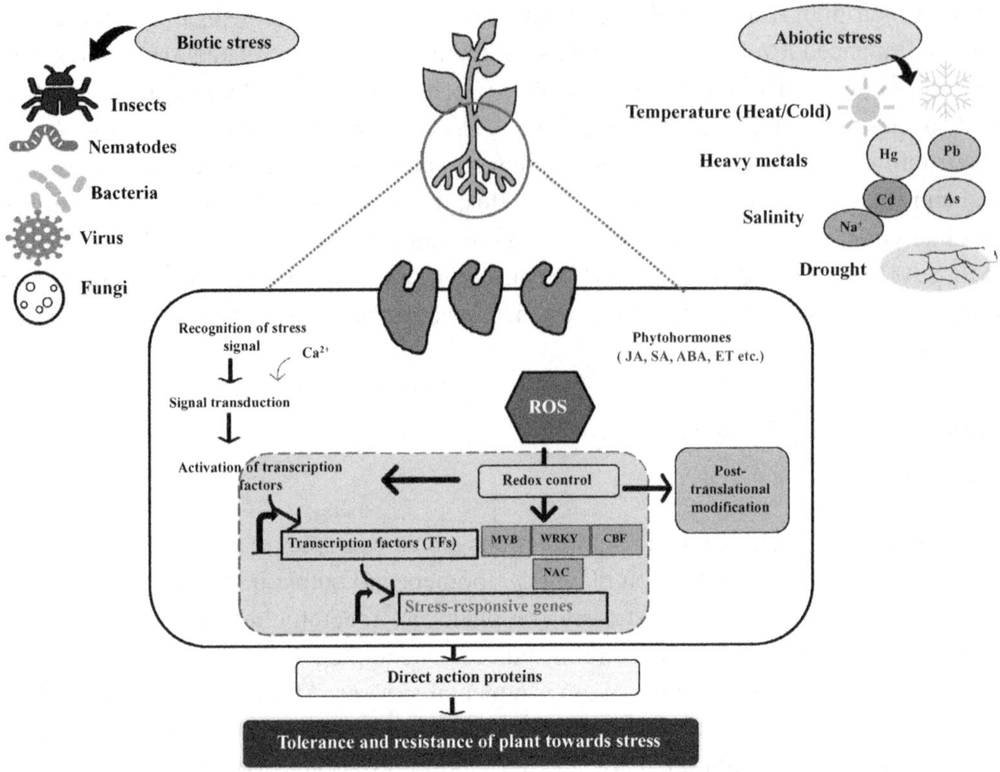

Fig. 5.1
Defense signaling in plants against various environmental stresses.

biotic conditions as well as combinations thereof. A biochemical reaction of plants to a combination of drought and heat stress was identified in 2002 and 2004, and it cannot be inferred directly from the responses of plants to either drought or heat stress when applied separately. Since then, several research studies have revealed how plants respond to various stressors such as droughts, salt, severe temperatures, heavy metals, UV-B, strong light, O_3, CO_2, and soil compaction (Pérez-López et al., 2013). In their natural environment, plants are constantly exposed to a variety of biotic and abiotic stressors. Plants have evolved complicated mechanisms to perceive external signals in order to respond optimally to environmental conditions in order to survive in such situations. Salicylic acid (SA), jasmonic acid (JA), abscisic acid (ABA), and ethylene (ET) are low-molecular-weight chemicals that predominantly regulate a plant's defensive mechanism against both biotic and abiotic challenges. Different yet overlapping sets of genes are regulated by biotic and abiotic stresses. The link between transcriptional regulation and environmental challenges, such as incompatible necrotrophic pathogen infection and heavy metal ($CuSO_4$) stress, for example, shows that biotic and abiotic stress responses are highly overlapping. According to these studies, reactive oxygen species (ROS) appears to be a common signal that causes downstream stress reactions. As a result, a growing amount of evidence reveals that plant signaling pathways are complex networks with regular interactions that enable plants to govern both abiotic stress tolerance and disease resistance (Narusaka et al., 2004).

Biotic stress

Plants are sessile, which means that they cannot escape the stress caused by interactions with insects and microbes including fungi, viruses, bacteria, and nematodes, which cause major yield losses to crops. When a plant detects a stress, a complex network of morphological, physiological, and biochemical systems is activated (Lamers et al., 2020). Infecting plants and inducing biotic stress is caused by a variety of pests, parasites, and pathogens. Fungal parasites can be classified into two groups, i.e., necrotrophic (that kill host cells by secreting poison) or biotrophic (that feed on living host cells). Plants can get leaf spots, vascular wilts, and cankers as a result of them (Sobiczewski et al., 2017). Nematodes are parasitic worms that feed on plant parts and cause soil-borne diseases such as nutrient deficiencies, growth retardation, and wilting. Viruses, too, are likely to cause systemic and local devastation, such as chlorosis and stunting (Pallas and García, 2011). Plants, on the other hand, are degraded by mites and insects that feed on them or lay eggs on them. Other viruses and bacteria could also be transmitted by insects. Plants have evolved sophisticated sensory systems to detect biotic invasion and combat the consequences in terms of growth, yield, and survival. Notwithstanding this, plants have evolved a variety of defense mechanisms that enable them to adapt, thrive, and reproduce under stressful environments. Biotic stress adaptation is tightly regulated, which is crucial for plant survival, as evidenced by the advent of omics technologies (viz., genomics, proteomics, metabolomics, etc.) and ongoing functional characterizations of individual genes. Reactive oxygen species (ROS) signaling, redox status changes, plant hormones, and inorganic ion fluxes, such as Ca^{2+}, are all crucial parts of the regulatory networks that underpin pathogen and microbial detection and defense (Osman et al., 2020).

Plants close their stomata and limit photosynthesis in response to biotic stresses. They also produce harmful toxins such as phytoalexins and reactive oxygen species in reaction to biotic agents as well as induce localized cell death. Many of these responses are coordinated by phytohormones. Plants mainly rely on jasmonic acid (JA), salicylic acid (SA), and ethylene signaling for biotic stress protection. Nevertheless, in biotic interactions, ABA frequently increases plant sensitivity and usually operates abrasively with SA. These signaling cues are passed on to transcription factors, which either activate or inhibit the expression of genes involved in immunological responses and metabolic functions. Although most plant immune responses are governed by transcriptome profiling techniques (such as RNA-seq and hybridization-based microarray), transcription can be used to track the stimulation or inhibition of specific regulatory and metabolic pathways at the genome level during plant–microbe interactions. Independent gene expression profiles can be incorporated into coexpression networks, in which genes are clustered based on pairwise gene expression relationships (Kissoudis et al., 2014; Suzuki et al., 2014; Li et al., 2020).

Impact of stresses on plant productivity

Abiotic stresses are defined as unpleasant environmental conditions that obstruct the physicochemical properties of plants, soils, and associated microbial inhabitants (Sharma et al., 2021). The major abiotic stresses include temperature, drought, salinity, nutrient deficiency, heavy metals, and others (Raza, 2021). Due to plants' sessile nature, they have a significant impact on plant development and output. Temperature stresses involving either heat or cold are detrimental to plant growth and productivity. They can affect the cellular physiology, photosynthesis rate, reproduction, osmolyte concentrations, and the metabolism of a plant. Moreover, both heat and cold stress conditions enhance ROS production and impose oxidative damage to plants, which may lead to cell death (Chandra et al., 2020; Chandra and Sharma, 2020; Jain et al., 2021). Temperature fluctuations in the environment also damage the oxygen-evolving complex of photosystem II. Furthermore, they are known to reduce Rubisco activity, affect the stomatal activity, and disrupt the thylakoid membrane. This indicates how badly temperature stress affects the photosynthetic machinery (Sharma et al., 2020).

Water deficit is another major constraint that affects plant growth and development. This situation occurs when the rate of transpiration exceeds the rate of water absorbed by the plant. It may occur due to poor moisture retention by soil particles, low groundwater levels, and inadequate precipitation. Drought stress may result in reduced cell division, lower leaf surface area, lesser root differentiation, and reduced agronomical parameters (Sarshad et al., 2021). Moreover, it alters the stomatal opening and transportation of minerals and water inside plants. The rate of secondary metabolite formation was found to increase under drought conditions that can be correlated with a plant's adaptation to the stress (Raza, 2021). An increment in the soil salt concentration is referred to as soil salinization. Salinity is a composite natural hazard disturbing the productivity of agricultural crops in barren and semi-barren regions worldwide. Salinization is likely to augment with an extreme change in the global environment and ever-escalating anthropogenic activities such as rapid urbanization and industrialization. Salinity stress affects plant growth and development in several ways, viz., reduced seed germination, reproductive failure, chaffy grains, and reduced agronomical parameters. They ultimately lead to lesser productivity and crop yield (Sharma et al., 2021).

According to reports, salt stress affects more than 44 million hectares of agricultural land worldwide, providing roughly one-third of all food crops (Mahanty et al., 2017; Giordano et al., 2021). The surplus concentration of salt ions in the soil system reduces agricultural productivity and converts productive fields into marginal fields and results in land abandonment. It is contemplated as the main hazard to traditional agricultural practices as it is observed as a prime factor of defacement and inhibition of plant health and development. Solutes containing different ions such as K^+, Cl^-, SO_4^{2-}, CO_3^{2-}, HCO_3^-, Na^+, Ca^{2+}, and Mg^{2+} in the soil system contribute to salinity (Giordano et al., 2021). Plant growth and development are influenced by the relative quantities of Mg^{2+}, K^+, Na^+, and Ca^{2+} in the soil system. Saline soil is defined as soil with an electrical conductivity (EC) of the saturation extract above 4 dS m^{-1} (ca. 40 mmol L^{-1} NaCl) at 25°C and a 15% exchangeable sodium percentage. The increase in soil salinity has impacted a significant percentage of arable lands (20% of all cultivated lands and 33% of all irrigated areas) over the years. Severe soil salinity is among the most damaging abiotic stresses to world agriculture, as it can cause arable soil degradation, especially in extensively irrigated areas (Feng et al., 2021).

Macro- and micronutrient deficiencies in soils are another major limiting factor for plant development. Phosphorus, potassium, and nitrogen are important plant nutrients that must be available in an adequate amount. Their deficiencies cause several diseases that can be identified through visible symptoms (Raza, 2021; Sharma et al., 2021). Nutritional deficiencies may lead to retarded growth, lesser productivity, and reduced crop yield. Moreover, in severe cases, these may lead to plant death. Plants are known to make several associations with microorganisms to overcome these problems.

Heavy metals are environmental pollutants that adversely affect soil and plant health. They may be classified into essential (Co, Ni, Cu, Fe, Zn, Mn, and Cr) and nonessential (Pb, Cd, Hg, As) categories. They may be present naturally in soils as colloids or as complex or soluble ions in extremely small concentrations. However, rapid urbanization, excessive use of agrochemicals, and other anthropogenic activities have increased their concentration enormously. These are extremely hazardous for living organisms and need to be taken care. Authorities have identified 53 heavy metals as global contaminants with densities higher than 5 g/cm^3 (Prieto et al., 2018). They are nonbiodegradable and can persist in the soil for hundreds of years (Singh et al., 2021). Heavy metals damage the physicochemical properties of soils, including pH, texture, and micronutrient concentration. This ultimately leads to the alteration of a plant's metabolism and retarded growth. Arsenic and cadmium are known to interfere with the electron transport chain and ATP synthesis in plants. Moreover, they alter the enzymatic functions involved in CO_2 fixation, phosphorus metabolism, and Calvin cycle. Thus, heavy metal stress is responsible for leaf epinasty, increased lipid peroxidation, chlorosis, oxidative damage, stunted growth, and metabolic alterations (Tiwari and Lata, 2018).

Plant approaches for adaptation and mitigation against stresses

Adaptations

Stresses are the most significant constraints to plants, especially crop yields. Crops must deal with detrimental external factors induced by climatic and edaphic conditions as well as biotic factors through their intrinsic underlying pathways. There is a growing body of knowledge in the biochemical, molecular, and physiological sciences that is opening the door to a better understanding of how cells work in concert. Plants are adapted to detecting, managing, maintaining, or evading changes in their surrounding environments. There is a metabolic cross talk among different biosynthetic channels and mechanisms in their perception of environmental stimuli and their responses to stress (Singh et al., 2020; Kumar et al., 2020). In order to grow and reproduce properly, plants require a variety of nutrients and elements. Plant growth and development are hampered by extreme circumstances (below or above the optimum ranges). An adverse environment characterized by dramatic temperature extremes, salinity, drought, and heat as well as biotic factors, such as fungi, pests, nematodes, insects, etc., produces a diverse array of stress factors. Plants are capable of sensing and responding to a variety of stresses in ways that are beneficial to their survival. Plants have a quick and effective reaction to stress stimuli, triggering a sophisticated stress-specific signaling cascade. To mount an appropriate defense system, phytohormones such as jasmonic acid, abscisic acid, salicylic acid, and ethylene are synthesized, phenolic acids and flavonoids are accumulated, different antioxidants and osmolytes are produced, and transcription factors (TFs) are activated along with the expression of stress-specific genes (Jiang et al., 2016).

Plants may be subjected to a variety of stresses in their environment, which increases the complexity of their reactions to multiple stresses as well as the amount of stresses that they are exposed to, as compared to a single stress. The issue lies in the activation of a specific gene expression, followed by biochemical programming in cells in response to a variety of stressors. Stress susceptibility, as well as its tolerance and defense systems, is a dynamic process that happens throughout a plant's life. Abiotic stress responses, rather than having an additional influence on plant growth, may reduce or enhance plant vulnerability to biotic stresses produced by pests or pathogens, depending on the circumstances (Rizhsky et al., 2004).

Mitigation

Understanding fine-scale molecular mechanisms and their connections that operate under stress conditions is the most important component in plants. This entails a thorough examination of a number of metabolic processes and their regulatory genes in various plant genotypes. To propose effective mitigation techniques, multigenic features connected to stress reactions, as well as the assessment of linked markers for such genes and the probability of pooling out significant genes, must be identified (Meena et al., 2017). Another technique for treating abiotic stressors in plants has been proposed: the use of different biomolecules of plant and microbial origin, whereas biotic stress can be effectively understood by analyzing pathogen interactions. These strategies allow scientists to explore new avenues for finding new ways to reduce abiotic stressors (Meena et al., 2017; Stallworth et al., 2020). Both biotic and abiotic stress mitigation techniques include activating processes that maintain membrane stability as well as inducing the mitogen-activated protein kinase (MAPK) and calcium-dependent protein kinase (CDPK) cascades, which are followed by transcription factor, hormone, and peptide production (Stallworth et al., 2020). Recent agricultural techniques have also demonstrated that PGPR do more than just help mitigate environmental challenges, as in they also boost the yield of a variety of crop plants such as maize, rice, soybeans, and barley (Suarez et al., 2015).

Stress-responsive genes for mitigating abiotic stress responses in plants

Drought

A drought is a major cause of crop yield reductions around the world. It will affect more areas in the future and will last longer due to climate change. An important scientific goal for increasing crop yields is to better understand the mechanisms that underlie a drought response (Singh and Laxmi, 2015). Physiological and molecular changes occur in plants in response to drought and other environmental stresses. These alterations assist plants in growing and resisting stress more efficiently. Many genes whose expression is believed to be important in stress response and tolerance are altered in response to drought stress. Many of these genes have already been discovered and studied (Todaka et al., 2015). Functional and regulatory genes are the two types of stress-inducible *Arabidopsis* genes. Chaperones, late embryogenesis abundant (LEA) proteins, detoxifying enzymes, reactive oxygen species, and proline biosynthetic enzymes fall under the first category of genes encoding proteins. Genes that code for signal transduction and gene expression, such as protein kinases, components of ABA signaling, enzymes involved in lipoprotein communication, or various transcription factors, fall under the first category (Todaka et al., 2015).

The plant hormone ABA, as previously mentioned, regulates a number of transcriptional networks in response to water shortage. Exogenous ABA, on the other hand, has no effect on the expression of several genes that are activated when there is a water shortage. Transcription factors are proteins that bind to the promoters of a group of genes and control their expression. The conversion of stress-induced signals into cellular responses is an important function of these enzymes. A single transcription factor can control the expression of a number of genes at the same time. Regulons are groups of genes that are regulated by the same regulatory protein. Plants activate a large number of regulons in response to drought and other stresses in order to promote optimum crop yield; some of these regulators have been well-studied in *Arabidopsis thaliana*. To overcome drought stress, the transcriptional response is predominantly regulated through both ABA-dependent and ABA-independent pathways by altering one or more regulons that are active when a plant is challenged by drought (Nakashima et al., 2009). There are a number of genes discovered, which are involved in drought stress tolerance such as aquaporins, bZIP, AP2/EREBP, MYB and LEA proteins, etc.

OsPIP proteins are aquaporin subfamilies that are separated into two groups: OsPIP1 and OsPIP2. Several members of the OsPIP1 and OsPIP2 subfamilies have been identified to be affected by drought and salt stress (Hadiarto and Tran, 2011). ARAG1 expression was found in rice roots, inflorescences, endosperms of germinating seeds, and embryos and was quickly upregulated by ABA and drought stress (Zhao et al., 2010).

Temperature stress

Heat stress

As the earth's temperature rises, plants experience stress, posing a serious threat to crop productivity and food security. Heat stress causes plants to activate complex molecular networks such as signal transduction, metabolite synthesis, and the activation of heat stress-related genes. Heat stress tolerance is a polygenic trait influenced by a number of genes, transcription factors, proteins, and hormones. As a result, many mechanisms and phenomena involved in the response to heat stress are necessary to improve heat stress tolerance (Kumar et al., 2015, 2016). Many technologies, such as genomics, next-generation sequencing techniques, genome sequence availability, and advanced biotechnological tools, are now part of a platform that discloses information about plant molecules and offers opportunities to enhance heat stress endurance in crop plants. The potential of several functional genomic approaches, such as gene expression profiling with microarrays, genome-wide association studies (GWASs), and suppression subtractive hybridization (SSH), can catalyze the discovery of novel genes associated with heat stress (Duan et al., 2019). Furthermore, reverse genetic approaches can help researchers better understand their expression patterns in response to heat stress. Using both reverse and forward genetic methods, new genes relevant to heat stress can be discovered and functionally validated. These validated genes can be used in plant breeding strategies and novel biotechnological tools, such as genome editing techniques, to improve heat stress tolerance in crop plants (Table 5.1).

As discussed above, regarding the gene identification and validation technologies, every technology has its merits and demerits, which are discussed below to get a clearer picture. Traditional breeding has become extremely tedious and laborious, whereas precision breeding can accelerate the breeding of a heat-tolerant variety more efficiently. Heat tolerance is a multigenic complex trait that is regulated by genotype-environment interactions. As a result, markers play an important role in precision breeding. Among the markers, SNPs are valued high due to their specific nature. The SNP sequence of a heat-shock protein (HSP16.9) between a heat-tolerant and a heat-susceptible wheat genotype was discovered in investigations (Garg et al., 2012). However, several SNPs linked to heat tolerance have been found in various food crops. Although heat tolerance is a multigenic trait, single SNP molecular markers do not have a significant influence on the progress of numerous SNPs linked to distinct QTLs that affect heat stress tolerance mechanisms (Paul et al., 2018). The

Table 5.1: Techniques used in crops during stress conditions.

Crop	Stress	Technique	Gene/TF	References
Wheat	Heat	Transcriptomics	*HSF* genes, *Hsp* genes, and genes encoding proteins involved in phytohormone biosynthesis/signaling, calcium and sugar signal pathways, RNA metabolism and ribosomes, and primary and secondary metabolism	Qin et al. (2008)
Barley	Heat	Transcriptomics	Heat stress on the central developmental functions of the caryopsis was studied at the transcriptional level	Mangelsen et al. (2011)
Wheat	Heat	Transcriptomics	Heat stress on grain development, which affects storage compounds and quality and yield of grains	Hurkman et al. (2013)
Wheat	Heat	Proteomics	Stress-associated active proteins (SAAPs)	Kumar et al. (2019)
Wheat	Heat	GWAS	QTL mapping for heat stress	Li et al. (2019)
Rice	Salt	GWAS	Transcription factor and transporter	Patishtan et al. (2018)
Maize	Heat	GWAS	QTL and SNP for thermotolerance	Gao et al. (2019)
Sorghum	Heat	GWAS	SNP for heat stress response	Chen et al. (2017)
Brassica napus L.	Heat	GWAS	SNP for heat stress traits and QTL for pollen sterility and pod number	Rahaman et al. (2018)
Tomato	Heat	Transcriptomics	HsfA1a TF heat-tolerant heat-shock protein	Scharf et al. (2012)

locus of those genes/markers/QTL areas related to heat tolerance can be used to improve thermotolerance because there is considerable genetic variability in the germplasms of various crops. Another technique, genome-wide association study (GWAS), has emerged as a powerful tool for identifying the genetic basis for complex phenotypic traits, and it provides a higher mapping resolution than traditional genetic mapping, which provides an exact idea for gene manipulation of forward and reverse genetics, which are used in the identification and validation of heat-responsive genes (Bahuguna et al., 2018; Kumar et al., 2017, 2019).

Growth has been observed in response to heat stress by triggering a variety of heat-responsive genes; hence, transcriptome investigations are useful for identifying prospective heat-responsive genes. Furthermore, new genome editing methods that contributed to the development of the tolerant variety of the ethylene response factor from the AP2/ERF superfamily improved crop plant abiotic stress tolerance (Debbarma et al., 2019).

In heat-tolerant development, recombinant technology is commonly used. Heat stress tolerance in transgenic *Arabidopsis* plants has been improved using AmDREB2C from *Ammopiptanthus mongolicus* (Yin et al., 2018). Furthermore, genome editing techniques such as zinc-finger nucleases (ZFNs), transcription activator-like effector nucleases (TALENs), and clustered regularly interspaced short palindromic repeats (CRISPR) can be used to determine the function of a gene and its validation. Several other technologies are used in gene validation, which is essential before incorporating into cultivars. Virus-induced gene silencing (VIGS), also known as transfer of T mutants, is a quick, efficient, and cost-effective posttranscriptional gene silencing (PTGS) technique used to investigate the function of target genes. Targeting Induced Local Lesions IN Genomes (TILLING) is another nontransgenic technique that permits the PCR-based detection of directed mutations in the target gene sequence, and the function of the target gene may be evaluated from the transformed phenotype owing to that mutation (Gilliham et al., 2017; Kumar et al., 2021).

Several heat-related genes have been discovered in a variety of crops using various methodologies. These genes boost the scavenging activity of reactive oxygen species (ROS) and are important to the response and acclimatization of crops to heat stress. Basically, ROS-related genes upregulate to cope with heat stress. *Hsp* gene expression is influenced by a heat stress transcription factor involved in heat sensing and signaling (Berz et al., 2019). TFs are involved in the control of HSR, since *TF* genes account for 92 of the genes upregulated by HS (bZIP, bHLH, MYB, ERF, and WRKY families) (Table 5.2).

Cold stress

Not only high temperatures but also low temperatures have deleterious effects on plants. As temperate-zone plants have the ability to tolerate low temperatures and their capacity changes from plant to plant, plants such as tomato, potato, cotton, rice, corn, and soybean are chilling-sensitive and cannot acclimatize to cold, whereas oats are chilling-tolerant but freezing-sensitive. However, wheat, barley, and rye are well-adapted to freezing temperatures. Without cold acclimation, winter wheat (*Triticum aestivum* L.) and barley (*Hordeum vulgare* L.) are unable to endure nonfreezing temperatures (Zhao et al., 2015). Different levels of cold exposures alter the quantity of cellular components in cells. A transcriptome analysis has revealed flavonoid biosynthesis; fructose and plant hormones for signal transduction were enriched at an extremely early stage of cold stress (Hao et al., 2018). In tomato plants, a low temperature decreases the ethylene signaling genes with no significant change in the protein level to maintain the ripening activity and cold acclimation (Mata et al., 2019). Plant glutamate-like receptors (GLRs) are implicated in the response to environmental stress. Under cold stress, *Arabidopsis* overexpresses two receptors, GLR1.2 and GLR1.3, which increase endogenous jasmonate levels and promote the downstream CBF/DREB1 signaling pathways. PIP2 aquaporin, MaPIP2–7, and ZjICE1 in banana and *Zoysia japonica* improve multilevel stresses, such as salt, cold, and drought stress. Under stress and recovery conditions,

Table 5.2: Functions of stress tolerance genes in crops.

Plant/crop species	Gene	Functions	Tolerance	Condition	References
Tobacco	*MtSAP1*	Overexpression of *MtSAP1* enhances stress tolerance	Temperature stress	Controlled condition	Charrier et al. (2013)
Sweet potato (*Ipomoea batatas*)	Chloroplastic *BADH* gene from *Spinacia oleracea* (*SoBADH*)	Glycine betaine accumulation, maintenance of cell membrane integrity, photosynthetic activity, reduced ROS production, and quick ROS scavenging by increased activity of free radical scavenging enzymes	Salinity and oxidative and low-temperature stress	Growth chamber	Fan et al. (2012)
Fescue plants	2-Cys peroxiredoxins (2-Cys Prx)	Reduction of lipid peroxidation and electrolyte leakage and maintenance of chlorophyll fluorescence	Heat stress	Growth chamber	Kim et al. (2010)
Arabidopsis	*AT4G18280* (GRPL1 protein)	Glycine-rich cell wall protein as a potential multi-stress-responsive gene	Salinity, drought, and high temperature	Transgenic plant	Tawab et al. (2020)
Arabidopsis	Thpl Thermal hysteresis proteins (antifreeze protein)	Transcription factor in plants An enhanced low-temperature tolerance in transgenic plants was observed by changes in the electrolyte leakage activity, malonyldialdehyde, and proline contents	Low temperature	Transgenic plant	Zhu et al. (2010)
Ammopiptanthus mongolicus	Cold-induced galactinol synthase gene (*AmGS*)	Increase cold tolerance	Low temperature	Transformed into *Photinia serrulata* tree by *Agrobacterium*-mediated transformation	Song et al. (2013)
Tomato	Carotenoid epsilon-ring hydroxylase gene (*LeLUT1*)	*LeLUT1* alleviated photoinhibition and photooxidation and decreased the sensitivity of photosynthesis to chilling stress	Chilling stress	Transgenic tobacco	Zhou et al. (2013)
Triticum aestivum	*bZIP60* (TF)	*TabZIP* TFs are induced by ABA and regulate the expression of stress-related genes in an ABA-dependent manner through interaction with specific ABA-responsive *cis*-acting elements (*ABRE*) in their promoter region	Drought, salt, and freezing tolerance	*Arabidopsis*	Zhang et al. (2015)
Lycospersicum esculentum	*LeAN2*	Produce a high level of antioxidant activity (nonenzymatic) and a low level of ROS accumulation	Heat stress	Tomato	Meng et al. (2015)

aquaporin MaPIP2–7 overexpression reduces malondialdehyde (MDA) and ion leakage (IL) but increases soluble sugar, proline, and abscisic acid (ABA) (Zheng et al., 2018).

C-repeat binding factor (CBF) genes have been found in a variety of plant species, including grasses and trees. CBF is a member of the APETALA2/ETHYLENE RESPONSE FACTOR (AP2/ERF) family that regulates the expression of the *COR* gene (Ritonga and Chen, 2020). C-repeat binding factors (CBFs) activate cold-responsive (*COR*) genes and cryoprotectants during cold acclimation, resulting in cold tolerance. CBF is a transcription factor that controls the expression of several *COR* genes, including *DREB*, *DEAR1*, *CZF2*, *ZF*, *AZF2* (*Arabidopsis* zinc-finger protein 2), and *ZAT10* proteins. Many other *COR* genes are regulated by ZAT12, HSFC1, and CZF1, which increase cold tolerance. Other transcription factors such as ICE1, SOC1, MYB15, and CAMTAs (calmodulin-binding transcription activators) regulate CBF expression. MYB15 inhibits CBF expression by binding directly to MYB recognition sites in the promoters of CBF1, CBF2, and CBF3. It is also plausible that MAPK signaling controls the MYB15 protein during cold acclimation, implying a regulatory network upstream of CBFs (Kim et al., 2017).

Apart from changes in genes, cold also imparts changes in membrane rigidity, protein kinase activation, and calcium influx induction and causes a balance between protein activation and degradation. These calcium influxes activate calcium/calmodulin-controlled receptors on the plasma membrane, such as kinases 1 and 2 (CRLK 1/2) (Zhao et al., 2017). These changes upregulate the CBF expression for cold tolerance. The circadian clock, photoperiod, and light quality all affect CBF expression. In *A. thaliana*, the phytochrome interacting factor 3/4/7 (PIF3/4/7) directly binds to CBF promoters and suppresses CBF expression, whereas circadian clock-associated 1 (CCA1) and late elongated hypocotyl (LHY) directly bind to CBF promoters and enhance CBF expression (Lee and Thomashow, 2012; Jiang et al., 2017). Plastid signals have been involved in the circadian regulation of CBF expression, according to a recent research. CBFs are repressed by the proteins LONG HYPOCOTYL 5 (HY5) and PRR5 (Norén et al., 2016).

Plants' cold tolerance is influenced by the ICE–CBF–COR pathways. Plant hormones such ABA, GA, brassinosteroids, jasmonates, auxins, cytokinins, melatonin, and polyamines play a role in CBF expression during stress. Ubiquitination-mediated proteolysis is a negative regulator of ICE1, whereas SUMO is a positive regulator (a small ubiquitin-related modifier) (Chinnusamy et al., 2007). Cold stress alters the physical and chemical compositions of a cell membrane for more tolerance. Lipids play an important role in resilience and have more tolerance capacity in plants (Barrero-Sicilia et al., 2017). HOS (high expression of osmotically responsive genes) aids plants in freezing cold and is vital for plant growth and development as well as for freezing tolerance via a CBF-independent mechanism. A member of the MADS-box gene family, which includes regulators of blooming time and the *ABCE* genes for floral organ identities, is also engaged in reproductive development under cold stress. Under chilling stress, the MADS-box genes are engaged in plant vegetative and reproductive development, perhaps through interactions with other MADS-box proteins via the K domain, thus controlling gene expression (Song and Chen, 2018).

In *Arabidopsis*, the myeloblastosis transcription factor assists in cold response. The R_2R_3 myeloblastosis protein has expressed its role in several physiological processes, hormonal signaling, and stress response. Several R_2R_3 myeloblastosis genes have been linked to the response to environmental stressors, such as cold stress. Cold acclimation requires osmotically responsive genes such *HOS10*, which encodes an R2R3-type myeloblastosis TF (Zhu et al., 2005). The zinc-finger protein family is involved in cold stress tolerance. *OsiSAP8*, *OsCOIN*, *OsbHLH1*, *OsISAP1*, *ROs-bZIP*, *OsDREB1*/CBF, *SNAC2*, and *OsNAC6* are zinc-finger genes that confer cold stress resistance on rice seedlings. The *Arabidopsis* FAD8 gene encodes a fatty acid desaturase that alters the lipid composition and thus contributes to freezing tolerance. Cold-responsive genes encoding molecular chaperones, such as *hsp70* in spinach and *hsp90* in *Brassica napus*, contribute to freezing tolerance by stabilizing proteins against denaturation caused by freezing. It has been shown that the *OsTPP1* gene, which encodes trehalose-6-phosphate phosphatase, confers cold stress tolerance on rice via promoting the expression of cold-responsive genes (Yadav, 2010) that probably function in stress tolerance, such as LEA protein, chaperones, mRNA-binding proteins, osmotin, antifreeze proteins, the key enzymes for osmolyte biosynthesis, such as water channel proteins, proline, sugar, and proline transporters, detoxification enzymes, proteinase inhibitors, enzymes for fatty acid metabolism, ferritin, and lipid-transfer proteins (Cushman and Bohnert, 2000).

Light and radiation stress

Light is considered an essential factor in terms of plant development, starting from the seed to flowering and fruiting. Sunlight provides light to plants for photosynthesis; generally, plants utilize only 1% of sunlight. However, plants possess the capacity to respond to a broad spectrum of light ranging from UV-B to far red light. Excessive light and ultraviolet (UV) may cause stress and damage at the molecular level. Generally, UV-B (280–315 nm) is more harmful than are UV-A and UV-C. Plant have memories at the cellular level for defense. Low doses of UV-B cause changes in the antioxidant status, such as regulation of glutathione pathways, phenylpropanoid, cinnamate, or flavonoid pathways, and pyridoxine biosynthesis pathways, all of which are linked to the UV RESISTANCE LOCUS 8 pathway, whereas high doses of UV-B cause increases in antioxidants, ROS, and, in severe cases, result in programmed cell death (Hideg et al., 2013). In studies it reflects that UV targets CHALCONE SYNTHASE (CHS), UV-B receptor UV RESISTANCE LOCUS 8 (UVR8) confirm the expedite nature of UV signaling. Generally, UV-B is perceived by the UVr8 protein in plants.

UVR8 is a seven-bladed propeller protein that interacts with CONSTITUTIVELY PHOTOMORPHOGENIC 1 (COP1), REPRESSOR OF UV-B PHOTOMORPHOGENESIS

RUP1, and RUP2 during UV-B signaling (Kaiserli and Jenkins, 2007; Christie et al., 2012). UV-B has an impact at the morphological, physiological, and molecular levels. In plants, UV-B induces the production of flavonol glycosides in the vacuoles of epidermal and subepidermal cells (Hectors et al., 2014). A set of genes encoding biosynthetic enzymes, including *CHALCONE, ISOMERASE* (*CHI*), *CHALCONESYNTHASE* (*CHS*), *FLAVONOL SYNTHASE* (*FLS*), *FLAVANONE 3-HYDROZYLASE* (*F3H*), *LEUCOANTHOCYANIDIN DIOZYGENASE*, and *DIHYDROFLAVONOL4-REDUCTASE* (*DFR*), regulate flavonoid biosynthesis (Tilbrook et al., 2013). The recombination endonuclease gene in *Arabidopsis* (*AtMRE11*) is involved in DNA damage detection and repair as well as male fertility (Stracker and Petrini, 2011).

Numerous genes linked to two phototropins (phot1 and phot2), two UV-A/blue light-absorbing cryptochromes (cry 1 and cry 2), three members of the Zeitlupe family (FKF, ZTL, and LKP2), and various phytochromes that absorb red and far red light have been identified (Schäfer and Nagy, 2006). Similarly, flavin-binding kelch repeat F-box (FKF) and the Lov Kelch repeat protein 2 (LKP2) perceive blue/UV-A light in the range 315–400 nm (Li et al., 2011). These photoreceptors transmit signals to modulate the expression of multiple genes, which leads to a physiological response. The transcriptional regulatory network plays crucial roles in balancing the activation and downregulation of the genes involved in light response. Chromatin immunoprecipitation (ChIP) assays have revealed that UVR8 can bind to the promoter region of its downstream target genes, such as basic leucine-zipper (bZIP) family transcription factor ELONGATED HYPOCOTYL 5 (HY5), which is a key regulator of photomorphogenesis (Brown et al., 2005). Regulation of UVR8-mediated signaling is also carried out through its interaction with WRKY36 and BIM1/BES1 (Yang et al., 2018), inducing histone modification in WRKY transcription gene factors. Mitogen-activated protein kinase phosphatase 1 (mkp 1)/MPK 6/MPK3-mediated stress signaling pathways are crucial to UV-B tolerance in plants (González Besteiro et al., 2011). Another type of high-dose UV-B response might be mediated by ROS, which leads to production of phytohormones, e.g., jasmonate, ethylene, and salicylic acid, and activates a general defense response to UV-B (Brosché and Strid, 2003). Even UV-C light activates the flowering transition in *Arabidopsis* through salicylic acid (SA), which assists in flower development (Martínez et al., 2004). The expression of UVR8 is also found in roots along with stems and leaves. MYB transcription factors, MYB73/77, regulate lateral root growth by interacting with UVR8 in roots in the presence of UV-B by inhibiting auxin signaling and hence integrating UV-B with auxin signaling pathways (Yang et al., 2019). Plants synthesize carotenoids to protect themselves from multiple stresses such as high light and heat stress.

Light stress is characterized by stress markers such as ZAT 12 and BAP1 and increased oxidative stress at night following an extended light period. Light stress induces oxidative bursts, increases apoplastic peroxidase, decreases catalase activity (Abuelsoud et al., 2020), lowers the expression of CIRCADIAN CLOCK ASSOCIATED1 (CCA1) and LATE

ELONGATED HYPOCOTYL (LHY), and reduces cytokinin content or signaling. Ultraviolet stresses damage the covalent bonds of DNA and inhibit transcription and replication. UV stress can be managed by reducing ROS, several transcription factors, and proteases (Nitschke et al., 2016).

Salinity stress

Extreme salinity is a damaging abiotic stress that affects agricultural yield all around the world. Whenever crops are subjected to high salt stress, various stress-responsive genes are triggered, with the ensuing by-products helping to protect the plants. Genetic engineering of salt-tolerant plants can be accomplished using genes that code for salt-tolerant molecules such as osmolytes, ion channels, receptors, calcium signaling components, or enzyme regulators. Multiple stress-responsive genes interact with other stress signal transduction pathway components, resulting in high-salinity stress tolerance or sensitivity in plants. High salinity damages cells primarily by upsetting their ionic and osmotic balance. Abscisic acid is an endogenous hormone that governs plant growth and development as well as the response to abiotic challenges such as salinity stress (Mahajan and Tuteja, 2005). It has been reported that ABA-deficient mutants exhibit poor performance under salinity stress, which confirms the role of ABA in salt stress. Under stressful conditions, the ABA level has been shown to rise due to the induction of genes for the enzymes involved in ABA biosynthesis. β-Carotene is converted to ABA via a variety of enzymatic genes (Xiong et al., 2001).

High salinity stress appears to influence the activation of ABA-sensitive genes such as zeaxanthin oxidase, 9-*cis*-epoxycarotenoid dioxygenase, and molybdenum cofactor sulfurase via a calcium-dependent phosphorylation pathway (Chinnusamy et al., 2004). If there is an excess of ABA, then ABA biosynthetic genes can be activated via the calcium signaling pathway and can degrade ABA through the activation of ABA catabolic enzymes. Osmotic stress-responsive genes are activated by salinity, and this activation can be transmitted via ABA-dependent or ABA-independent pathways (Chinnusamy et al., 2004). However, RD29A gene's transcript accumulation has been found to be ABA-dependent as well as ABA-independent and they also contribute in plant proline accumulation (Zhu, 2002).

In response to salinity stress, the transcripts of pea DNA helicase 45 (PDH45) were upregulated via an ABA-dependent pathway (Sanan-Mishra et al., 2005), whereas pea CBL and CIPK were upregulated via an ABA-independent pathway (Mahajan et al., 2006). Knight et al. (1997) discovered that calcium plays a role in the ABA-dependent induction of the *P5CS* gene under salinity stress. Proline biosynthesis in *Arabidopsis* may also be negatively regulated by phospholipase D, ABA, and calcium (Thiery et al., 2004). Some upstream transcriptional factors regulate stress-induced gene promoters, which contain *cis*-regulatory elements like DRE/CRT, ABRE, and the MYC recognition sequence (also known as MYCRS) and the MYB recognition sequence (also known as MYBRS) (Mahajan et al., 2006). To activate the stress-responsive gene, ABA-dependent salinity stress signaling activates the AREB transcription factors, which bind to the ABRE element (RD29A). Osmotic stress genes have transcription factors such as DREB2A and DREB2B that transactivate the DRE, which helps keep the cell's osmotic balance in check (Mahajan and Tuteja, 2005).

Heavy metal stress

Plants have developed a number of mechanisms to deal with biotic and abiotic stress signals that they receive from their environment. Studies show that a wide range of proteins, genes, and secondary metabolites are produced to help plants cope with a wide range of stresses, such as salt, drought, and waterlogging. These stressors provide plants with an extra layer of protection that they can use to fight back. Here, we have attempted to summarize the key molecules that provide plants with metal toxicity responses and their roles in various plant species.

Exposure to toxic heavy metals is the major cause of adverse health effects in both animals and plants. They can have a significant impact on the developmental processes, aging, and energy synthesis due to their high reactivity. Plants have a remarkable ability to alter their metabolism quickly in response to environmental changes. They have complex processes like perception, transduction, and stress stimuli transmission for this purpose (Turner et al., 2002). Heavy metals have emerged as key abiotic stress agents for living organisms due to their increasing use in developing industries and agrotechnologies as well as their high bioaccumulation and toxicity. These toxins have a significant effect on plants, as they inhibit both above- and below-ground growth while decreasing the activity of the photosynthetic apparatus. These effects are commonly seen as plants age and become more fragile (Krupa et al., 1993).

Excessive exposure to heavy metals can have intriguing effects such as growth inhibition and accelerated aging, and understanding the mechanisms behind these effects is critical in fields such as ecophysiology and medicine. Several genes have been investigated in relation to various heavy metal stresses. For example, jasmonate induces subtilisin-like serine proteases; At-SLP2, a 70 kDa protein equivalent to Hsp70 (heat-shock protein 70), and studied in relation to the stress caused by Cd toxicity in plants (Golldack et al., 2003), and gsh, gsh2, and gr1-glutathione metabolic genes respond to stresses caused by Cu and Cd metal toxicity. When it came to dealing with the toxicity of Cu, Cd, and Hg, both the OsAOC-allene oxide cyclase gene and the OsAOS-allene oxide synthase gene showed positive responses (Agrawal et al., 2003). Another connection with signaling responses can be traced back to heavy metals' strong influence on oxidative processes. The increased production of ROS caused by heavy metals usually damages cellular components such as membranes, nucleic acids, and chloroplast pigments. It is possible that heavy metal-induced oxidative stress causes lipid peroxide formation, which can act as an activation signal for plant defense genes by increasing octadecanoid pathways (Gratão et al., 2005).

Stress-responsive genes for mitigating biotic stress responses in plants

Microbes and pathogens

Living organisms, such as viruses, bacteria, fungi, nematodes, insects, arachnids, and weeds, induce biotic stress in plants. In contrast to abiotic stress, which is caused by environmental factors such as drought and heat, biotic stress agents deprive their hosts of nutrients, resulting in reduced plant vigor and, in extreme cases, death. Pathogen infection causes various changes in the plant physiology, including loss of biomass, early flowering, reduced seed set, formation of defensive metabolites, and more (Cipollini et al., 2014). Nematodes devour plant cells and cause harm to all plant parts; they also allow soil-borne diseases to enter the root system, causing nutrient deficiency symptoms such as stunting and wilting. Viruses, too, cause systemic and local diseases in plants, causing malformations, stunting, and chlorosis in various parts of the plant, even when their hosts are not destroyed. On the other hand, insects and mites should be addressed since they destroy plants by laying eggs or feeding on them. Using their stylets, piercing-sucking insects can transmit diseases to plants (Hilker and Fatouros, 2015). All agricultural systems require effective and long-term management of bacteria, fungi, oomycetes, viruses, nematodes, and insects. Despite the continuing emergence of novel resistant cultivars, pathogen-causing global output losses are substantial. Plant diseases reduce crop yields and crop quality by generating chemicals that are harmful to human health. Pathogens are also becoming increasingly resistant to current resistance genes and insecticides. As a result, some other disease control approaches are required (Baker et al., 2010).

Molecular marker techniques and biotechnology are being used to enhance crop improvement programs based on plant disease resistance genes. As a result, plant resistance genes must be thoroughly investigated in order to address the current problem of pests and diseases as well as abiotic challenges (Chisholm et al., 2006). Pathogen resistance induced by the R protein is usually race-specific, meaning that it only acts against pathogen strains that produce the Avr protein. Such resistance is frequently linked to a hypersensitive response (HR), which manifests as the invasion cell and, in some situations, a few surrounding cells dying quickly. Plant resistance genes and R gene loci need to be structurally and functionally analyzed in order to successfully assemble multiple resistance sources and develop new disease resistance methods in agriculture. Apart from that, to attain the aforementioned objectives, it is critical to comprehend the plant–pathogen interactions (Petit-Houdenot and Fudal, 2017).

Plant resistance genes imparting resistance against a bacterial attack have been examined; however, the genetics of sensitivity for the bulk of plant diseases is less evident. The type-III secretion system (T3SS), which is also known as the injectisome, permits Gram-negative bacteria to release and inject pathogenicity proteins into the cytosol of eukaryotic host cells and has long been recognized to transfer virulence proteins into the host cytoplasm of both plants and animals (Desvaux et al., 2004). The type-III secretion system is encoded by the *hrp* and *hrc* genes, which are related to pathogenicity and HR. On mutations, these genes prevent bacterial pathogenicity in susceptible host plants and have the potential to elicit HRs in the nonhost or cultivar-specific resistant plants. Several type-III secretion system effector proteins have been demonstrated to be reliant on molecular chaperones to keep them partially unfolded in the bacterial cytoplasm. The emerging evidence on their role in pathogenesis suggests that they act as molecular double agents in some interactions, disclosing the pathogen to plant defenses while lowering host responses in others (Puhar and Sansonetti, 2014).

In practically all significant crops, fungal infections are the most important cause of output losses. Several fungus resistance genes have been identified and are currently being explored in crop development efforts. However, sequence diversity within the central LRR domain and variation in the gene's LRR copy number play a key role in defining recognition specificity. For example, sequence differences in the tomato Cf-*4* and Cf-*9* genes play a key role in determining recognition specificity, which confers resistance on the biotrophic leaf mold disease caused by *Cladosporium* and triggers a hypersensitive response (HR) when the fungus's Avr4 and Avr9 peptides are recognized. Ve has a role in the *Verticillium* species' infection resistance in tomatoes. Ve1-mediated resistance signaling only partially overlaps with Cf protein-mediated resistance signaling. A virus-induced gene silencing technique has been recently used to characterize Ve-mediated signaling, and it was discovered that two genes, *EDS1* (Enhanced Disease Susceptibility1) and *NDR1* (non-race-specific disease resistance 1), are required for the signaling cascade downstream of Ve1 (Gayoso et al., 2010).

Nematodes

Plants are constantly exposed to biotic stress, which causes physiological damage and changes in plant metabolism, resulting in a decrease in yield (Gimenez et al., 2018). Due to the constant and rapid evolution of biotic agents such as viruses, bacteria, fungi, insects, and nematodes, biotic stress has become a serious problem. Plants have evolved to mitigate stress responses by expressing or regulating stress-responsive genes and transcription factors because of biotic interactions. Plants respond to nematode stress in a variety of biochemical, physiological, and molecular ways (Chauhan et al., 2017). Plant parasitic nematodes feed on the contents of plant cells and can feed on all sections of the plant, but they predominantly cause soil-borne infections and affect the root system. Plant parasitic nematodes that are sedentary form long-term associations with their hosts. They cause wilting and stunting, which are signs of nutritional inadequacy. Nematodes feed for more than a month on root vascular cells, which are converted into enormous multinucleate feeding cells. It has been determined that nematodes affect several elements of plant development, like transport of auxins and pathways related to plant cell differentiation. These findings also show that effectors play a role in the establishment of feeding cells by inhibiting plant stress and defensive responses (Gheysen and Mitchum, 2011).

An antagonism between the hormones jasmonic acid, salicylic acid, and ethylene defines the response to biotic stimuli. ABA, on the other hand, can serve as a global stress regulator by inhibiting biotic stress signaling pathways. This could enable fine-tuning plant stress responses to focus more easily on the most significant threat, and it could have both positive and negative implications for pathogen defense (Ton et al., 2009). A molecular analysis was principally conducted in *Arabidopsis thaliana*, *Solanum lycopersicum*, *Medicago* spp., and *Nicotiana tabacum* to study the responses of the plants and their interaction with nematodes (de Almeida Engler et al., 2005). Several other crop species are found to have nematode resistance genes, such as *Hs1pro-1*, a gene from a wild variety related to sugar beet, provides resistance to *Heterodera schachtii*; the beet cyst nematode is the first nematode resistance gene to be cloned. Other than this, the *Mi* gene of tomato, which is singly dominant and encodes a protein with CC-NBS-LR motifs, has been found to provide resistance to three major root-knot nematodes, namely, *Meloidogyne arenaria*, *Meloidogyne incognita*, and *Meloidogyne javanica*. The salicylic acid (SA) signaling pathway and mitogen-activated protein kinase (MAPK) cascades are required for Mi-1 resistance (Gururani et al., 2012). AZI1, a lipid transfer protein that is critical for systemic immune priming in distal tissues, is locally elevated after pathogen infection. This pathogen response gene *AZI1* was downregulated when water deficiency and nematode stress were combined (Parker, 2009).

Long noncoding RNAs (lncRNAs) are essential determinants of posttranscriptional and transcriptional processes in a variety of biological regulations, which include development of plants and their responses to stress. A total of 565 long noncoding RNAs (lncRNAs) were found to be differentially expressed in nematode stress responses (Li et al., 2018). Transcription factors (TFs) play a key role in nematode infection responses in the host plant. By releasing miRNAs inside the host plant, nematodes cause plant immunity to be impaired (Zhang et al., 2016). Many target TFs react to nematode stress in either a negative or positive manner. SPL, MYB, CKX, ARF, DCL1, NAM, WRKY, TCP, NAC, ZIP, NFYA, and TOE are among them. These transcription factors promote plant nematode stress tolerance by regulating gene expression, activating miRNAs, and improving plant nematode stress tolerance. The WRKY factors are important components of a plant's innate immune system, and they play an important role in host plant responses to nematode invasion. WRKY11 and WRKY17 were found to be positive regulators of nematode defense and negatively regulated gene expression in *Arabidopsis* roots in several studies (Ali et al., 2014). Furthermore, WRKY2 is a pathogen-inducible defense signaling protein that is preferentially activated during pepper–pathogen interactions (Oh et al., 2006).

Insects

In the natural environment, plants are subjected to a variety of abiotic and biotic stresses. Because of their abundance and diversity, insects are one of the most serious threats to plant survival. Insect herbivores have exerted enormous selection pressure on plants over millions of years, and, as a result, plants are constantly looking for novel methods to escape insect infestations. This has resulted in the evolution of highly sophisticated plant defenses. Plant–insect interactions have been extensively researched in terms of their ecology and evolution (War et al., 2018). Herbivory causes plants to activate signal transduction pathways, which causes changes in the expression of defense-related genes and, eventually, in the activation of biosynthetic pathways. Herbivores evoke plant defense elicitors, and plants respond by activating kinase networks and phytohormones, which start the defensive process (Pieterse et al., 2012).

It is necessary to recognize herbivores and translate that recognition into defense signaling, to reprogram cellular functions in order to have an effective defense response (Erb and Reymond, 2019). The ability of plants to withstand an insect/pest attack is determined by resistance genes (R genes), whereas susceptible genes (S genes) cause them to surrender to stress (Rathinam et al., 2019). It is believed that floras distinguish herbivorous insects by binding HAMPs and DAMPs (herbivore- and damage-associated molecular patterns, respectively) to a specific pattern recognition receptor (PRR) protein, which is consistent with the canonical model of plant herbivore perception. HAMPs have been isolated from a wide variety of insect herbivores (Stahl et al., 2018), and information about their interactions with PRRs is beginning to be gathered.

After an invasion, plants evolve resistance mechanisms that are frequently governed by dominant resistance genes, the products of which directly or indirectly detect certain pathogen effectors and trigger effective defense reactions. The resistance to these insect pathogens activated by the R protein is usually race-specific and only works against pathogen strains that produce the cognate effector protein (the Avr protein) identified by the R protein. This resistance is frequently linked to a hypersensitive response (HR), which manifests as the invaded cell and, in some situations, a few surrounding cells dying quickly (Takahashi et al., 2003). The phytohormones jasmonic acid (JA), salicylic acid (SA), and ethylene (ET) are significant plant signaling phytohormones. Chewing insects and phloem-feeding insects both have signaling pathways mediated by jasmonic acid and SA, respectively. Resistance to the leaf folder *Cnaphalocrocis medinalis* is mediated through the SA and ET signaling pathways in rice plants (Wang et al., 2011). The radiolabeled volicitin (17-hydroxylinolenoyl-L-Gln) characterized from *Spodoptera exigua* secretions (OSs) binds with high affinity to a potential PRR on the plasma membrane of the maize plant. Inhibition of rice defense responses by the C-terminal region of a mucin-like protein receptor produced by a planthopper (BPH), *Nilaparvata lugens*, has been demonstrated (Shangguan et al., 2018). It was reported by

Liu et al. (2015) that rice resistance to BPH is conferred by a cluster of three G-type lectin receptor kinases (LecRKs), but it is unclear whether these proteins bind to the mucin-like protein or to another ligand in this case. Recently, it has been discovered that the rice LRR-RK (leucine-rich repeat receptor kinase protein) is obligatory for awareness and defense against the striped stem borer (SSB) *Chilo suppressalis* in plants (Hu et al., 2018). The nature of the ligands for these orphan PRRs that are found on the cell surface may be revealed by further investigation of these orphan PRRs.

Defense activation by HAMPs may take place independently of PRRs in some cases. It has been discovered that different caterpillar species have different levels of glucose oxidase in their saliva (Acevedo et al., 2015). According to Louis et al. (2013), it encourages defenses in tomato plants, whereas suppresses them in *Nicotiana attenuate* (Diezel et al., 2009). Systemin, an 18 amino acid polypeptide cleaved from prosystemin (an inactive form) in tomato in response to injury, was the first damage-related peptide signal to be reported. Systemin stimulates the production of JA and the expression of genes that code for insecticide-like proteinase inhibitors.

Only a few insect resistance genes from the NBS-LRR plant resistance gene family have been cloned and described to date. Some examples are as follows: the *Mi-1*gene in tomatoes provides resistance to the potato aphid (*Macrosiphum euphorbiae*) and whitefly (*Bemisia tabaci*) and the *Nr* gene in lettuce provides resistance to a single aphid species (*Nasonovia ribisnigri*). In apple and melon, the *Sd1* gene provides resistance to the rosy leaf-curling aphid (*Dysaphis devecta*). An HR to the blue-green aphid and pea aphid is triggered by a single gene called *AIN*. However, even though the HR is elicited in both situations, the resistance is only imparted to the blue-green aphid. Regardless of whether the HR is present, reduced phloem feeding appears to be a common mechanism of the *R* gene-mediated resistance to piercing-sucking insects (Gururani et al., 2012). Plant development and stress response are two examples of biological pathways in which long noncoding RNAs (lncRNAs) play critical roles in posttranscriptional and transcriptional regulation and are particularly abundant (Li et al., 2018). In addition to inflicting crop damage, several insects and pests have developed resistance to chemicals. In pests, a greater activity of complex multigene enzymes such as glutathione-*S*-transferase, esterases, and cytochrome P450 has been frequently observed (Bass and Field, 2011).

Conclusions

As sessile organisms, plants have evolved many stress-responsive genes that allow them to mitigate not just a single stress but also multiple stresses. Understanding the complexities of the plant defense mechanisms, stress signaling, and stress-responsive genes are critical for reducing the impact of stresses on plant growth, development, and production while also extending plant longevity. Studies on plant immunological responses to single and multiple biotic and abiotic stresses have recently revealed the complicated nature and the high degree of signaling cross talk in response to various assailants. Despite remarkable advancements in high-throughput sequencing technologies and potent bioinformatic tools for genomic, proteomic, and transcriptomic studies, stress condition study is still limited. Recent research has shown that biotic and abiotic environmental stressors induce the formation of reactive oxygen species (ROS) and hormone signaling in a nonspecific manner and that these pathways play critical roles in the immune system's response to biotic and abiotic challenges. Different genes also play a significant complex role in plant defense against biotic and abiotic stresses. Thus, identifying multifunctional genes and the genes responsible for biotic and abiotic stress resistance would aid in the development of resistant crop plants through precise genome editing using new technologies like transcription activator-like effector nucleases (TALENs) and CRISPR/Cas9 systems.

References

Abuelsoud, W., Cortleven, A., Schmülling, T., 2020. Photoperiod stress alters the cellular redox status and is associated with an increased peroxidase and decreased catalase activity. bioRxiv. https://doi.org/10.1101/2020.03.05.978270.

Acevedo, F.E., Rivera-Vega, L.J., Chung, S.H., Ray, S., Felton, G.W., 2015. Cues from chewing insects—the intersection of DAMPs, HAMPs, MAMPs and effectors. Curr. Opin. Plant Biol. 26, 80–86.

Agrawal, G.K., Jwa, N.S., Agrawal, S.K., Tamogami, S., Iwahashi, H., Rakwal, R., 2003. Cloning of novel rice allene oxide cyclase (OsAOC): mRNA expression and comparative analysis with allene oxide synthase (OsAOS) gene provides insight into the transcriptional regulation of octadecanoid pathway biosynthetic genes in rice. Plant Sci. 164 (6), 979–992.

Ali, M.A., Wieczorek, K., Kreil, D.P., Bohlmann, H., 2014. The beet cyst nematode *Heterodera schachtii* modulates the expression of WRKY transcription factors in syncytia to favour its development in *Arabidopsis* roots. PLoS One 9 (7), e102360.

Atkinson, N.J., Urwin, P.E., 2012. The interaction of plant biotic and abiotic stresses: from genes to the field. J. Exp. Bot. 63, 3523–3543.

Bahuguna, R.N., Gupta, P., Bagri, J., Singh, D., Dewi, A.K., Tao, L., Islam, M., Sarsu, F., Singla-Pareek, S.L., Pareek, A., 2018. Forward and reverse genetics approaches for combined stress tolerance in rice. Indian J. Plant Physiol. 23 (4), 630–646. https://doi.org/10.1007/s40502-018-0418-0.

Bailey-Serres, J., Voesenek, L.A., 2008. Flooding stress: acclimations and genetic diversity. Annu. Rev. Plant Biol. 59, 313–339.

Baker, C.M., Chitrakar, R., Obulareddy, N., Panchal, S., Williams, P., Melotto, M., 2010. Molecular battles between plant and pathogenic bacteria in the phyllosphere. Braz. J. Med. Biol. Res. 43, 698–704.

Barrero-Sicilia, C., Silvestre, S., Haslam, R.P., Michaelson, L.V., 2017. Lipid remodelling: unravelling the response to cold stress in Arabidopsis and its extremophile relative *Eutrema salsugineum*. Plant Sci. 263, 194–200.

Bass, C., Field, L.M., 2011. Gene amplification and insecticide resistance. Pest Manag. Sci. 67 (8), 886–890.

Berz, J., Simm, S., Schuster, S., Scharf, K.D., Schleiff, E., Ebersberger, I., 2019. HEATSTER: a database and web server for identification and classification of heat stress transcription factors in plants. Bioinf. Biol. Insights 13. 1177932218821365.

Brosché, M., Strid, Å., 2003. Molecular events following perception of ultraviolet-B radiation by plants. Physiol. Plant. 117, 1–10.

Brown, B.A., Cloix, C., Jiang, G.H., Kaiserli, E., Herzyk, P., Kliebenstein, D.J., Jenkins, G.I., 2005. A UV-B-specific signaling component orchestrates plant UV protection. Proc. Natl. Acad. Sci. U. S. A. 102 (50), 18225–18230.

Chandra, D., Sharma, A.K., 2020. Field evaluation of consortium of bacterial inoculants producing ACC deaminase on growth, nutrients and yield components of rice and wheat. J. Crop. Sci. Biotechnol. 24 (3), 293–305.

Chandra, D., Srivastava, R., Glick, B.R., Sharma, A.K., 2018a. Drought-tolerant *Pseudomonas* spp. improve the growth performance of finger millet (*Eleusine coracana* (L.) Gaertn.) under non-stressed and drought-stressed conditions. Pedosphere 28, 227–240.

Chandra, D., Srivastava, R., Sharma, A., 2018b. Influence of IAA and ACC deaminase producing fluorescent pseudomonads in alleviating drought stress in wheat (*Triticum aestivum*). Agric. Res. 7, 290–299.

Chandra, D., Srivastava, R., Gupta, V.V., Franco, C.M., Sharma, A.K., 2019a. Evaluation of ACC-deaminase-producing rhizobacteria to alleviate water-stress impacts in wheat (*Triticum aestivum* L.) plants. Can. J. Microbiol. 65 (5), 387–403.

Chandra, D., Srivastava, R., Glick, B.R., Sharma, A.K., 2020. Rhizobacteria producing ACC deaminase mitigate water-stress response in finger millet (*Eleusine coracana* (L.) Gaertn.). 3 Biotech 10, 65.

Chandra, D., Srivastava, R., Gupta, V.V., Franco, C.M., Paasricha, N., Saifi, S.K., Tuteja, N., Sharma, A.K., 2019b. Field performance of bacterial inoculants to alleviate water stress effects in wheat (*Triticum aestivum* L.). Plant Soil 441 (1), 261–281.

Charrier, A., Lelievre, R., Limami, A.M., Planchet, E., 2013. *Medicago truncatula* stress associated protein 1 gene (MtSAP1) overexpression confers tolerance to abiotic stress and impacts proline accumulation in transgenic tobacco. J. Plant Physiol. 170, 874–877.

Chauhan, S., Yogindran, S., Rajam, M.V., 2017. Role of miRNAs in biotic stress reactions in plants. Indian J. Plant Physiol. 22 (4), 514–529.

Chen, J., Chopra, R., Hayes, C., Morris, G., Marla, S., Burke, J., Xin, Z., Burow, G., 2017. Genome-wide association study of developing leaves' heat tolerance during vegetative growth stages in a *Sorghum* association panel. Plant Genome 10 (2). https://doi.org/10.3835/plantgenome2016.09.0091.

Chinnusamy, V., Schumaker, K., Zhu, J.K., 2004. Molecular genetic perspectives on cross-talk and specificity in abiotic stress signalling in plants. J. Exp. Bot. 55 (395), 225–236.

Chinnusamy, V., Zhu, J., Zhu, J.K., 2007. Cold stress regulation of gene expression in plants. Trends Plant Sci. 12 (10), 1360–1385.

Chisholm, S.T., Coaker, G., Day, B., Staskawicz, B.J., 2006. Host-microbe interactions: shaping the evolution of the plant immune response. Cell 124 (4), 803–814.

Christie, J.M., Arvai, A.S., Baxter, K.J., Heilmann, M., Pratt, A.J., O'Hara, A., Kelly, S.M., Hothorn, M., Smith, B.O., Hitomi, K., Jenkins, G.I., 2012. Plant UVR8 photoreceptor senses UV-B by tryptophan-mediated disruption of cross-dimer salt bridges. Science 335 (6075), 1492–1496.

Cipollini, D., Walters, D., Voelckel, C., 2014. Costs of resistance in plants: from theory to evidence. Annu. Plant Rev. (47), 263–307. https://doi.org/10.1002/9781119312994.apr0512.

Cushman, J.C., Bohnert, H.J., 2000. Genome approaches to plant stress tolerance. Curr. Opin. Plant Biol. 3, 117–124.

Dalvi, A.A., Bhalerao, S.A., 2013. Response of plants towards heavy metal toxicity: an overview of avoidance, tolerance and uptake mechanism. Ann. Plant Sci. 2 (9), 362–368.

de Almeida Engler, J., Favery, B., Engler, G., Abad, P., 2005. Loss of susceptibility as an alternative for nematode resistance. Curr. Opin. Biotechnol. 16 (2), 112–117.

Debbarma, J., Sarki, Y.N., Saikia, B., Boruah, H.P.D., Singha, D.L., Chikkaputtaiah, C., 2019. Ethylene response factor (ERF) family proteins in abiotic stresses and CRISPR–Cas9 genome editing of ERFs for multiple abiotic stress tolerance in crop plants: a review. Mol. Biotechnol. 61 (2), 153–172.

Desvaux, M., Parham, N.J., Scott-Tucker, A., Henderson, I.R., 2004. The general secretory pathway: a general misnomer? Trends Microbiol. 12, 306–309.

Diezel, C., von Dahl, C.C., Gaquerel, E., Baldwin, I.T., 2009. Different lepidopteran elicitors account for cross-talk in herbivory-induced phytohormone signaling. Plant Physiol. 150 (3), 1576–1586.

Duan, S., Liu, B., Zhang, Y., Li, G., Guo, X., 2019. Genome-wide identification and abiotic stress-responsive pattern of heat shock transcription factor family in *Triticum aestivum* L. BMC Genomics 20 (1), 257.

Erb, M., Reymond, P., 2019. Molecular interactions between plants and insect herbivores. Annu. Rev. Plant Biol. 70, 527–557.

Fan, W., Zhang, M., Zhang, H., Zhang, P., 2012. Improved tolerance to various abiotic stresses in transgenic sweet potato (*Ipomoea batatas*) expressing spinach betaine aldehyde dehydrogenase. PLoS One 7 (5), e37344.

Feng, G., Zhu, C., Wu, Q., Wang, C., Zhang, Z., Mwiya, R.M., Zhang, L., 2021. Evaluating the impacts of saline water irrigation on soil water-salt and summer maize yield in subsurface drainage condition using coupled HYDRUS and EPIC model. Agric. Water Manag. 258, 107175.

Fraire-Velázquez, S., Rodríguez-Guerra, R., Sánchez-Calderón, L., 2011. Abiotic and biotic stress response crosstalk in plants. In: Abiotic Stress Response in Plants—Physiological, Biochemical and Genetic Perspectives. InTech, pp. 3–26.

Gao, J., Wang, S., Zhou, Z., Wang, S., Dong, C., Mu, C., Song, Y., Ma, P., Li, C., Wang, Z., He, K., 2019. Linkage mapping and genome-wide association reveal candidate genes conferring thermotolerance of seed-set in maize. J. Exp. Bot. 70 (18), 4849–4864.

Garg, D., Sareen, S., Dalal, S., Tiwari, R., Singh, R., 2012. Heat shock protein based SNP marker for terminal heat stress in wheat (*Triticum aestivum* L.). Aust. J. Crop. Sci. 6 (11), 1516–1521.

Gayoso, C., Pomar, F., Novo-Uzal, E., Merino, F., de Ilárduya, Ó.M., 2010. The Ve-mediated resistance response of the tomato to *Verticillium dahliae* involves H_2O_2, peroxidase and lignins and drives PAL gene expression. BMC Plant Biol. 10 (1), 1–19.

Gheysen, G., Mitchum, M.G., 2011. How nematodes manipulate plant development pathways for infection. Curr. Opin. Plant Biol. 14 (4), 415–421.

Gilliham, M., Able, J.A., Roy, S.J., 2017. Translating knowledge about abiotic stress tolerance to breeding programmes. Plant J. 90 (5), 898–917.

Gimenez, E., Salinas, M., Manzano-Agugliaro, F., 2018. Worldwide research on plant defense against biotic stresses as improvement for sustainable agriculture. Sustainability 10 (2), 391.

Giordano, M., Petropoulos, S.A., Rouphael, Y., 2021. Response and defence mechanisms of vegetable crops against drought, heat and salinity stress. Agriculture 11 (5), 463.

Golldack, D., Vera, P., Dietz, K.J., 2003. Expression of subtilisin-like serine proteases in *Arabidopsis thaliana* is cell-specific and responds to jasmonic acid and heavy metals with developmental differences. Physiol. Plant. 118 (1), 64–73.

González Besteiro, M.A., Bartels, S., Albert, A., Ulm, R., 2011. *Arabidopsis* MAP kinase phosphatase 1 and its target MAP kinases 3 and 6 antagonistically determine UV-B stress tolerance, independent of the UVR8 photoreceptor pathway. Plant J. 68 (4), 727–737.

Gratão, P.L., Polle, A., Lea, P.J., Azevedo, R.A., 2005. Making the life of heavy metal-stressed plants a little easier. Funct. Plant Biol. 32 (6), 481–494.

Grayson, M., 2013. Agriculture and drought. Nature 501, S1. https://doi.org/10.1038/501S1a.

Gururani, M.A., Venkatesh, J., Upadhyaya, C.P., Nookaraju, A., Pandey, S.K., Park, S.W., 2012. Plant disease resistance genes: current status and future directions. Physiol. Mol. Plant Pathol. 78, 51–65.

Hadiarto, T., Tran, L.S.P., 2011. Progress studies of drought-responsive genes in rice. Plant Cell Rep. 30 (3), 297–310.

Hao, X., Wang, B., Wang, L., Zeng, J., Yang, Y., Wang, X., 2018. Comprehensive transcriptome analysis reveals common and specific genes and pathways involved in cold acclimation and cold stress in tea plant leaves. Sci. Hortic. 240, 354–368.

Hectors, K., Van Oevelen, S., Geuns, J., Guisez, Y., Jansen, M.A., Prinsen, E., 2014. Dynamic changes in plant secondary metabolites during UV acclimation in *Arabidopsis thaliana*. Physiol. Plant. 152 (2), 219–230.

Hideg, É., Jansen, M.A., Strid, A., 2013. UV-B exposure, ROS, and stress: inseparable companions or loosely linked associates? Trends Plant Sci. 18 (2), 107–115.

Hilker, M., Fatouros, N.E., 2015. Plant responses to insect egg deposition. Annu. Rev. Entomol. 60, 493–515.

Hu, L., Ye, M., Kuai, P., Ye, M., Erb, M., Lou, Y., 2018. OsLRR-RLK1, an early responsive leucine-rich repeat receptor-like kinase, initiates rice defense responses against a chewing herbivore. New Phytol. 219 (3), 1097–1111.

Hurkman, W.J., Tanaka, C.K., Vensel, W.H., Thilmony, R., Altenbach, S.B., 2013. Comparative proteomic analysis of the effect of temperature and fertilizer on gliadin and glutenin accumulation in the developing endosperm and flour from *Triticum aestivum* L. cv. Butte 86. Proteome Sci. 11 (1), 1–15.

Jain, P., Pandey, B., Singh, P., Singh, R., Singh, S.P., Sonkar, S., Gupta, R., Rathore, S.S., Singh, A.K., 2021. Plant performance and defensive role of glycine betaine under environmental stress. In: Husen, A. (Ed.), Plant Performance under Environmental Stress. Springer, Cham, pp. 225–248.

Jiang, Q.Y., Zhuo, F., Long, S.H., Zhao, H.D., Yang, D.J., Ye, Z.H., Li, S.S., Jing, Y.X., 2016. Can arbuscular mycorrhizal fungi reduce Cd uptake and alleviate Cd toxicity of Lonicera japonica grown in Cd-added soils? Sci. Rep. 6 (1), 1–9.

Jiang, B., Shi, Y., Zhang, X., Xin, X., Qi, L., Guo, H., Li, J., Yang, S., 2017. PIF3 is a negative regulator of the CBF pathway and freezing tolerance in *Arabidopsis*. Proc. Natl. Acad. Sci. U. S. A. 114 (32), E6695–E6702.

Kaiserli, E., Jenkins, G.I., 2007. UV-B promotes rapid nuclear translocation of the Arabidopsis UV-B–specific signaling component UVR8 and activates its function in the nucleus. Plant Cell 19, 2662–2673. https://doi.org/10.1105/tpc.107.053330.

Kim, K.H., Alam, I., Lee, K.W., Sharmin, S.A., Kwak, S.S., Lee, S.Y., Lee, B.H., 2010. Enhanced tolerance of transgenic tall fescue plants overexpressing 2-Cys peroxiredoxin against methyl viologen and heat stresses. Biotechnol. Lett. 32 (4), 571–576.

Kim, S.H., Kim, H.S., Bahk, S., An, J., Yoo, Y., Kim, J.Y., Chung, W.S., 2017. Phosphorylation of the transcriptional repressor MYB15 by mitogen-activated protein kinase 6 is required for freezing tolerance in Arabidopsis. Nucleic Acids Res. 45 (11), 6613–6627.

Kissoudis, C., van de Wiel, C., Visser, R.G., van der Linden, G., 2014. Enhancing crop resilience to combined abiotic and biotic stress through the dissection of physiological and molecular crosstalk. Front. Plant Sci. 5, 207.

Knight, H., Trewavas, A.J., Knight, M.R., 1997. Calcium signalling in *Arabidopsis thaliana* responding to drought and salinity. Plant J. 12 (5), 1067–1078.

Krupa, Z., Siedlecka, A., Maksymiec, W., Baszyński, T., 1993. In vivo response of photosynthetic apparatus of *Phaseolus vulgaris* L. to nickel toxicity. J. Plant Physiol. 142 (6), 664–668.

Kumar, N., Kumar, N., Shukla, A., Shankhdhar, S.C., Shankhdhar, D., 2015. Impact of terminal heat stress on pollen viability and yield attributes of rice (*Oryza sativa* L.). Cereal Res. Commun. 43 (4), 616–626.

Kumar, N., Shankhdhar, S.C., Shankhdhar, D., 2016. Impact of elevated temperature on antioxidant activity and membrane stability in different genotypes of rice (*Oryza sativa* L.). Indian J. Plant Physiol. 21 (1), 37–43. https://doi.org/10.1007/s40502-015-014-z.

Kumar, N., Suyal, D.C., Sharma, I.P., Verma, A., Singh, H., 2017. Elucidating stress proteins in rice (*Oryza sativa* L.) genotype under elevated temperature: a proteomic approach to understand heat stress response. 3 Biotech 7, 205.

Kumar, N., Jeena, N., Singh, H., 2019. Elevated temperature modulates rice pollen structure: a study from foothill Himalayan Agro-ecosystem in India. 3 Biotech 9, 175. https://doi.org/10.1007/s13205-019-1700-1.

Kumar, A., Kumar, P., Singh, H., Kumar, N., 2020. Adaptation and mitigation potential of roadside trees with bio-extraction of heavy metals under vehicular emissions and their impact on physiological traits during seasonal regimes. Urban For. Urban Green. 58. https://doi.org/10.1016/j.ufug.2020.126900, 126900.

Kumar, N., Jeena, N., Kumar, A., Khairakpam, R., Singh, H., 2021. Comparative response of rice cultivars to elevated air temperature in Bhabar region of Indian Himalaya: status on yield attributes. Heliyon 7, e07474.

Lamaoui, M., Jemo, M., Datla, R., Bekkaoui, F., 2018. Heat and drought stresses in crops and approaches for their mitigation. Front. Chem. 6, 26.

Lamers, J., Van Der Meer, T., Testerink, C., 2020. How plants sense and respond to stressful environments. Plant Physiol. 182 (4), 1624–1635.

Lee, C.M., Thomashow, M.F., 2012. Photoperiodic regulation of the C-repeat binding factor (CBF) cold acclimation pathway and freezing tolerance in *Arabidopsis thaliana*. Proc. Natl. Acad. Sci. U. S. A. 109 (37), 15054–15059.

Li, J., Li, G., Wang, H., Deng, X.W., 2011. Phytochrome signaling mechanisms. Arabidopsis Book 9.

Li, X., Xing, X., Xu, S., Zhang, M., Wang, Y., Wu, H., Yang, T., 2018. Genome-wide identification and functional prediction of tobacco lncRNAs responsive to root-knot nematode stress. PLoS One 13 (11), e0204506.

Li, L., Mao, X., Wang, J., Chang, X., Reynolds, M., Jing, R., 2019. Genetic dissection of drought and heat-responsive agronomic traits in wheat. Plant Cell Environ. 42 (9), 2540–2553.

Li, X.L., Xiang, L., Wang, X.H., Qin, P.E.N.G., Zhang, M.S., Ren, M.J., 2020. Biotic and abiotic stress-responsive genes are stimulated to resist drought stress in purple wheat. J. Integr. Agric. 19 (1), 33–50.

Liu, Y., Wu, H., Chen, H., Liu, Y., He, J., Kang, H., Wan, J., 2015. A gene cluster encoding lectin receptor kinases confers broad-spectrum and durable insect resistance in rice. Nat. Biotechnol. 33 (3), 301–305.

Louis, J., Luthe, D.S., Felton, G.W., 2013. Salivary signals of European corn borer induce indirect defenses in tomato. Plant Signal. Behav. 8 (11), e27318.

Mahajan, S., Tuteja, N., 2005. Cold, salinity and drought stresses: an overview. Arch. Biochem. Biophys. 444 (2), 139–158.

Mahajan, S., Sopory, S.K., Tuteja, N., 2006. CBL-CIPK paradigm: role in calcium and stress signaling in plants. Proc. Indian Natl. Sci. Acad. 72 (2), 63.

Mahanty, T., Bhattacharjee, S., Goswami, M., Bhattacharyya, P., Das, B., Ghosh, A., Tribedi, P., 2017. Biofertilizers: a potential approach for sustainable agriculture development. Environ. Sci. Pollut. Res. 24 (4), 3315–3335.

Mangelsen, E., Kilian, J., Harter, K., Jansson, C., Wanke, D., Sundberg, E., 2011. Transcriptome analysis of high-temperature stress in developing barley caryopses: early stress responses and effects on storage compound biosynthesis. Mol. Plant 4, 97–115.

Martínez, C., Pons, E., Prats, G., León, J., 2004. Salicylic acid regulates flowering time and links defence responses and reproductive development. Plant J. 37 (2), 209–217.

Mata, C.I., Hertog, M.L., Van Raemdonck, G., Baggerman, G., Tran, D., Nicolai, B.M., 2019. Omics analysis of the ethylene signal transduction in tomato as a function of storage temperature. Postharvest Biol. Technol. 155, 1–10.

Meena, K.K., Sorty, A.M., Bitla, U.M., Choudhary, K., Gupta, P., Pareek, A., Minhas, P.S., 2017. Abiotic stress responses and microbe-mediated mitigation in plants: the omics strategies. Front. Plant Sci. 8, 172.

Meng, X., Wang, J.R., Wang, G.D., Liang, X.Q., Li, X.D., Meng, Q.W., 2015. An R2R3-MYB gene, LeAN2, positively regulated the thermo-tolerance in transgenic tomato. J. Plant Physiol. 175, 1–8. https://doi.org/10.1016/j.jplph.2014.09.018.

Mickelbart, M.V., Hasegawa, P.M., Bailey-Serres, J., 2015. Genetic mechanisms of abiotic stress tolerance that translate to crop yield stability. Nat. Rev. Genet. 16 (4), 237–251.

Mittler, R., Blumwald, E., 2010. Genetic engineering for modern agriculture: challenges and perspectives. Annu. Rev. Plant Biol. 61, 443–462.

Nakashima, K., Ito, Y., Yamaguchi-Shinozaki, K., 2009. Transcriptional regulatory networks in response to abiotic stresses in *Arabidopsis* and grasses. Plant Physiol. 149 (1), 88–95.

Narusaka, M., Seki, M., Umezawa, T., Ishida, J., Nakajima, M., Enju, A., Shinozaki, K., 2004. Crosstalk in the responses to abiotic and biotic stresses in Arabidopsis: analysis of gene expression in cytochrome P450 gene superfamily by cDNA microarray. Plant Mol. Biol. 55 (3), 327–342.

Nicol, J.M., Turner, S.J., Coyne, D.L., den Nijs, L.J.M.F., Hockland, S., Maafi, Z.T., 2011. Current nematode threats to world agriculture. In: Fenoll, C., et al. (Eds.), Genomics and Molecular Genetics of Plant-Nematode Interactions. Springer, Dordrecht, pp. 21–43.

Nitschke, S., Cortleven, A., Iven, T., Feussner, I., Havaux, M., Riefler, M., Schmülling, T., 2016. Circadian stress regimes affect the circadian clock and cause jasmonic acid-dependent cell death in cytokinin-deficient *Arabidopsis* plants. Plant Cell 28 (7), 1616–1639.

Norén, L., Kindgren, P., Stachula, P., Rühl, M., Eriksson, M.E., Hurry, V., Strand, A., 2016. Circadian and plastid signaling pathways are integrated to ensure correct expression of the CBF and COR genes during photoperiodic growth. Plant Physiol. 171 (2), 1392–1406.

Oh, S.K., Yi, S.Y., Yu, S.H., Moon, J.S., Park, J.M., Choi, D., 2006. CaWRKY2, a chili pepper transcription factor, is rapidly induced by incompatible plant pathogens. Mol. Cells 22 (1) (Springer Science & Business Media BV).

Osman, H.A., Ameen, H.H., Mohamed, M., Elkelany, U.S., 2020. Efficacy of integrated microorganisms in controlling root-knot nematode *Meloidogyne javanica* infecting peanut plants under field conditions. Bull. Natl. Res. Cent. 44 (1), 1–10.

Pallas, V., García, J.A., 2011. How do plant viruses induce disease? Interactions and interference with host components. J. Gen. Virol. 92 (12), 2691–2705.

Parker, J.E., 2009. The quest for long-distance signals in plant systemic immunity. Sci. Signal. 2 (70), pe31.

Patishtan, J., Hartley, T.N., Fonseca de Carvalho, R., Maathuis, F.J., 2018. Genome-wide association studies to identify rice salt-tolerance markers. Plant Cell Environ. 41 (5), 970–982.

Paul, P.J., Samineni, S., Thudi, M., Sajja, S.B., Rathore, A., Das, R.R., Khan, A.W., Chaturvedi, S.K., Lavanya, G.R., Varshney, R., Gaur, P.M., 2018. Molecular mapping of QTLs for heat tolerance in chickpea. Int. J. Mol. Sci. 19 (8), 2166.

Pérez-López, U., Miranda-Apodaca, J., Muñoz-Rueda, A., Mena-Petite, A., 2013. Lettuce production and antioxidant capacity are differentially modified by salt stress and light intensity under ambient and elevated CO_2. J. Plant Physiol. 170 (17), 1517–1525.

Petit-Houdenot, Y., Fudal, I., 2017. Complex interactions between fungal avirulence genes and their corresponding plant resistance genes and consequences for disease resistance management. Front. Plant Sci. 8, 1072.

Pieterse, C.M., Van der Does, D., Zamioudis, C., Leon-Reyes, A., Van Wees, S.C., 2012. Hormonal modulation of plant immunity. Annu. Rev. Cell Dev. Biol. 28, 489–521.

Prieto, M.J., Acevedo, S.O.A., Prieto, G.F., 2018. Phytoremediation of soils contaminated with heavy metals. Biodivers. Int. J. 2 (4), 362–376.

Puhar, A., Sansonetti, P.J., 2014. Type III secretion system. Curr. Biol. 24 (17), R784–R791.

Qin, D., Wu, H., Peng, H., Yao, Y., Ni, Z., Li, Z., Zhou, C., Sun, Q., 2008. Heat stress-responsive transcriptome analysis in heat susceptible and tolerant wheat (*Triticum aestivum* L.) by using wheat genome array. BMC Genomics 9 (1), 1–19.

Rahaman, M., Mamidi, S., Rahman, M., 2018. Genome-wide association study of heat stress-tolerance traits in spring-type *Brassica napus* L. under controlled conditions. Crop J. 6 (2), 115–125.

Rathinam, M., Mishra, P., Mahato, A.K., Singh, N.K., Rao, U., Sreevathsa, R., 2019. Comparative transcriptome analyses provide novel insights into the differential response of Pigeonpea (*Cajanus cajan* L.) and its wild relative (*Cajanus platycarpus* (Benth.) Maesen) to herbivory by *Helicoverpa armigera* (Hübner). Plant Mol. Biol. 101 (1), 163–182.

Raza, A., 2021. Eco-physiological and biochemical responses of rapeseed (*Brassica napus* L.) to abiotic stresses: consequences and mitigation strategies. J. Plant Growth Regul. 40, 1368–1388.

Ritonga, N.F., Chen, S., 2020. Physiological and molecular mechanism involved in cold stress tolerance in plants. Plan. Theory 9, 560. https://doi.org/10.3390/plants9050560.

Rizhsky, L., Liang, H., Shuman, J., Shulaev, V., Davletova, S., Mittler, R., 2004. When defense pathways collide. The response of Arabidopsis to a combination of drought and heat stress. Plant Physiol. 134 (4), 1683–1696.

Sanan-Mishra, N., Pham, X.H., Sopory, S.K., Tuteja, N., 2005. Pea DNA helicase 45 overexpression in tobacco confers high salinity tolerance without affecting yield. Proc. Natl. Acad. Sci. U. S. A. 102 (2), 509–514.

Sarshad, A., Talei, D., Torabi, M., Rafiei, F., Nejatkhah, P., 2021. Morphological and biochemical responses of *Sorghum bicolor* (L.) Moench under drought stress. SN Appl. Sci. 3 (1), 1–12.

Schäfer, E., Nagy, F., 2006. Photomorphogenesis in Plants and bacteria: Function and Signal Transduction Mechanisms. Springer Science & Business Media.

Scharf, K.D., Berberich, T., Ebersberger, I., Nover, L., 2012. The plant heat stress transcription factor (Hsf) family: structure, function and evolution. Biochim. Biophys. Acta Gene Regul. Mech. 1819 (2), 104–119.

Shangguan, X., Zhang, J., Liu, B., Zhao, Y., Wang, H., Wang, Z., Guo, J., Rao, W., Jing, S., Guan, W., Ma, Y., 2018. A mucin-like protein of planthopper is required for feeding and induces immunity response in plants. Plant Physiol. 176 (1), 552–565.

Sharma, A., Kumar, V., Shahzad, B., Ramakrishnan, M., Singh Sidhu, G.P., Bali, A.S., Handa, N., Kapoor, D., Yadav, P., Khanna, K., Bakshi, P., 2020. Photosynthetic response of plants under different abiotic stresses: a review. J. Plant Growth Regul. 39 (2), 509–531.

Sharma, S., Chandra, D., Sharma, A.K., 2021. Rhizosphere plant–microbe interactions under abiotic stress. In: Gupta, V.V.S.R., Sharma, A.K. (Eds.), Rhizosphere Biology: Interactions between Microbes and Plants. Rhizosphere Biology, Springer, Singapore, https://doi.org/10.1007/978-981-15-6125-2_10.

Singh, D., Laxmi, A., 2015. Transcriptional regulation of drought response: a tortuous network of transcriptional factors. Front. Plant Sci. 6, 895.

Singh, H., Yadav, M., Kumar, N., Kumar, A., Kumar, M., 2020. Assessing adaptation and mitigation potential of roadside trees under the influence of vehicular emissions: a case study of *Grevillea robusta* and *Mangifera indica* planted in an urban city of India. PLoS One 15 (1). https://doi.org/10.1371/journal.pone.0227380, e0227380.

Singh, M., Singh, D., Rai, P.K., Suyal, D.C., Saurabh, S., Soni, R., Giri, K., Yadav, A.N., 2021. Fungi in remediation of hazardous wastes: current status and future outlook. In: Yadav, A.N. (Ed.), Recent Trends in Mycological Research. Springer, Cham, pp. 195–224.

Sobiczewski, P., Iakimova, E.T., Mikiciński, A., Węgrzynowicz-Lesiak, E., Dyki, B., 2017. Necrotrophic behaviour of *Erwinia amylovora* in apple and tobacco leaf tissue. Plant Pathol. 66 (5), 842–855.

Song, G.Q., Chen, Q., 2018. Overexpression of the MADS-box gene K-domain increases the yield potential of blueberry. Plant Sci. 276, 22–31.

Song, J., Liu, J., Weng, M., Huang, Y., Luo, L., Cao, P., Sun, H., Liu, J., Zhao, J., Feng, D., Wang, B., 2013. Cloning of galactinol synthase gene from *Ammopiptanthus mongolicus* and its expression in transgenic *Photinia serrulata* plants. Gene 513 (1), 118–127.

Stahl, E., Hilfiker, O., Reymond, P., 2018. Plant–arthropod interactions: who is the winner? Plant J. 93 (4), 703–728.

Stallworth, S., Schumaker, B., Fuller, M.G., Tseng, T.M., 2020. Consequences and mitigation strategies of biotic and abiotic stress in rice (*Oryza sativa* L.). In: Hossain, A. (Ed.), Plant Stress Physiology. IntechOpen, pp. 101–112.

Stracker, T.H., Petrini, J.H., 2011. The MRE11 complex: starting from the ends. Nat. Rev. Mol. Cell Biol. 12 (2), 90–103.

Suarez, C., Cardinale, M., Ratering, S., Steffens, D., Jung, S., Montoya, A.M.Z., Geissler-Plaum, R., Schnell, S., 2015. Plant growth-promoting effects of *Hartmannibacter diazotrophicus* on summer barley (*Hordeum vulgare* L.) under salt stress. Appl. Soil Ecol. 95, 23–30.

Suzuki, N., Rivero, R.M., Shulaev, V., Blumwald, E., Mittler, R., 2014. Abiotic and biotic stress combinations. New Phytol. 203 (1), 32–43.

Takahashi, A., Casais, C., Ichimura, K., Shirasu, K., 2003. HSP90 interacts with RAR1 and SGT1 and is essential for RPS2-mediated disease resistance in *Arabidopsis*. Proc. Natl. Acad. Sci. U. S. A. 100 (20), 11777–11782.

Tawab, F., Munir, I., Nasim, Z., Khan, M.S., Tawab, S., Nasim, A., Iqbal, A., Ahmad, M.A., Ali, W., Munir, R., Asim, N., 2020. Identification and characterization of a novel multi-stress responsive gene in *Arabidopsis*. PLoS One 15 (12), e0244030.

Thiery, L., Leprince, A.S., Lefebvre, D., Ghars, M.A., Debarbieux, E., Savouré, A., 2004. Phospholipase D is a negative regulator of proline biosynthesis in *Arabidopsis thaliana*. J. Biol. Chem. 279 (15), 14812–14818.

Tilbrook, K., Arongaus, A.B., Binkert, M., Heijde, M., Yin, R., Ulm, R., 2013. The UVR8 UV-B photoreceptor: perception, signaling and response. Arabidopsis Book 11. https://doi.org/10.1199/tab.0164, e0164.

Tiwari, S., Lata, C., 2018. Heavy metal stress, signaling, and tolerance due to plant-associated microbes: an overview. Front. Plant Sci. 9, 452.

Todaka, D., Shinozaki, K., Yamaguchi-Shinozaki, K., 2015. Recent advances in the dissection of drought-stress regulatory networks and strategies for development of drought-tolerant transgenic rice plants. Front. Plant Sci. 6, 84.

Ton, J., Flors, V., Mauch-Mani, B., 2009. The multifaceted role of ABA in disease resistance. Trends Plant Sci. 14 (6), 310–317.

Turner, J.G., Ellis, C., Devoto, A., 2002. The jasmonate signal pathway. Plant Cell 14 (suppl_1), S153–S164.

Wang, X., Hu, L., Zhou, G., Cheng, J., Lou, Y., 2011. Salicylic acid and ethylene signaling pathways are involved in production of rice trypsin proteinase inhibitors induced by the leaf folder *Cnaphalocrocis medinalis* (Guenée). Chin. Sci. Bull. 56 (22), 2351–2358.

War, A.R., Taggar, G.K., Hussain, B., Taggar, M.S., Nair, R.M., Sharma, H.C., 2018. Plant defence against herbivory and insect adaptations. AoB Plants 10, ply037.

Xiong, L., Ishitani, M., Lee, H., Zhu, J.K., 2001. The *Arabidopsis* LOS5/ABA3 locus encodes a molybdenum cofactor sulfurase and modulates cold stress–and osmotic stress–responsive gene expression. Plant Cell 13 (9), 2063–2083.

Yadav, S.K., 2010. Cold stress tolerance mechanisms in plants. A review. Agron. Sustain. Dev. 30 (3), 515–527.

Yang, Y., Liang, T., Zhang, L., Shao, K., Gu, X., Shang, R., Shi, N., Li, X., Zhang, P., Liu, H., 2018. UVR8 interacts with WRKY36 to regulate HY5 transcription and hypocotyl elongation in Arabidopsis. Nat. Plants 4 (2), 98–107.

Yang, Y., Zhang, L., Chen, P., Liang, T., Li, X., Liu, H., 2019. UV-B photoreceptor UVR8 interacts with MYB73/MYB77 to regulate auxin responses and lateral root development. EMBO J. 39 (2), e101928.

Yin, Y., Jiang, X., Ren, M., Xue, M., Nan, D., Wang, Z., Xing, Y., Wang, M., 2018. AmDREB2C, from *Ammopiptanthus mongolicus*, enhances abiotic stress tolerance and regulates fatty acid composition in transgenic Arabidopsis. Plant Physiol. Biochem. 130, 517–528.

Zhang, L., Zhang, L., Xia, C., Zhao, G., Liu, J., Jia, J., Kong, X., 2015. A novel wheat bZIP transcription factor, TabZIP60, confers multiple abiotic stress tolerances in transgenic *Arabidopsis*. Physiol. Plant. 153 (4), 538–554.

Zhang, Y., Wang, Y., Xie, F., Li, C., Zhang, B., Nichols, R.L., Pan, X., 2016. Identification and characterization of microRNAs in the plant parasitic root-knot nematode *Meloidogyne incognita* using deep sequencing. Funct. Integr. Genomics 16 (2), 127–142.

Zhao, L., Hu, Y., Chong, K., Wang, T., 2010. ARAG1, an ABA-responsive DREB gene, plays a role in seed germination and drought tolerance of rice. Ann. Bot. 105 (3), 401–409.

Zhao, C., Lang, Z., Zhu, J.K., 2015. Cold responsive gene transcription becomes more complex. Trends Plant Sci. 20 (8), 466–468.

Zhao, C., Wang, P., Si, T., Hsu, C.C., Wang, L., Zayed, O., Yu, Z., Zhu, Y., Dong, J., Tao, W.A., Zhu, J.K., 2017. MAP kinase cascades regulate the cold response by modulating ICE1 protein stability. Dev. Cell 43 (5), 618–629.

Zheng, Y., Luo, L., Wei, J., Chen, Q., Yang, Y., Hu, X., Kong, X., 2018. The glutamate receptors AtGLR1.2 and AtGLR1.3 increase cold tolerance by regulating jasmonate signaling in *Arabidopsis thaliana*. Biochem. Biophys. Res. Commun. 506 (4), 895–900.

Zhou, B., Deng, Y.S., Kong, F.Y., Li, B., Meng, Q.W., 2013. Overexpression of a tomato carotenoid ε-hydroxylase gene alleviates sensitivity to chilling stress in transgenic tobacco. Plant Physiol. Biochem. 70, 235–245.

Zhu, J.K., 2002. Salt and drought stress signal transduction in plants. Annu. Rev. Plant Biol. 53 (1), 247–273.

Zhu, J., Verslues, P.E., Zheng, X., Lee, B.H., Zhan, X., Manabe, Y., Sokolchik, I., Zhu, Y., Dong, C.H., Zhu, J.K., Hasegawa, P.M., 2005. HOS10 encodes an R2R3-type MYB transcription factor essential for cold acclimation in plants. Proc. Natl. Acad. Sci. U. S. A. 102 (28), 9966–9971.

Zhu, B., Xiong, A.S., Peng, R.H., Xu, J., Jin, X.F., Meng, X.R., Yao, Q.H., 2010. Over-expression of *ThpI* from *Choristoneura fumiferana* enhances tolerance to cold in *Arabidopsis*. Mol. Biol. Rep. 37 (2), 961–966.

Further reading

Shinozaki, K., Yamaguchi-Shinozaki, K., 1997. Gene expression and signal transduction in water-stress response. Plant Physiol. 115 (2), 327.

CHAPTER 6

Microbial diversity and root exudates as an important facet in the rhizosphere ecosystem

Kunal Kumar
Amity Institute of Biotechnology, Amity University, Ranchi, Jharkhand, India

Chapter outline

Introduction

The plant root and soil interfacial region, rhizosphere, is occupied by exceptional populations of microorganisms influenced by released biomolecules, phytochemicals, and other plant exudates. The term 'rhizosphere' was coined more than a century ago by the German plant physiologist Lorenz Hiltner; this has always ignited research enthusiasm among earlier plant agronomists and botanists right from the time the term was coined in 1904 to present-day plant biotechnologists. Since then, it has been postulated that the rhizosphere widely accommodates three zones within its periphery, which are defined based on their relative closeness to roots and are influenced by their proximity to roots (Bashir et al., 2016). The apoplastic space between the region of the cortex and endodermis is the endorhizosphere, which is inhabited by microbes and metal ions. The outermost zone is the ectorhizosphere, which ranges from the rhizoplane out to the main soil. The rhizoplane is the zone between the endorhizosphere and the ectorhizosphere and is directly neighboring the roots including the mucilage and root epidermis (Fig. 6.1).

Unravelling Plant-Microbe Synergy. https://doi.org/10.1016/B978-0-323-99896-3.00014-X

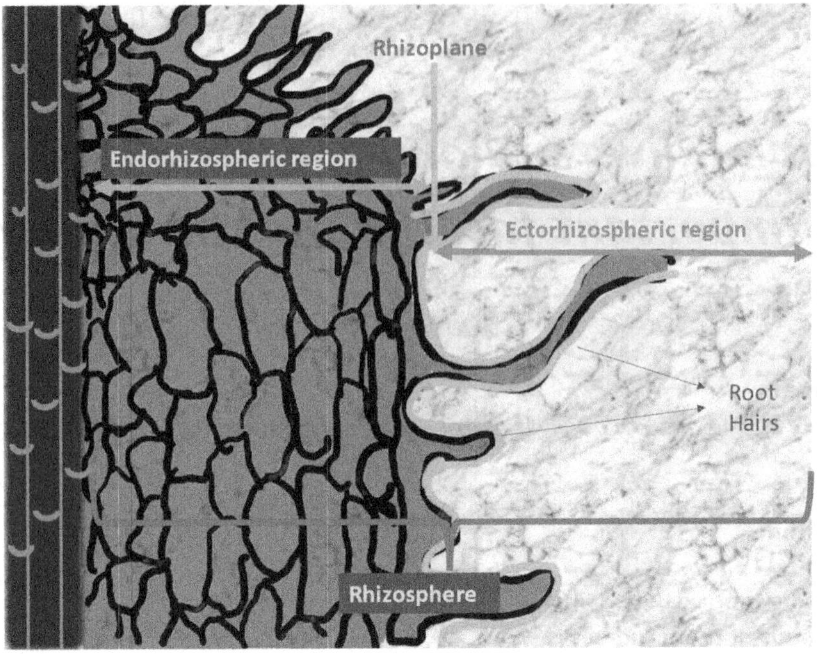

Fig. 6.1
Schematic diagram of the rhizosphere ecosystem (Bashir et al., 2016).

The rhizosphere comprises diverse microbial populations as elucidated in metagenomic studies (Leveau, 2007; Sørensen et al., 2009; Hirsch and Mauchline, 2012) conducted on rhizosphere microbiomes. The profile of the major microbial proportion remains untapped from the vicinity of the rhizosphere (Amann et al., 1995; Kent and Triplett, 2002; Doornbos et al., 2012), and it is expected to harbor a plethora of useful microbes with immense commercial importance. The zones of the soil around the plant root comprise much diverse populations of microorganisms (Foster et al., 1983) as compared to soils deprived of any vegetation. The studies on the complete set of RNA transcripts of these microbiota have unveiled microbial activities in these intricate and complicated environments (Urich et al., 2008; Gosalbes et al., 2012; Jansson et al., 2012), and these would generate the complete details of microbial metabolites being synthesized and the kinds of survival and adaptability strategies employed by the microbial population. Plant genotypes, soil environment, and plant developmental stages influence the microbial diversity of the rhizosphere. The interactions and communication among microorganisms that coexist in the rhizosphere are dynamic, and variations of metabolites exuded by plant roots help outline the configuration of the root microbiome (Bever et al., 2012).

The pattern and chemical composition of root exudates influence microbial activity and diversity, which ultimately have an impact on the invertebrate population including species of the phyla *Nematoda* and *Arthropoda* that share a niche with the microbial population. The symbiotic and parasitic associations existing between rhizosphere organisms and plants, in turn, modify root function and plant growth (Kennedy and de Luna, 2005). The scope of rhizosphere organisms is not restricted to influencing plant health alone, as these do have the ability to form balanced aggregation of soil components for the bioremediation of contaminated soils. With better understanding of the ecosystem and complete knowledge of the components of the rhizosphere, plant productivity and environmental quality can be improved with better soil health, biotic activity, and interactions (Kennedy and de Luna, 2005). The intricate and unclear features of the rhizosphere will continue to motivate scientists interested in the ecology of the plant–soil interface and its associated biota to unveil the intricate developments associated with rhizosphere interactions (Broeckling et al., 2008). Recent findings in rhizosphere biology have highlighted the event of soil microbiomes being invaded by potential human pathogens. Thus, this ecology harbors hazards that may possibly affect the community strength of both producers and consumers. These findings inspire the possibilities of the availability of human health-promoting rhizobacteria, an intriguing thought for future investigation (Dazzo and Ganter, 2009).

The rhizosphere, which is the area near a plant root and the soil adjacent to the roots hairs, and the rhizoplane, which is the plant root surface, both harbor microbes rich in diversity as previously stated. The microbiome of the rhizoplane is influenced by plant root exudates. Plant root exudates are a group of substances in the rhizosphere that are secreted by the root tips of living plants. Root exudates are a complex mixture of polar and semipolar organic substances comprising a range of sugars, amino acids, organic acids, enzymes, and species-specific substances (Vives-Peris et al., 2020).

Plants releasing root exudates

Root exudates include both secretions and diffusates, which are secreted along as well as against the concentration gradients from the roots to the bulk soil. These could also be lysates released by cell disruption of the cortical as well as the epidermal region. The chemical nature of exudates from the plant roots has been identified as an attractant in the soil root ecosystem (Bais et al., 2001; Estabrook & Yoder, 1998). They influence the microbial population of the rhizosphere as well as help in plant water uptake, plant defense, and stimulation. Numerous experiments have discovered that the metabolite profile of the compounds exuded into the rhizosphere is shaped by different internal and external factors. Internal factors such as species type, growth form, and functional characteristics of plants influence and modify the exudate composition in the rhizosphere. Rhizospheric microbial communities, herbivores, and plants growing nearby also define the metabolite cocktail of the rhizosphere ecosystem as exogenous biotic factors (Dietz et al., 2020). The diverse group of chemical compounds released from undamaged and healthy roots comprise a range of carbohydrates, amino acids, small peptides, organic acids, growth factors, enzymes, nucleotides, fungal inhabitants, stimulants, and some other compounds that help in plant water uptake, plant defense, and stimulation (Vives-Peris et al., 2020).

Types and forms of root exudates

Among amino acids, deoxymugineic is the dominant phytosiderophore released from the roots of rice and wheat plants (Prasad et al., 2014). Phytosiderophores (PS) are organic substances produced by plants under iron- and zinc-deficient conditions, which can chelate Fe or Zn and increase their uptake by plants. Many of the phytosiderophores secreting perennial grasses exhibit similar behavior in response to the deficiencies of iron (Fe). *Lolium perenne cv. Tove* secreted 3-epihydroxy-2′-deoxymugineic acid in addition to 2′-deoxymugineic acid (DMA) (Ueno et al., 2007). A nonconventional amino acid, α-aminoadipic, was found to be released from the roots of sweet pea (Kuo et al., 1982). α-Aminoadipic acid plays a critical role in the synthesis of penicillin N by the *Cephalosporium* species (Warren et al., 1967). This could explain the kind of mutual benefit that both plants and microbes can get in a close association. Interestingly γ-aminobutyric, another nonconventional amino acid, has been found to be released in plants under phosphorus-deficient conditions (Carvalhais et al., 2011).

Organic acid production in plants is an integral part of normal metabolism. The synthesis of organic acids in few cases is because of the defense against the toxic effects of intracellular metal–ion concentration (Singh et al., 2016). These acids are chelating agents and scavenge the harmful high concentration of intracellular metal ions. However, more recently, a plethora of organic acids have been reported to be excreted from plant roots. The conventional organic acids found in root exudates are malonic, malic, acetic, fumaric, citric, succinic, oxalic, tartaric, and lactic acids, among others (Badri and Vivanco, 2009). Organic acids of root exudates appear to have a significant role in the solubilization of mineral nutrients of the soil for better uptake by the plant roots. Among nonconventional organic acids, adonic acid has been found to more specifically exudated directly by the roots of N_2-fixing legumes, which serve as the major signals to the *Rhizobiaceae* bacteria responsible for root nodulation, in which N_2 is reduced to ammonia, suggesting the role of aldonic acid in inducing rhizobial *nod* genes (Dakora, 2003).

Another nonconventional organic phenolic acid, caffeic acid in tobacco root exudates, provides protection of tobacco plants from *Ralstonia Solanacearum* infection and defends against tobacco bacterial wilt disease (Li et al., 2021). Other phenolic acids like vanillic acid, *p*-coumaric acid, 4-hydroxybenzoic acid, ferulic acid, and cinnamic acid have been found to affect the soil fungal and bacterial populations either by supporting and enhancing their growth or by inhibiting a few of them and preventing pathogenesis (Zhou and Wu, 2012; Zhang et al., 2020). These reports suggest that root-derived organic inputs influence the diversity of the soil microbial population. Among sterols, stigmasterol and β-sitosterol released from the plant roots of *Spirodela polyrhiza* promote nitrogen removal by the denitrifying rhizosphere microorganisms. The unconventional sterols intensify nitrogen removal by targeting nitrite reductase in bacteria, thereby affecting the 'rhizobioaugmention' in duckweed *Spirodela polyrhiza* and could help in enhancing wastewater purification by promoting plant growth (Lu et al., 2021).

Strigolactones, among many derived carbohydrates, appear in the exudates of vegetation from taxonomic groups of the entire plant kingdom. It has been suggested that they are essentially compounds responsible for the formation of an arbuscular mycorrhizal symbiotic association (Steinkellner et al., 2007). The chemotropic effects of plant peroxidase exudated from stressed plants have been observed to promote the direct growth of the pathogenic fungus *Fusarium oxysporium* toward the roots (Turrà et al., 2015), and, at the same time, stress-induced peroxidase behaves as an attractant for the beneficial *Trichoderma* species and the roots of such plants harbor this beneficial organism (Lombardi et al., 2018).

Peroxidase can utilize phenolic acids exudated by these plants as a source of hydrogen donors for the substrate H_2O_2, thereby influencing the qualitative and quantitative profiles of the phenolic acid exudated by the plants in the rhizoplane (Vaughan et al., 1994). Root exudates of *Medicago sativa* L. and *Phaseolus mungo* L., commonly called lucerne and urid, respectively, upon inoculation with *Rhizobium*, exhibited polygalacturonase activity along with the release of amino-nitrogen, ortho-dihydroxy phenols, and nonreducing sugars (Rao, 1976).

Variations in the metabolite profile among growth forms and within species

The richness of root exudates of the rhizosphere ecosystem has been found to be diverse among growth forms and within species. However, production of few classes of metabolites remains conserved within species and is exudated in the rhizoplane and does not show any qualitative variation. The work of Dietz et al. (2020) revealed that the polar metabolites released by forbs and grasses were mostly identical in quality among the species and less influenced by the growth form. However, grasses and forbs release partial polar metabolites of diverse compositions. The high diversity in the metabolite profiles that has been observed among forbs is because of the recent identification of new species and exudation of more species-specific metabolites than grasses. The phylogenetic distance between the growth forms also contributes to differences in exudate composition. Similarly, the results of Herz et al. (2018), while working on European grassland communities, corroborated similar assumptions.

The plant species of a wetland ecosystem were found to produce several organic compounds including esters, alkanes, amines, alcohols, benzene, and organic acids such as benzoic acid, phthalic acid, cyclopentasiloxane, and cycloheptasiloxane as root exudates of the wetland plants. The wetland plants in response to the presence of fertilizers and chemicals in the water bodies produced root exudates of diverse compositions and varied concentrations. A research work carried by Duan et al. (2020) revealed that *Pistia stratiotes*, *Cyperus alternifolius*, and black algae under various phosphorus concentrations exudated organic compounds. The relative contents of cyclohexasiloxane, benzene dicarboxylic acid, and phthalate in the root exudates during the study under the influence of varied phosphorus concentrations reduced after a slight increase in their content.

Microbial diversity in response to root exudates

The root–microbe interaction of mutualists targets beneficial microbes and suppresses plant pathogens (Marschner et al., 2004; Berg & Smalla, 2009). The roots of plants harboring beneficial microbes in their vicinity and dominantly on the rhizoplane promote their growth under the influence of root exudates. These roots have inhibitory or stimulatory influence on soil microorganisms, which supports their community structure development as they expand their competition for food and resources required for survival (Cesco et al., 2010). Few of the overabundant carbon substrates behave as signaling compounds as they are produced by plant roots. These substrates are categorized into a class called flavonoids, which are accountable for unique host–microbe interactions. More than 4000 different flavonoids are known to mediate host specificity (Birt and Jeffery, 2013; Pascale et al., 2020).

Interestingly, in a host pathogen study among *Fusarium* and host plant species, flavonoids have been found to promote micro- and macroconidia germination but could not exert any influence on the growth of hyphae throughout infections. Specifically, legumes have been identified to release flavonoids into their root exudates. Certain isoflavonoid compounds like genistein and daidzein in the root exudates of legume *Glycine* max effectively and specifically target *Bradyrhizobium japonicum*, a species known to influence root nodulation in legumes and are therefore known as nitrogen-fixing microsymbiotic bacteria. The *nod* gene of the microbes is stimulated by the isoflavones of *Glycine max* (Sugiyama, 2019). The same sets of isoflavones have negative impacts on another nitrogen-fixing bacterium *Sinorhizobium meliloti* and exhibit the phenomenon of inhibition of the *nod* gene (Zhang et al., 2009). On the other hand, strigolactones have been recently identified as significant signaling molecules in the arbuscular mycorrhizal fungi–plant interaction. Plants of the *Brassicaceae* family stimulate ectomycorrhizal fungi for symbiotic association (Zeng et al., 2003). There occurs a specific, interesting interaction between rhizobia and legume, allowing only selective rhizobial strains to nodulate with specific host legumes. *Sinorhizobium meliloti* promotes root nodulation in the genera *Melilotus*, *Medicago* and *Trigonella*, whereas the *Vicia*, *Pisum*, *Lathyrus*, and *Lens* genera are nodulated by *Rhizobium leguminosarum* bv. *viciae* (Bais et al., 2006).

The developmental stage of plants influences root exudation and impacts the microbes sheltering around roots. Variations in sugar levels in response to different developmental stages of the model plant *Arabidopsis* accordingly influence the sugar and secondary metabolism of the microbes harboring in the roots. It has been reported that *Arabidopsis* plants throughout the initial and final stages of their growth can impact the richness of *Bacteroidetes*, *Actinobacteria*, and *Cyanobacteria* as well as their microbial activity (Chaparro et al., 2014). Bacterial species with close genetic similarity are all attracted to common root exudates.

Among all, phenolics appear to act as a major attractant for these bacterial units, followed by carbohydrates and amino acids. A major role of phenolics is witnessed in the *Arabidopsis* ABC transporter mutant (*abcg30*) in attracting beneficial microbes. Phenolics released from the *Arabidopsis* ABC transporter mutant (*abcg30*) have abundant plant growth-promoting rhizobacteria and bacterial species in the vicinity of their roots, thus playing a significant role in the remediation of heavy metals as compared to nonmutant-type plants (Badri et al., 2009). A recent work on phenolics has demonstrated that coumarins display a differential level of toxicity against beneficial and pathogenic microbes and influence the rhizosphere microbiome (Stringlis et al., 2018, Stringlis et al., 2019; Voges et al., 2019). The microenvironment of roots and their associated microbiomes is balanced by benzoxazinoids (Hu et al., 2018; Cotton et al., 2019), triterpenes, and camalexin (Pascale et al., 2020). Arabinogalactan proteins, one among the many secreted molecules found in the entire plant kingdom, regulate *Agrobacterium* and *Rhizobium* attachment on roots (Pascale et al., 2020).

Plant–plant interaction

The growth, survival, development, and reproduction of one organism is influenced by biochemicals of another, a common biological phenomenon called allelopathy. The biochemicals called allelochemicals have beneficial as well as detrimental effects on target organisms. Plant root exudates in plant allelopathy exert positive and negative effects on both receptor and donor plants, thereby playing a significant constructive role in managing agricultural output and in weed control and crop protection. This natural phenomenon also causes autotoxicity, soil sickness, or biological invasion. Reports suggest that the agricultural areas of northwest United States with a widespread presence of the introduced weed diffuse knapweed (*Centaurea diffusa* Lam.) have lost their potential to grow crops of all types. This deleterious effect is a classic example of the plant–plant allelopathy phenomenon, and the phytotoxins of knapweed account for this destructive behavior. The fraction of shoot extracts containing the phytotoxin sesquiterpene caused seed germination and seedling growth inhibition as evident in the studies carried out by Muir et al. (1987). In a different dimension, lactones released by *Oryza sativa* do not allow the growth of other grass species and neighboring lettuce plants (Xu et al., 2014). The incident of intraspecies competition is prevented by catechin, an allelochemical from the noxious weed *Centaurea maculosa*. This is achieved by delayed germination of seeds (Biedrzycki et al., 2010). However, not many experiments examining the deleterious effects of plants on the exudate forms of neighboring plants in a natural habitat have been reported so far.

The exudate sorgoleone released from the roots of Sorghum (*Sorghum bicolor* L.) is a classic example of allelochemicals that inhibit the rise and growth of various weed species (Muzell Trezzi et al., 2016). Infestation of the world's worst weed, nutsedge (*Cyperus rotundus* L.), has been reported to be reduced by sesame plants (Hussain et al., 2017). Numerous other annual crop species are known for their allelochemical synthesis and release, including wheat (*Triticum aestivum* L.), alfalfa (*Medicago sativa* L.), rye (*Secale cereale* L.), barley (*Hordeum vulgare* L.), sunflower (*Helianthus annuus* L.), and rice (*Oryza sativa* L.) (de Albuquerque et al., 2011). Crop rotation and intercropping systems have been employed to control biomass density and weed population through allelochemical-mediated interactions (Cheng and Cheng, 2015). Interesting results were obtained with overall better benefits in the intercropping of sesame (*Sesamum indicum* L.), sorghum (*Sorghum bicolor* L.), and soybean (*Glycine max* L.) in a cotton (*Gossypium hirsutum* L.) field. Such experiments were fruitful in inhibiting purple nutsedge (*Cyperus rotundus* L.) in comparison with fields growing cotton alone (Iqbal et al., 2007). Similarly, intercropping of eggplant and garlic (Wang et al., 2015) contributed to a good yield and better eggplant growth, and, thus, testified intercropping could be an advantageous cultivation practice. Allelochemicals from plants play a dominant role in soil nitrification management with a nitrification-inhibiting property. Soil microbes causing soil nitrification and contributing to pollution with the emission of N_2O can be inhibited by wheat allelochemicals, ferulic acid, hydroxamic acid, and *p*-hydroxybenzoic acid. This prevents loss of nitrogen fertilizers and contributes to their better consumption rate (MA, 2005). Similarly, the allelopathic property of plantain (*Plantago lanceolata* L.) is evident in the reduction of soil nitrogen leaching with mineralization of the soil nitrogen (Dietz et al., 2013), suggesting the beneficial effects of the plant (Table 6.1).

The mechanism of transport of root exudates into the rhizoplane

Passive and active transport are few of the several mechanisms of transport of root exudates into the rhizoplane. Naturally, the release of root exudates has been considered along the concentration gradient (passive transport), facilitated by diffusion through the root membrane, transport mediated by vesicles, and transport involving ionic channels (Baetz and Martinoia, 2014). Carbohydrates, carboxylic acids, amino acids, and phenolics having low molecular weight are transported through diffusion (Fig. 6.2). The driving force behind this transport is the gradient created by the diverse concentrations between the cytoplasm of root cells and the rhizosphere influenced by the nature of the molecules to be exuded, the integrity of the root cells, and root membrane permeability (Badri and Vivanco, 2009).

Table 6.1: List of unique root exudates from major groups of biomolecules.

Root exudates	Chemical group	Host plant	Role in the rhizosphere	References
Deoxymugineic	Amino acids	Roots of rice and wheat plants	Phytosiderophores	Prasad et al. (2014)
3-Epihydroxy-2′-deoxymugineic acid	Amino acids	*Lolium perenne*	Phytosiderophores	Ueno et al. (2007)
α-Aminoadipic	Amino acids	Sweet pea	Synthesis of penicillin N by the *Cephalosporium* species	Kuo et al. (1982)
Adonic acids	Organic acids	Legumes	Inducing rhizobial *nod* genes	Dakora (2003)
Phenolic acid and caffeic acid	Organic acids	Tobacco root	Protecting tobacco plants from *Ralstonia solanacearum* infection and defending against tobacco bacterial wilt disease	Li et al. (2021)
Vanillic acid, ferulic acid, *p*-coumaric acid, cinnamic acid, and 4-hydroxybenzoic acid	Phenolic acids	Diverse group of plants including peanut and tobacco	Affecting the fungal and bacterial populations of the soil	Zhou and Wu (2012), Zhang et al. (2020)
Stigmasterol and β-sitosterol	Sterols	*Spirodela polyrhiza*	Promoting nitrogen removal by the denitrifying rhizosphere microorganism	Lu et al. (2021)
Strigolactones	Derived carbohydrates	Plants from diverse taxa	Essential signaling molecules exhibiting arbuscular mycorrhizal symbiosis	Steinkellner et al. (2007)
Peroxidase	Enzymes	All plants under stress	Attractant for the beneficial *Trichoderma* species	Lombardi et al. (2018)
Polygalacturonase	Enzymes	Urid (*Phaseolus mungo* L.) and lucerne (*Medicago sativa* L.)	Exhibiting polygalacturonase activity along with the release of amino-nitrogen, ortho-dihydroxy phenols, and nonreducing sugars	Rao (1976)
Genistein and daidzein	Isoflavonoid	*Glycine max*	The *nod* gene of the microbe is stimulated by isoflavones of *Glycine max*	Zhang et al. (2009)

Continued

Table 6.1 List of unique root exudates from major groups of biomolecules—cont'd

Root exudates	Chemical group	Host plant	Role in the rhizosphere	References
Benzoxazinoids	Tryptophan-derived secondary metabolites	Maize	Contributing to the balance between roots and the microbiome	Cotton et al. (2019)
Sorgoleone	Allelochemicals	Sorghum	Preventing the rise and development of many weed species	Muzell Trezzi et al. (2016)
Camalexin	Indole alkaloid	Soybean	Providing resistance to nematodes in soybean	Youssef et al. (2013)
Lactones	Cyclic carboxylic esters	Rice	Rice plants prevent the growth of grass species and neighboring lettuce plants	Xu et al. (2014)
Sesquiterpene	Terpenes	Diffuse knapweed	Causing seed germination and seedling growth inhibition	Muir et al. (1987)
Catechin	Phenols	*Centaurea maculosa*	Preventing intraspecies competition and delaying germination of seeds	Biedrzycki et al. (2010)
Ferulic acid and *p*-hydroxybenzoic acid	Phenolic derivatives	Wheat	Wheat allelochemicals prevent loss of nitrogen fertilizers and contribute to their better consumption rate	Ma (2005)

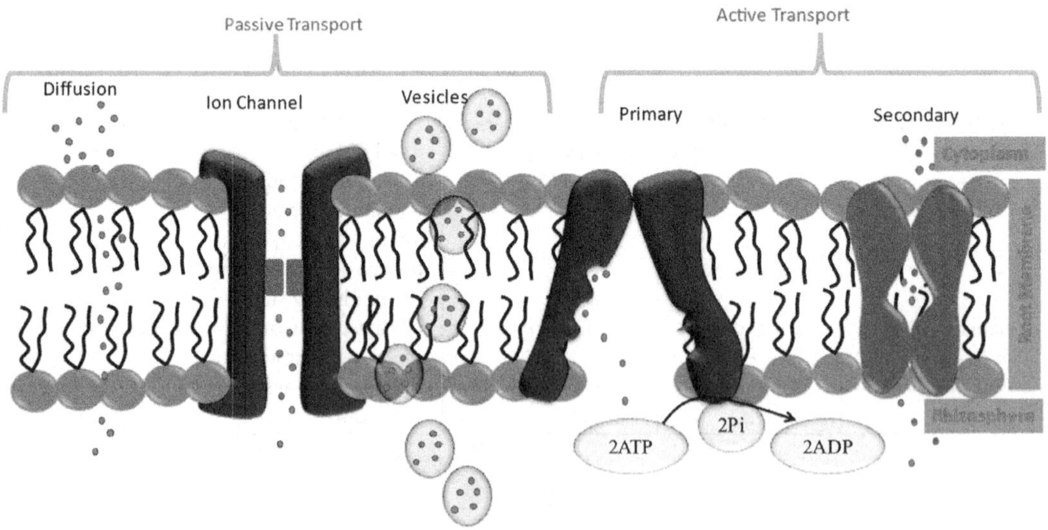

Fig. 6.2
Mechanism of transport of root exudates into the rhizoplane (Vives-Peris et al., 2020).

The secretion of sugars and specific carboxylates such as oxalate and malate (exuded in high quantities) is dependent on ionic channels facilitated by proteins. Two different anionic channels have been described: QUACs (QUick Anion Channels), originally named R-type (Rapid-type), which can be activated in a few milliseconds (Dreyer et al., 2012), and SLACs (SLow Anion Channels), originally named S-type (Slow-type), which need several seconds to be activated.

Conclusions

Transcriptomic and metabolite profiling studies have revealed that the rhizosphere is a hotspot and an intriguing center, imparting scope for conducting ample research in the future. The microbial load of the microbiomes and the plethora of plant exudates released into the microenvironment of the rhizosphere carry immense potential to address the challenges of life. Studies on the genetic makeup of microbes and the plants influenced during the long association could encourage researchers worldwide to isolate novel microbial strains from the vicinity of root hairs and culture them to acquire novel compounds of industrial importance.

References

Amann, R.I., Ludwig, W., Schleifer, K.H., 1995. Phylogenetic identification and in situ detection of individual microbial cells without cultivation. Microbiol. Rev. 59 (1), 143–169. https://doi.org/10.1128/mr.59.1.143-169.1995.

Badri, D.V., Quintana, N., el Kassis, E.G., Kim, H.K., Choi, Y.H., Sugiyama, A., Verpoorte, R., Martinoia, E., Manter, D.K., Vivanco, J.M., 2009. An ABC transporter mutation alters root exudation of phytochemicals that provoke an overhaul of natural soil microbiota. Plant Physiol. 151 (4), 2006–2017. https://doi.org/10.1104/pp.109.147462.

Badri, D.V., Vivanco, J.M., 2009. Regulation and function of root exudates. Plant Cell Environ. 32 (6), 666–681. https://doi.org/10.1111/j.1365-3040.2009.01926.x.

Baetz, U., Martinoia, E., 2014. Root exudates: the hidden part of plant defense. Trends Plant Sci. 19 (2), 90–98. https://doi.org/10.1016/j.tplants.2013.11.006.

Bais, H.P., Loyola-Vargas, V.M., Flores, H.E., Vivanco, J.M., 2001. Root-specific metabolism: the biology and biochemistry of underground organs. In Vitro Cell Dev. Biol. Plant 37 (6), 730–741. https://doi.org/10.1007/s11627-001-0122-y.

Bais, H.P., Weir, T.L., Perry, L.G., Gilroy, S., Vivanco, J.M., 2006. The role of root exudates in rhizosphere interactions with plants and other organisms. Annu. Rev. Plant Biol. 57, 233–266. https://doi.org/10.1146/annurev.arplant.57.032905.105159.

Bashir, O., Khan, K., Hakeem, K.R., Mir, N.A., Rather, G.H., Mohiuddin, R., 2016. Soil microbe diversity and root exudates as important aspects of rhizosphere ecosystem. In: Hakeem, K., Akhtar, M. (Eds.), Plant, Soil and Microbes. Springer, Cham, pp. 337–357, https://doi.org/10.1007/978-3-319-29573-2_15.

Berg, G., Smalla, K., 2009. Plant species and soil type cooperatively shape the structure and function of microbial communities in the rhizosphere. FEMS Microbiol. Ecol. 68 (1), 1–13. https://doi.org/10.1111/j.1574-6941.2009.00654.x.

Bever, J.D., Platt, T.G., Morton, E.R., 2012. Microbial population and community dynamics on plant roots and their feedbacks on plant communities. Annu. Rev. Microbiol. 66 (1), 265–283. https://doi.org/10.1146/annurev-micro-092611-150107.

Biedrzycki, M.L., Jilany, T.A., Dudley, S.A., Bais, H.P., 2010. Root exudates mediate kin recognition in plants. Commun. Integr. Biol. 3 (1), 28–35. https://doi.org/10.4161/cib.3.1.10118.

Birt, D.F., Jeffery, E., 2013. Flavonoids. Adv. Nutr. 4 (5), 576–577. https://doi.org/10.3945/an.113.004465.

Broeckling, C.D., Manter, D.K., Paschke, M.W., Vivanco, J.M., 2008. Rhizosphere ecology. In: Jørgensen, S.E., Fath, B.D. (Eds.), Encyclopedia of Ecology. Academic Press, pp. 3030–3035, https://doi.org/10.1016/B978-008045405-4.00540-1.

Carvalhais, L.C., Dennis, P.G., Fedoseyenko, D., Hajirezaei, M.-R., Borriss, R., von Wirén, N., 2011. Root exudation of sugars, amino acids, and organic acids by maize as affected by nitrogen, phosphorus, potassium, and iron deficiency. J. Plant Nutr. Soil Sci. 174 (1), 3–11. https://doi.org/10.1002/jpln.201000085.

Cesco, S., Neumann, G., Tomasi, N., Pinton, R., Weisskopf, L., 2010. Release of plant-borne flavonoids into the rhizosphere and their role in plant nutrition. Plant Soil 329 (1), 1–25. https://doi.org/10.1007/s11104-009-0266-9.

Chaparro, J.M., Badri, D.V., Vivanco, J.M., 2014. Rhizosphere microbiome assemblage is affected by plant development. ISME J. 8 (4), 790–803. https://doi.org/10.1038/ismej.2013.196.

Cheng, F., Cheng, Z., 2015. Research progress on the use of plant allelopathy in agriculture and the physiological and ecological mechanisms of allelopathy. Front. Plant Sci. 6, 1020. https://doi.org/10.3389/fpls.2015.01020.

Cotton, T.E.A., Pétriacq, P., Cameron, D.D., Almeselmani, M., Schwarzenbacher, R., Rolfe, S.A., Ton, J., 2019. Metabolic regulation of the maize rhizobiome by benzoxazinoids. ISME J. 13 (7), 1647–1658. https://doi.org/10.1038/s41396-019-0375-2.

Dakora, F.D., 2003. Defining new roles for plant and rhizobial molecules in sole and mixed plant cultures involving symbiotic legumes. New Phytol. 158 (1), 39–49. http://www.jstor.org/stable/1514078.

Dazzo, F.B., Ganter, S., 2009. Rhizosphere. In: Schaechter, M. (Ed.), Encyclopedia of Microbiology (Third Edition). Academic Press, pp. 335–349, https://doi.org/10.1016/B978-012373944-5.00287-X.

de Albuquerque, M.B., dos Santos, R.C., Lima, L.M., de Melo Filho, P.A., Nogueira, R.J.M.C., da Câmara, C.A.G., de Rezende Ramos, A., 2011. Allelopathy, an alternative tool to improve cropping systems. A review. Agron. Sustain. Dev. 31 (2), 379–395. https://doi.org/10.1051/agro/2010031.

Dietz, M., Machill, S., Hoffmann, H.C., Schmidtke, K., 2013. Inhibitory effects of *Plantago lanceolata* L. on soil N mineralization. Plant Soil 368 (1), 445–458. https://doi.org/10.1007/s11104-012-1524-9.

Dietz, S., Herz, K., Gorzolka, K., Jandt, U., Bruelheide, H., Scheel, D., 2020. Root exudate composition of grass and forb species in natural grasslands. Sci. Rep. 10 (1), 10691. https://doi.org/10.1038/s41598-019-54309-5.

Doornbos, R.F., van Loon, L.C., Bakker, P., 2012. Impact of root exudates and plant defense signaling on bacterial communities in the rhizosphere. A review. Agron. Sustain. Dev. 32 (1), 227–243. https://doi.org/10.1007/s13593-011-0028-y.

Dreyer, I., Gomez-Porras, J., Riaño Pachón, D.M., Hedrich, R., Geiger, D., 2012. Molecular evolution of slow and quick anion channels (SLACs and QUACs/ALMTs). Front. Plant Sci. 3, 263. https://doi.org/10.3389/fpls.2012.00263.

Duan, X., Zhao, Y., Zhang, J., 2020. Characteristics of the root exudate release system of typical plants in plateau lakeside wetland under phosphorus stress conditions. Open Chem. 18 (1), 808–821. https://doi.org/10.1515/chem-2020-0059.

Estabrook, E.M., Yoder, J.I., 1998. Plant-plant communications: rhizosphere signaling between parasitic angiosperms and their hosts. Plant Physiol. 116 (1), 1–7. https://www.ncbi.nlm.nih.gov/pmc/articles/PMC1539169/.

Foster, R.C., Rovira, A.D., Cock, T.W., 1983. Ultrastructure of the Root-Soil Interface. American Phytopathological Society.

Gosalbes, M.J., Abellan, J.J., Durban, A., Perez-Cobas, A.E., Latorre, A., Moya, A., 2012. Metagenomics of human microbiome: beyond 16s rDNA. Clin. Microbiol. Infect. 18, 47–49. https://doi.org/10.1111/j.1469-0691.2012.03865.x.

Herz, K., Dietz, S., Gorzolka, K., Haider, S., Jandt, U., Scheel, D., Bruelheide, H., 2018. Linking root exudates to functional plant traits. PLoS One 13 (10). https://doi.org/10.1371/journal.pone.0204128, e0204128.

Hirsch, P.R., Mauchline, T.H., 2012. Who's who in the plant root microbiome? Nat. Biotechnol. 30 (10), 961–962. https://doi.org/10.1038/nbt.2387.

Hu, L., Robert, C.A.M., Cadot, S., Zhang, X., Ye, M., Li, B., Manzo, D., Chervet, N., Steinger, T., van der Heijden, M.G.A., Schlaeppi, K., Erb, M., 2018. Root exudate metabolites drive plant-soil feedbacks on growth and defense by shaping the rhizosphere microbiota. Nat. Commun. 9 (1), 1–13. https://doi.org/10.1038/s41467-018-05122-7.

Hussain, I., Singh, N.B., Singh, A., Singh, H., 2017. Allelopathic potential of sesame plant leachate against *Cyperus rotundus* L. Ann. Agrar. Sci. 15 (1), 141–147. https://doi.org/10.1016/j.aasci.2016.10.003.

Iqbal, J., Cheema, Z.A., An, M., 2007. Intercropping of field crops in cotton for the management of purple nutsedge (*Cyperus rotundus* L.). Plant Soil 300 (1–2). https://doi.org/10.1007/s11104-007-9400-8.

Jansson, J.K., Neufeld, J.D., Moran, M.A., Gilbert, J.A., 2012. Omics for understanding microbial functional dynamics. Environ. Microbiol. 14 (1), 1–3. https://doi.org/10.1111/j.1462-2920.2011.02518.x.

Kennedy, A.C., de Luna, L.Z., 2005. Rhizosphere. In: Hillel, D. (Ed.), Encyclopedia of Soils in the Environment. Elsevier, pp. 399–406, https://doi.org/10.1016/B0-12-348530-4/00163-6.

Kent, A.D., Triplett, E.W., 2002. Microbial communities and their interactions in soil and rhizosphere ecosystems. Annu. Rev. Microbiol. 56 (1), 211–236. https://doi.org/10.1146/annurev.micro.56.012302.161120.

Kuo, Y.-H., Lambein, F., Ikegami, F., van Parijs, R., 1982. Isoxazolin-5-ones and amino acids in root exudates of pea and sweet pea seedlings. Plant Physiol. 70 (5), 1283–1289. https://doi.org/10.1104/pp.70.5.1283.

Leveau, J.H.J., 2007. The magic and menace of metagenomics: prospects for the study of plant growth-promoting rhizobacteria. Eur. J. Plant Pathol. 119 (3), 279–300. https://doi.org/10.1007/s10658-007-9186-9.

Li, S., Pi, J., Zhu, H., Yang, L., Zhang, X., Ding, W., 2021. Caffeic acid in tobacco root exudate defends tobacco plants from infection by *Ralstonia solanacearum*. Front. Plant Sci. 12, 690586. https://doi.org/10.3389/fpls.2021.690586.

Lombardi, N., Vitale, S., Turrà, D., Reverberi, M., Fanelli, C., Vinale, F., Marra, R., Ruocco, M., Pascale, A., d'Errico, G., Woo, S.L., Lorito, M., 2018. Root exudates of stressed plants stimulate and attract *Trichoderma* soil Fungi. Mol. Plant-Microbe Interact. 31 (10), 982–994. https://doi.org/10.1094/MPMI-12-17-0310-R.

Lu, Y., Kronzucker, H.J., Shi, W., 2021. Stigmasterol root exudation arising from Pseudomonas inoculation of the duckweed rhizosphere enhances nitrogen removal from polluted waters. Environ. Pollut. 287. https://doi.org/10.1016/j.envpol.2021.117587.

Ma, Y., 2005. Allelopathic studies of common wheat (*Triticum aestivum* L.). Weed Biol. Manag. 5 (3), 93–104. https://doi.org/10.1111/j.1445-6664.2005.00164.x.

Marschner, P., Crowley, D., Yang, C.H., 2004. Development of specific rhizosphere bacterial communities in relation to plant species, nutrition and soil type. Plant Soil 261 (1), 199–208. https://doi.org/10.1023/B:PLSO.0000035569.80747.c5.

Muir, A.D., Majak, W., Balza, F., Towers, G.H.N., 1987. A search for the allelopathic agents in diffuse knapweed. In: Allelochemicals: Role in Agriculture and Forestry. vol. 330. American Chemical Society, pp. 238–246, https://doi.org/10.1021/bk-1987-0330.ch022.

Muzell Trezzi, M., Vidal, R.A., Balbinot Junior, A.A., von Hertwig Bittencourt, H., da Silva Souza Filho, A.P., 2016. Allelopathy: driving mechanisms governing its activity in agriculture. J. Plant Interact. 11 (1), 53–60. https://doi.org/10.1080/17429145.2016.1159342.

Pascale, A., Proietti, S., Pantelides, I.S., Stringlis, I.A., 2020. Modulation of the root microbiome by plant molecules: the basis for targeted disease suppression and plant growth promotion. Front. Plant Sci. 10, 1741. https://doi.org/10.3389/fpls.2019.01741.

Prasad, R., Shivay, Y.S., Kumar, D., 2014. Chapter two—agronomic biofortification of cereal grains with iron and zinc. In: Sparks, D.L. (Ed.), Advances in Agronomy. vol. 125. Academic Press, pp. 55–91, https://doi.org/10.1016/B978-0-12-800137-0.00002-9.

Rao, A.V., 1976. Root exudation in relation to inoculation with rhizobia. Zentralbl. Bakteriol. Parasitenkd. Infektkrankh. Hyg. 131 (1), 79–82. https://doi.org/10.1016/S0044-4057(76)80010-X.

Singh, S., Parihar, P., Singh, R., Singh, V.P., Prasad, S.M., 2016. Heavy metal tolerance in plants: role of transcriptomics, proteomics, metabolomics, and ionomics. Front. Plant Sci. 6, 1143. https://doi.org/10.3389/fpls.2015.01143.

Sørensen, J., Haubjerg Nicolaisen, M., Ron, E., Simonet, P., 2009. Molecular tools in rhizosphere microbiology—from single-cell to whole-community analysis. Plant Soil 321 (1), 483–512. https://doi.org/10.1007/s11104-009-9946-8.

Steinkellner, S., Lendzemo, V., Langer, I., Schweiger, P., Khaosaad, T., Toussaint, J.P., Vierheilig, H., 2007. Flavonoids and strigolactones in root exudates as signals in symbiotic and pathogenic plant-fungus interactions. Molecules 12 (7), 1290–1306. https://doi.org/10.3390/12071290.

Stringlis, I.A., de Jonge, R., Pieterse, C.M.J., 2019. The age of coumarins in plant–microbe interactions. Plant Cell Physiol. 60 (7), 1405–1419. https://doi.org/10.1093/pcp/pcz076.

Stringlis, I.A., Yu, K., Feussner, K., de Jonge, R., van Bentum, S., van Verk, M.C., Berendsen, R.L., Bakker, P.A.H.M., Feussner, I., Pieterse, C.M.J., 2018. MYB72-dependent coumarin exudation shapes root microbiome assembly to promote plant health. Proc. Natl. Acad. Sci. 115 (22), E5213–E5222. https://doi.org/10.1073/pnas.1722335115.

Sugiyama, A., 2019. The soybean rhizosphere: metabolites, microbes, and beyond—a review. J. Adv. Res. 19, 67–73. https://doi.org/10.1016/j.jare.2019.03.005.

Turrà, D., el Ghalid, M., Rossi, F., di Pietro, A., 2015. Fungal pathogen uses sex pheromone receptor for chemotropic sensing of host plant signals. Nature 527 (7579), 521–524. https://doi.org/10.1038/nature15516.

Ueno, D., Rombolà, A.D., Iwashita, T., Nomoto, K., Ma, J.F., 2007. Identification of two novel phytosiderophores secreted by perennial grasses. New Phytol. 174 (2), 304–310. https://doi.org/10.1111/j.1469-8137.2007.02056.x.

Urich, T., Lanzén, A., Qi, J., Huson, D.H., Schleper, C., Schuster, S.C., 2008. Simultaneous assessment of soil microbial community structure and function through analysis of the meta-transcriptome. PLoS One 3 (6). https://doi.org/10.1371/journal.pone.0002527, e2527.

Vaughan, D., Cheshire, M.V., Ord, B.G., 1994. Exudation of peroxidase from roots of Festuca rubra and its effects on exuded phenolic acids. Plant Soil 160 (1), 153–155. http://www.jstor.org/stable/42939459.

Vives-Peris, V., de Ollas, C., Gómez-Cadenas, A., Pérez-Clemente, R.M., 2020. Root exudates: from plant to rhizosphere and beyond. Plant Cell Rep. 39 (1), 3–17. https://doi.org/10.1007/s00299-019-02447-5.

Voges, M.J.E.E.E., Bai, Y., Schulze-Lefert, P., Sattely, E.S., 2019. Plant-derived coumarins shape the composition of an *Arabidopsis* synthetic root microbiome. Proc. Natl. Acad. Sci. U. S. A. 116 (25), 12558–12565. https://doi.org/10.1073/pnas.1820691116.

Wang, M., Wu, C., Cheng, Z., Meng, H., 2015. Growth and physiological changes in continuously cropped eggplant (*Solanum melongena* L.) upon relay intercropping with garlic (*Allium sativum* L.). Front. Plant Sci. 6, 262. https://doi.org/10.3389/fpls.2015.00262.

Warren, S., Newton, G., Abraham, E., 1967. Use of α-aminoadipic acid for the biosynthesis of penicillin N and cephalosporin C by a *Cephalosporium* sp. Biochem. J. 103 (3), 891–901. https://doi.org/10.1042/bj1030891.

Xu, J., Jiang, C., Zhang, Z., Ma, W., Guo, Y., 2014. Recent progress regarding the bioactivities, biosynthesis and synthesis of naturally occurring resorcinolic macrolides. Acta Pharmacol. Sin. 35 (3), 316–330. https://doi.org/10.1038/aps.2013.155.

Youssef, R.M., MacDonald, M.H., Brewer, E.P., Bauchan, G.R., Kim, K.H., Matthews, B.F., 2013. Ectopic expression of AtPAD4 broadens resistance of soybean to soybean cyst and root-knot nematodes. BMC Plant Biol. 13 (1), 67. https://doi.org/10.1186/1471-2229-13-67.

Zeng, R.S., Mallik, A.U., Setliff, E., 2003. Growth stimulation of ectomycorrhizal fungi by root exudates of brassicaceae plants: role of degraded compounds of indole glucosinolates. J. Chem. Ecol. 29 (6), 1337–1355. https://doi.org/10.1023/A:1024257218558.

Zhang, H., Yang, Y., Mei, X., Li, Y., Wu, J., Li, Y., Wang, H., Huang, H., Yang, M., He, X., Zhu, S., Liu, Y., 2020. Phenolic acids released in maize rhizosphere during maize-soybean intercropping inhibit *Phytophthora* blight of soybean. Front. Plant Sci. 11, 886. https://doi.org/10.3389/fpls.2020.00886.

Zhang, J., Subramanian, S., Stacey, G., Yu, O., 2009. Flavones and flavonols play distinct critical roles during nodulation of *Medicago truncatula* by *Sinorhizobium meliloti*. Plant J. 57 (1), 171–183. https://doi.org/10.1111/j.1365-313X.2008.03676.x.

Zhou, X., Wu, F., 2012. Dynamics of the diversity of fungal and *Fusarium* communities during continuous cropping of cucumber in the greenhouse. FEMS Microbiol. Ecol. 80 (2), 469–478. https://doi.org/10.1111/j.1574-6941.2012.01312.x.

CHAPTER 7

Advantages of using halotolerant/halophilic bacteria in agriculture

Furkan Orhan[a], **Derya Efe**[b], **and Arzu Gormez**[c]
[a]*Department of Molecular Biology and Genetics, Agri Ibrahim Cecen University, Agri, Turkey,* [b]*Department of Medicinal and Aromatic Plants, Giresun University, Giresun, Turkey,* [c]*Department of Molecular Biology and Genetics, Erzurum Technical University, Erzurum, Turkey*

Chapter outline

Introduction

The population of the world is estimated to reach approximately 9 billion by 2050. As soil productivity of agricultural lands has decreased due to human activities, the yield obtained from these lands has dramatically decreased (Wood, 2001). On the other hand, by 2030, it is estimated (Food and Agriculture Organization) that the total demand for agricultural products will increase by 60% compared to the current demand due to the expected increase in population. Consequently, one of the major problems, which human beings will face in the future, is malnutrition, as the world population and agricultural production are not increasing in parallel. Each country, for its own sake, must develop agricultural strategies to meet the food demands of its growing population (Chandra et al., 2020; Chandra and Sharma, 2020). The low yield of agricultural production is generally attributed to different abiotic stresses, including drought, cold, heat, and high salinity (Ahmad et al., 2012; Chandra et al., 2018a, b, 2019a, b). The term 'salinity', in its broad sense, is used to describe the presence of an

Unravelling Plant-Microbe Synergy. https://doi.org/10.1016/B978-0-323-99896-3.00011-4

excessive amount of soluble salts in soil and water. Salinity is one of the most important factors restricting the utilization of arable land. The reduction in the utilization of arable lands as a result of salinity is about 1%–2% every year. More specifically, approximately 7% of the land on Earth has been adversely affected by salinity (FAO, 2005; Shrivastava and Kumar, 2015). Changes in climate are expected to make the situation even worse, and, by 2050, it is predicted that approximately 50% of cultivated agricultural soils would be affected by salinity (Othman et al., 2006; Jamil et al., 2011). Under moderate saline conditions, the yield loss of 30 crop species is about 50%–80%, which contributes to about 90% of plant-derived food demands (Panta et al., 2014). Soil salinity is one of the most serious problems, reducing agricultural productivity in arid and semiarid areas as these regions receive insufficient rainfall to leach the excess salts from the plant root zones (Munns, 2005). Natural or primary salinity occurs due to salt accumulation in the soil and/or groundwater because of natural processes. In these natural processes, soluble salts (chlorides of sodium, magnesium, calcium, carbonates, and sulfates) are released due to the weathering of the parent rock containing soluble salts and the accumulating oceanic salts, mainly sodium chloride, carried through wind and rain (Safdar et al., 2019). Human-induced or secondary salinity occurs due to anthropic activities such as irrigation using salt-rich water or insufficient drainage.

Salinity results in 'salt stress', which is described as the adverse effect of excess minerals such as Na^+ and/or Cl^- on plants. As known, osmotic stress and ionic toxicity are the adverse effects of salinity that damage plants. Osmotic stress first exerts detrimental hyperosmotic effects on plants via nutritional imbalance and low osmotic potential of soil solution. Second, ionic toxicity causes oxidative damages to plants (Zhu, 2001). Salinity significantly decreases crop yields because of its effects on water uptake and ion homeostasis, photosynthesis, respiration, nutrient assimilation, and hormonal imbalance. Moreover, salinity generates reactive oxygen species (ROS) and damages macromolecules including proteins, lipids, and nucleic acids in stressed plants and hence constrains crop productivity (Safdar et al., 2019). Saline soils can be remediated using several different approaches including application of chemicals or utilization of halophytic plants. While the former approach has the potential to deteriorate soils as more chemicals (including salts) are added, the latter will be insufficient to meet the food needs of plants, as the number of halophytic plants is limited (Qadir et al., 2007; Kiliç et al., 2008; Shaygan et al., 2018). The utilization of PGP bacteria is an exceedingly popular approach and provides promising opportunities. In fact, the effects of bacteria on plant growth have long been known, since from the time the term 'PGPR' was first coined in the late 1970s, inoculation of plants with bacteria has been used to increase agricultural yields. It is certain that bacteria have positive impacts on plant growth, but there are also situations including saline conditions in which their utilization is limited. The growth of a huge number of agricultural plants under salinity stress has been shown to be stimulated with various halophilic/halotolerant PGP bacterial inoculations (Etesami and Alikhani, 2019). In light of this information, we aimed to evaluate the proven potential of halophilic/halotolerant PGP bacteria in soils under salinity stress.

Halophilic/halotolerant bacteria and their importance in agricultural applications

Halophiles have been identified as extremophilic microorganisms that exhibit optimal growth in saline environments (Gupta et al., 2014). Halophiles grow optimally in environments containing 0.5–5.2 M NaCl (Sarwar et al., 2015). According to their response to salt concentration, these organisms can be divided into four categories: nonhalophiles, slight halophiles, moderate halophiles, and extreme halophiles. Organisms growing optimally in an environment containing < 0.2 M NaCl are termed 'nonhalophiles'. On the other hand, organisms surviving in an NaCl concentration between 0.2 and 0.5 M and those growing best in an NaCl concentration between 0.2 and 0.5 M are described as 'slight halophiles' and 'moderate halophiles', respectively. Finally, extreme halophiles are defined as those that optimally grow in hypersaline environments (1.5–5.2 M NaCl) (Kumar and Khare, 2012). Hypersaline environments contain a great diversity of microorganisms including *Eukarya*, *Bacteria*, and *Archaea*. When a nonhalophilic microorganism inhabits a hypersaline environment, its cells lose water, shrink, and, subsequently, die because of fatal loss of cellular structure as well as function (Oren, 1999). Therefore, any microorganism thriving in hypersaline conditions is expected to keep its cytoplasm at least isosmotic or hyperosmotic with the extracellular environment (Oren, 2002). These organisms provide osmotic balance between the cell cytoplasm and the hypersaline medium using two main strategies, defined as high-salt-in and low-salt-in (production of a compatible solute) (Gunde-Cimerman et al., 2018). Osmotic balance, a strategy used to deal with high salt concentrations, is a result of intracellularly agglomerating high-intensity concentrations of potassium and chloride ions.

The high-salt-in strategy is based on KCl rather than on NaCl as the main intracellular salt, excluding Na^+ ions as much as possible from cells in all three domains of life (Oren, 2008). The other strategy is identified as the low-salt-in or compatible solute strategy. In this strategy, organic osmotic molecules are either synthesized or taken up from the environment and then accumulated in cells (Oren, 1999). The compatible solutes used in the low-salt-in strategy are osmoprotectants with a neutral charge or zwitterion at the physiological pH level and with and low molecular weight. These compounds, even at extremely high concentrations, allow enzymes to perform their function in cells. Therefore, halophilic microorganisms that provide osmotic balance with the compatible solute strategy do not need enzymes that require extra adaptations (Galinski, 1995; Gunde-Cimerman et al., 2018; Jebbar et al., 2020).

The main mechanisms of halophilic PGP bacteria, which help plants cope with salt stress, are as follows: (i) alteration of plant hormone levels, (ii) nitrogen fixation, (iii) siderophore production, (iv) phosphate solubilization, (v) antagonistic activity, and (vi) EPS production (Fig. 7.1).

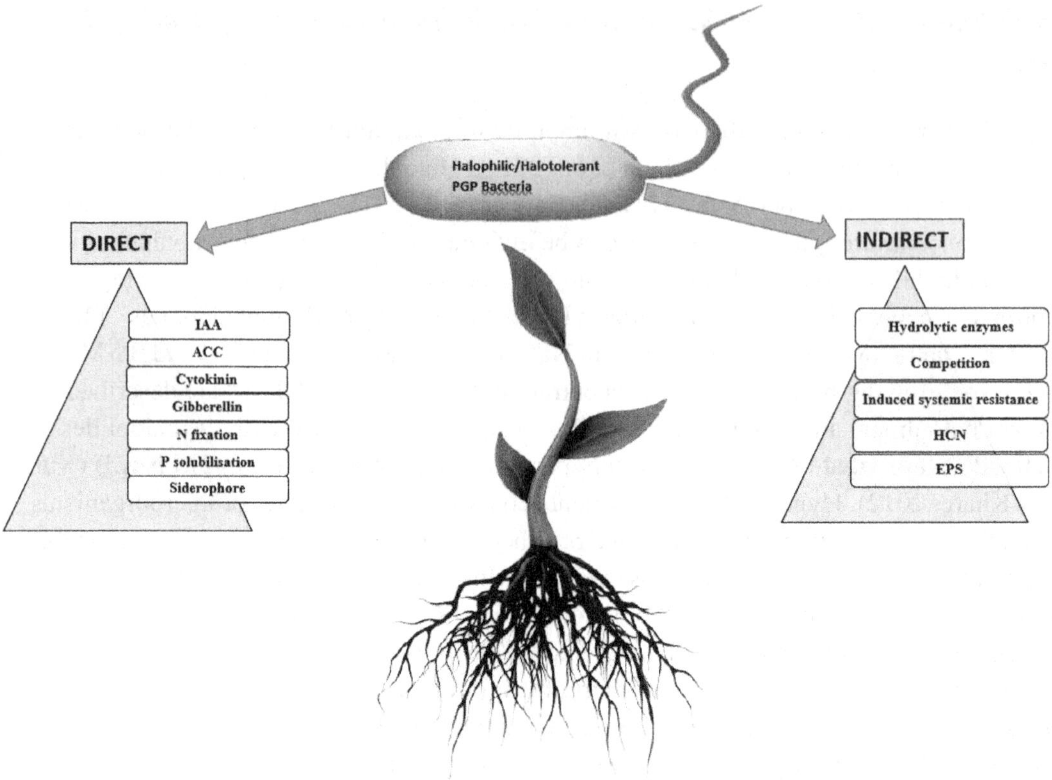

Fig. 7.1
Plant growth-promoting (PGP) mechanisms of halotolerant/halophilic bacteria.

Alteration of the levels of plant hormones

Plant hormones, which are natural organic compounds and vital growth regulators, have important roles in plant metabolism and amelioration of abiotic stresses in all of a plant's life cycles (Egamberdieva et al., 2017). On the other hand, abiotic stresses change the endogenous levels of phytohormones including gibberellins, auxins, abscisic acid (ABA), salicylic acid (SA), and jasmonic acid (JA). Changes in phytohormone levels result in plant growth perturbations (Khan et al., 2014).

Salt stress has been documented to be one of the most effective factors that alter phytohormone concentrations in plant tissues (Tao et al., 2015). Plants have evolved various alteration mechanisms to survive under salt stress at the biochemical, physiological, and molecular levels. These alterations depend on the soil characteristics, plant growth stage, and various phytohormone levels of plants (Bernstein, 2019). Plants sometimes cannot produce the required hormones for their use. In this case, plant-associated microbes produce these phytohormones (Chaiharn and Lumyong, 2011). These microorganisms have the ability to support plants by preventing the production of plant hormone such as ethylene or by producing plant hormones such as auxins, gibberellins, cytokinins, and ABA (Sgroy et al., 2009; Khan et al., 2014; Kudoyarova et al., 2014; Pereira et al., 2016; Etesami and Glick, 2020).

Ethylene, the only gaseous hormone, is found in all plants. It is involved in many vital metabolic processes either as a stimulator or as an inhibitor. Ethylene plays important roles in seed germination, tissue differentiation, initiation of flowering, flower blooming, ripening of fruit, anthocyanin and volatile compound production, hydrolysis of storage products, shedding of leaf and fruit, and senescence. In addition, ethylene also plays a role in coping with biotic and abiotic stresses in plants (Ryu and Cho, 2015). Plants increase ethylene synthesis considerably under environmental stresses, especially salt stress. Although ethylene is a significant hormone for plant growth, a high level of ethylene synthesis is called 'ethylene stress', which has detrimental effects on plants (Tao et al., 2015). Ethylene stress affects the root and shoot development and also N_2 fixation in a negative manner (Gamalero and Glick, 2015). PGP bacteria have the capacity to produce the 1-aminocyclopropane-1-carboxylic acid (ACC) deaminase enzyme. This enzyme is able to decrease ethylene level by affecting the gene expression, which is responsible in the ethylene production enzymes and protect plants against salt stress. Ethylene levels are also regulated based on the synthesis of indole-3-acetic acid (IAA). IAA stimulates the production of enzymes (such as ACC deaminase). The central molecule of ethylene biosynthesis, ACC, acts as a direct precursor to ethylene. When ACC deaminase reduces the level of ACC, ethylene stress is mitigated due to the high level of IAA (Misra and Chauhan, 2020). Moreover, IAA promotes many different development processes including elongation, division, and differentiation of cells (Khalid and Aftab, 2020). The processes of IAA (endogenous synthesis, transport, metabolism, and activity) can be negatively affected by salt stress (Ljung, 2013). In salinity stress, the PGP activity of nonhalotolerant bacteria can be reduced or inhibited (production of ACC deaminase or IAA) (Etesami and Glick, 2020). However, halophilic PGP bacteria are not affected by salinity stress due to their salt adaptation mechanisms and thus can enhance the growth of plants under salt stress (Karimzadeh et al., 2021).

Cytokinin (CK) and gibberellic acid (GA) are the phytohormones negatively affected by salt stress. There are many studies reporting that CK- and GA-producing PGP bacteria are involved in increasing plants' tolerance to salinity (Kang et al., 2014; Shahzad et al., 2016). Abscisic, jasmonic, and salicylic acids are well-known to participate in improving the stress responses and adaptation of plants, although their endogenous levels alter under abiotic stresses just as those of other phytohormones (Egamberdieva et al., 2017). Abscisic acid (ABA), a natural sequiterpenoid, regulates the gene expression of stress responses, which are induced by ABA and signaling, and leads to better tolerance responses (Sah et al., 2016). Moreover, ABA controls plant growth, water content, nitrogen fixation, and nutrient uptake under salt and drought stresses (Khadri et al., 2006). Exogenous ABA application has been proposed as a promising tool to mitigate the effects of stress conditions. ABA has the ability to stabilize the osmotic adjustment of plants by closing their stomata, decreasing water loss, activities of ion exchange, and accumulating various proteins and osmoprotectants under salt stress (Ryu and Cho, 2015). On the other hand, an abrupt increase in the ABA level inhibits plant growth (Asgher et al., 2015). Many studies have reported that ABA-producing and ABA-metabolizing PGP bacteria can play different roles in modulating the plant ABA status and thus can increase plant response to salt stress (Cohen et al., 2015; Shahzad et al., 2017). PGP bacterial application can increase or decrease the ABA level in plants according to the stress responses of plants (Cohen et al., 2015).

Salicylic acid (SA) plays a significant role in the stress endurance of plants by regulating antioxidant enzymes due to its phenolic nature and by reducing the endogenous ethylene levels of plants (Khan et al., 2014; Silva et al., 2017). Moreover, SA has the ability to reprogram gas exchange, stabilize the membrane, modify the ionic imbalance and water relation of plants, accumulate organic osmolytes (proline and free amino acids), decrease endogenous ethylene levels, increase the photosynthesis rate, and decrease electrolyte leakage (Janda et al., 2012; Jayakannan et al., 2015; Egamberdieva et al., 2018). Under stress conditions, PGP bacteria have the ability to produce SA and hence support plant growth by modulating SA levels in plants (Khan et al., 2017; Shahzad et al., 2017). Jasmonic acid (JA) facilitates salt stress by reducing the amount of sodium ions in plants (Ahmad et al., 2016). As mentioned above, under salt stress, the PGP activity of nonhalotolerant bacteria can be reduced or inhibited (Etesami and Glick, 2020); therefore, the use of halophilic/halotolerant PGP bacteria is a substantial tool for increasing JA levels in salt-stressed plants (Shahzad et al., 2016).

Nitrogen fixation

It has been reported that the most essential nutrient required for plants, which affects plants' growth and reproduction potential, is nitrogen (N). Atmospheric nitrogen can be used by plants only after nitrogen fixation (Kour et al., 2019). In traditional agricultural applications, farmers use chemical nitrogenous fertilizers to overcome N deficiency in soils. Fertilizer requirement in saline soils is generally higher than that in nonsaline soils. However, overutilization of fertilizers impairs the structure and composition of soils. Besides, it destroys soil microorganism compositions and increases soil salinity (Akhavan-Kharazian et al., 1991).

Since salinity interferes with nitrogen nutrition and reduces nitrogen nutrient concentration in plant tissues, the yield of agricultural crops, even halophytes, is reduced under salinity stress (Yadav, 2017). In addition, salinity (NaCl) represses both assimilation and uptake of nitrate (NO_3^-) and ammonium (NH_4^+) ions because of the high concentrations of chlorine. The uptake of potassium (K^+) can also be reduced due to the high intake of sodium ions by plants under salt stress (Ullrich, 2002). Diazotrophic bacteria, which have the ability to utilize dinitrogen (N_2) from the atmosphere as a nitrogen source for their metabolic activities, play a significant role in nitrogen fixation (Patle et al., 2018). In the N_2 fixation process, the conversion of atmospheric nitrogen into ammonia is carried out by the nitrogenase enzyme (Franche et al., 2009). Diazotrophic PGP bacterial application is one of the promising alternatives to nitrogenous fertilizer application for sustainable agriculture without environmental damage (Kour et al., 2017). On the other hand, the N_2-fixing process (enzymes) of nonhalophilic bacteria is known to be oversensitive to salts; thus, the formation and functioning of nodules are affected in a negative manner because of salts. This situation leads to low N_2-fixing activity of microorganisms because of toxic ions and high osmotic stress. Yet, halophilic diazotrophic PGP bacteria can deal with high osmotic stress owing to the low-salt-in strategy used by these microorganisms under salt stress (Yan et al., 2015; Kumar et al., 2019).

Siderophore production

Iron (Fe) is a requisite microelement for virtually all organisms as it is a significant cofactor for many enzymatic reactions (Suman et al., 2016). Fe is involved in the processes of photosynthesis, respiration, and molecular nitrogen fixation (Kobayashi and Nishizawa, 2012). Although the availability of Fe is high on Earth, its accessibility is quite limited. The limitation of Fe is due to its ferric state (Fe^{3+}), which converts into highly insoluble oxides, hydroxides, and oxyhydroxides under aerobic conditions. The insoluble forms of Fe are accumulated in soils; thus, this circumstance makes the utilization of Fe by plants and microorganisms more difficult (Rajkumar et al., 2010). The low accessibility of Fe can be dealt with by the use of bacteria-producing siderophores. The function of these compounds is to chelate Fe, which is then accessible for plant use. These molecules bind to Fe^{3+} with a high affinity and are efficiently taken into the bacterial cells (Rabhi et al., 2007). In addition, siderophores function as iron scavengers and can thus alleviate stress caused by excess heavy metal concentrations in soils. Plants have the ability to take up Fe from siderophores secreted by bacteria through special mechanisms including chelation and ligand exchange (Čolo et al., 2014). Consequently, PGP bacteria have great significance in supplying the required Fe for plant use. On the other hand, under salt stress, only halophilic/halotolerant PGP bacteria can maintain their function (siderophore production) (Slama et al., 2019; Mukhtar et al., 2020).

Phosphate solubilization

Phosphorus, a main restrictive macroelement, is as significant as nitrogen for the growth of plants (Yadav et al., 2020). Phosphorus is found in both inorganic and organic forms. However, plants have the ability to utilize phosphorus in only monobasic and dibasic ion forms, and the chemical and structural complexities of these phosphorus forms can significantly decrease the availability of this macroelement to plants (Behera et al., 2014). Besides, phosphate is deficit in most of the agricultural soils because of its strong interaction with soil particles and its rapid conversion into insoluble complexes (Bhattacharyya and Jha, 2012). So far, fertilizers have been commonly used in order to obviate the phosphorus deficiency in agricultural lands.

However, excessive usage of chemical phosphate fertilizers leads to many serious side effects on the soil and environment. Due to the side effects of chemical fertilizers, the utilization of PGP bacteria has drawn significant attention, especially because of their phosphate-solubilizing ability. PGP bacteria can solubilize phosphate owing to the production of various acids, chelating agents, enzymes (phosphatase), and ion-exchange reactions. Salt stress disturbs the diversity, density, and activity of microorganisms; consequently, nutrient and phosphate availability and crop yields are reduced (Sharma et al., 2013). If chemical phosphorus fertilizers are extensively applied to soils, then the salinity of the soils significantly increases. At this point, the application of halophilic PGP bacteria is an effective approach to protect both the microbial structure and nutrient balance of the soils (Mukhtar et al., 2020). Halophilic PGP bacteria (*Oceanobacillus*, *Staphylococcus*, *Azospirillum*, *Vibrio*, *Bacillus*, *Phyllobacterium*, *Achromobacter*, *Kocuria*, *Aeromonas*, *Arthrobacter*, and *Pseudomonas*) can dissolve different types of minerals including tricalcium phosphate, ferric phosphate, aluminum phosphate, and rock phosphate, thus promoting plant development under the stress caused by salts (Upadhyay and Singh, 2015; Etesami and Alikhani, 2019).

Antagonistic activity of halophilic bacteria

Phytopathogenic diseases, especially fungal diseases, negatively affect crop yields. These diseases diminish the annual yield between 10% and 20% (Etesami and Glick, 2020). Moreover, soil salinity increases plant susceptibility to pathogens because of disrupting plant physiology and morphology (Etesami and Beattie, 2018). PGP bacteria can indirectly support plant growth by limiting various stages (both colonization and development) in pathogenic organisms. This action mechanism performed by PGP bacteria is defined as an antagonistic activity (Kour et al., 2019). PGP bacteria exhibit their antagonistic activity via different mechanisms including induced systemic resistance (ISR) and production of various metabolites such as lytic enzymes and antibiotics (Kour et al., 2019; Yadav et al., 2019). PGP bacteria also compete with phytopathogens for nutrients and binding sites (ecological niches) on plant roots (Olanrewaju et al., 2017). The generation of hydrogen cyanide (HCN) by PGP bacteria is another significant factor for suppressing fungal pathogens. HCN, a volatile substance, is an inhibitor of most metalloenzymes. This compound inhibits the metalloenzymes of the respiratory system and ATP production of pathogens (Etesami and Glick, 2020). Induced systemic resistance (ISR), especially against fungal diseases, is one of the major plant defense mechanisms of PGP bacteria. Exposure to PGP bacteria results in faster and stronger response of plants to a specific phytopathogen due to the signaling pathway regulation of PGP bacteria (bacterial phytohormone) (Etesami, 2019). Since the effectiveness of PGP bacteria changes according to the conditions (salinity, pH, drought, temperature, etc.) in which microorganisms thrive, the use of bacteria that are adapted to environmental conditions can effectively overcome phytopathogen-related diseases (Etesami and Glick, 2020). There are many reports about the antagonistic activity of different halophilic/halotolerant PGP bacterial genera including *Bacillus*, *Streptomyces*, and *Halomonas* (Bibi et al., 2018; Slama et al., 2019). Consequently, the application of salt-loving PGP bacteria that exhibit antagonistic activity is the best option to remediate soils instead of using chemical fungicides for agricultural use (Etesami and Beattie, 2018).

EPS production

Halophilic and halotolerant PGP bacteria can produce exopolysaccharides (EPSs) differently from their nonhalophilic/halotolerant counterparts. EPSs are compounds that are able to form an EPS–cation complex, thus neutralizing the excess Na ions present in the soil and keeping plant roots away from the uptake of Na ions (Ashraf et al., 2004; Grover et al., 2011). The plant growth-promoting potential of several halophilic/halotolerant bacterial isolates (*Bacillus pumilus*, *Bacillus subtilis*, *Pseudomonas* sp., *Bradyrhizobium japonicum*, *Pantoea agglomerans*, *Serratia proteamaculans*, *Alcaligenes* sp., etc.) producing EPS has been previously documented (Han and Lee, 2005; Kasotia et al., 2016; Bhise and Dandge, 2019; Fatima et al., 2020; Kumar et al., 2021). This beneficial effect of EPSs to alleviate salt stress (restriction of Na^+) has been demonstrated in various plant genera such as wheat, maize, rice, and soybean (Han and Lee, 2005; Arora et al., 2010; Kasotia et al., 2016; Atouei et al., 2019).

Application of halophilic/halotolerant bacteria as plant growth-promoting agents

Even though non-salt-tolerant PGP bacteria have positive impacts on plant growth, there are still many limitations to their usage. These bacteria (non-salt-tolerant PGP) may exhibit their beneficial activity, particularly when their resilience enables them to tolerate stresses in different soil types with different soil properties (physicochemical). For non-salt-tolerant PGP bacteria, salts can affect their plant enhancement potential under salt conditions. It has been documented that salt stress (200 mM) can inhibit both nitrogen fixation and nitrogenase activity in *Azospirillum brasilense* Sp7 (Tripathi et al., 2002). Similarly, in salt-stressed conditions, a reduction in phosphate solubilization and production of IAA, gibberellic acid, and siderophores has been reported in various phosphate-solubilizing fungal strains and *Bacillus* species (Upadhyay et al., 2009; Xiao et al., 2011). In light of this information, it is a fact that the utilization of halotolerant/halophilic PGP bacteria produces noteworthy and advantageous outcomes for increasing crop production in saline soils, since they can survive in extreme conditions due to their adaptation mechanisms. Recently, halophilic microorganisms have drawn significant attention due to their valuable applications including those in the production of retinal pigments, stabilizing compounds (compatible solutes), biopolymers, and carotenoid pigments, and their plant growth-promoting potential (Corral et al., 2020). As known, microorganisms as plant growth-promoting agents can significantly increase many plants yields (Orhan and Demirci, 2020). Many research studies have documented the beneficial effects of halotolerant/halophilic bacterial applications for enhancing the yield of agricultural products in salt-stressed conditions (Shukla et al., 2012; Ullah and Bano, 2015; Mukhtar et al., 2020).

In a recent report, the growth of a plant (rice) has been improved with the inoculation of a halophilic PGP bacterium (*Halomonas* sp.) under salt stress conditions (Mukherjee et al., 2019). Inoculations of wheat with halotolerant PGP bacteria (*Pseudomonas stutzeri, Bacillus thuringiensis, Enterobacter asburiae*, and *Moraxella pluranimalium*) increased plant growth by more than 50% compared to uninoculated plants under salt stress (Raheem and Ali, 2015). The growth (total fresh weight and the lengths of the root and shoot) of wheat inoculated with halophilic/halotolerant PGP bacteria (*Bacillus* sp., *Staphylococcus succinus, Zhihengliuella halotolerans, Bacillus gibsonii, Thalassobacillus* sp., *Oceanobacillus oncorhynchi, Halobacillus* sp., and *Halomonas* sp.) has been reported to be significantly increased (62.2%–78.1%) in the presence of 200 mM NaCl (Orhan, 2016). In a study conducted by Sorty et al. (2016), halotolerant PGP bacteria (*Pantoea, Acinetobacter, Bacillus, Marinobacterium, Sinorhizobium, Enterobacter, Rhizobium*, and *Pseudomonas*) promoted wheat yield under saline conditions. Similarly, plant (tomato) growth enhancement has been shown with the application of the *Enterobacter* sp. strain EJ01 under salt stress (Kim et al., 2014).

Barley growth has also been shown to be improved (69.22%–81.86%) with the inoculation of halotolerant/halophilic PGP bacteria (*Zhihengliuella salsuginis, Marinibacillus* sp., *Promicromonospora* sp., and *Planococcus* sp.) under salt (200 mM NaCl) stress (Orhan and Demirci, 2020). Similar to the plant growth in wheat and barley, the plant growth of maize with inoculation of halotolerant/halophilic PGP bacteria has been reported by several researchers (Hameeda et al., 2008; Ullah and Bano, 2015; Mukhtar et al., 2020). For example, the inoculation of maize with halotolerant/halophilic bacteria (*Bacillus* sp. and *Arthrobacter pascens*) under salt stress (100 mM NaCl) has been shown to induce vegetation processes (length of the shoot and root, chlorophyll and sugar content) (Ullah and Bano, 2015). In field conditions, *Pseudomonas* sp. and *Serratia marcescens* have increased the maize yield at a rate of 64% and 85%, respectively, compared to that of the control group (Hameeda et al., 2008). In a new report, maize inoculated with halotolerant PGP bacteria (*Bacillus safensis, Aeromonas veronii, Kocuria rosea, Bacillus pumilus*, and *Pseudomonas fluorescens*) also showed significant development in plant dry weights compared with those of the uninoculated group (Mukhtar et al., 2020). Another plant whose growth (biomass, length, and weight) has been shown to be enhanced by the inoculation of halotolerant PGP bacteria (*Klebsiella* sp. and *Ochrobactrum anthropi*) is peanut (Sharma et al., 2016). The inoculation of *Beta vulgaris* L. with halotolerant/halophilic PGP bacteria (*Micrococcus yunnanensis*, *Variovorax paradoxus*, and *Planococcus rifietoensis*) is another example of a plant whose growth (seed germination and biomass) has been prompted under salt stress (50–125 mM NaCl) (Zhou et al., 2017). The effects of halotolerant/halophilic PGP bacterial isolates have also been evaluated on the plant genus *Salicornia*. A member of this genus, *Salicornia europaea*, was inoculated with a halotolerant PGP bacterium (*Pseudomonas pseudoalcaligenes*) and it was shown that the plant growth increased under salt stress (Ozawa et al., 2007). In a recent study, the growth of the same plant (*S. europaea*) has been shown to be enhanced by several other halotolerant/halophilic PGP bacteria (*Variovorax paradoxus*, *Bacillus endophyticus*, *Arthrobacter agilis*, and *Bacillus tequilensis*) under salt stress (50, 250, and 500 mM NaCl, respectively) (Zhao et al., 2016). Two other halotolerant/halophilic PGP bacteria (*Brachybacterium saurashtrense* and *Pseudomonas* sp.) have been reported to increase the length of the root (53%) and shoot (36%), fresh weight (93%), and germination (100%) of a different member of this genus (*Salicornia brachiate*) under salt stress conditions (250 and 500 mM NaCl, respectively) (Jha et al., 2012). In a more recent study, two different rhizogenic and endophytic strains of *Staphylococcus* sp. have been reported to enhance *Salicornia* sp.'s salt tolerance and growth in the presence of high salt concentrations (Komaresofla et al., 2019).

Another study showing the outcomes of salt-tolerant PGP bacterial inoculation was conducted on *Arthrocnemum macrostachyum*. A simultaneous inoculation of this plant with three halophilic PGP bacteria (*Gracilibacillus saliphilus*, *Bacillus thuringiensis*, and *Bacillus alcalophilus*) was conducted to enhance the ratio of germination and the growth of the plant in a salt stress situation (1030 mM NaCl) (Navarro-Torre et al., 2017). Similarly, the inoculation of a folk medicine plant (*Limonium sinense*) with halotolerant/halophilic PGP bacteria (*Bacillus flexus*, *Isoptericola dokdonensis*, *Arthrobacter soli*, and *Streptomyces pactum*) under salt stress (250 mM NaCl) increased the lengths of the shoot (3%–68%) and root (52%–137%) (Walia et al., 2014). In another study, the use of halotolerant PGP bacteria (*Brachybacterium saurashtrense*, *Brevibacterium casei*, and *Haererohalobacter*) was reported to boost the salinity adaptation of the plant (*Arachis hypogaea*) (Shukla et al., 2012). Similarly, the application of 14 halophilic/halotolerant PGP bacteria (*Brevibacterium epidermidis*, *Oceanimonas smirnovii*, *Brevibacterium iodinum*, *Planococcus rifietoensis*, *Bacillus aryabhattai*, *Exiguobacterium acetylicum*, *Zhihengliuella alba*, *Bacillus stratosphericus*, *Micrococcus yunnanensis*, *Bacillus licheniformis*, *Bacillus stratosphericus*, *Corynebacterium variabile*, *Arthrobacter nicotianae*, and *Bacillus aryabhattai*) increased plant (canola) growth (root length: 5.2%–47.8% and dry weight: 16.2%–43%) under salt stress conditions (150 mM NaCl) (Siddikee et al., 2010). In a more recent study, halotolerant/halophilic bacteria have been shown to affect plant growth differently (significant prompting, slight prompting, or inhibiting the plant growth). In that report, several bacteria (*Bacillus licheniformis*, *Bacillus subtilis* strain GB03, *Pseudomonas* species, and *Kushneria*) slightly prompted the growth of the plant (alfalfa), whereas *Halomonas* sp. and *Bacillus* sp. significantly stimulated the plant growth (Kearl et al., 2019). In addition, the growth of the *Suaeda maritima* plant has been shown to be prompted by inoculation of halotolerant/halophilic PGP bacteria (*Brevibacterium* sp., *Bacillus subtilis*, *Sanguibacter* sp., *Erwinia persicina*, *Zhihengliuella alba*, *Zhihengliuella halotolerans*, and *Brachybacterium* sp.) (Alishahi et al., 2020).

Conclusions

Plants are equipped to tolerate many stress factors such as heavy metals, droughts, and salinity. However, there are some situations that they cannot tolerate, and thus they need help to deal with such situations. One of the ways to enhance plant growth is by the application of chemical fertilizers. Initially, the use of chemical fertilizers can improve the soil structure and quality, but the use of chemical fertilizers will increase the existing salinity after a while. Another way to enhance plant growth in saline soils is by the application of halophilic/halotolerant PGP bacteria. As known, the most important feature that distinguishes halophilic/halotolerant PGP bacteria from nonhalophilic/halotolerant PGP is their salt adaptation. Although nonhalophilic/halotolerant PGP promote plant growth by altering phytohormones, producing osmolytes and siderophores, fixing nitrogen, and increasing nutrient accessibility, the growth of these microorganisms will be restricted due to salt stress in saline soils. Therefore, in such a situation, halophilic/halotolerant PGP bacteria are of great importance to enable plants to grow even in extreme saline conditions.

References

Ahmad, P., Kumar, A., Ashraf, M., Akram, N.A., 2012. Salt-induced changes in photosynthetic activity and oxidative defense system of three cultivars of mustard (*Brassica juncea* L.). Afr. J. Biotechnol. 11 (11), 2694–2703.

Ahmad, P., Rasool, S., Gul, A., Sheikh, S.A., Akram, N.A., Ashraf, M., Kazi, A.M., Gucel, S., 2016. Jasmonates: multifunctional roles in stress tolerance. Front. Plant Sci. 7, 813.

Akhavan-Kharazian, M., Campbell, W., Jurinak, J., Dudley, L.M., 1991. Effects of $CaSO_4$, $CaCl_2$, and NaCl on leaf nitrogen, nodule weight, and acetylene reduction activity in *Phaseolus vulgaris* L. Arid Land Res. Manag. 5 (2), 97–103.

Alishahi, F., Alikhani, H.A., Khoshkholgh-Sima, N.A., Etesami, H., 2020. Mining the roots of various species of the halophyte *Suaeda* for halotolerant nitrogen-fixing endophytic bacteria with the potential for promoting plant growth. Int. Microbiol. 23 (3), 415–427.

Arora, M., Kaushik, A., Rani, N., Kaushik, C., 2010. Effect of cyanobacterial exopolysaccharides on salt stress alleviation and seed germination. J. Environ. Biol. 31 (5), 701–704.

Asgher, M., Khan, M.I.R., Anjum, N.A., Khan, N.A., 2015. Minimising toxicity of cadmium in plants—role of plant growth regulators. Protoplasma 252 (2), 399–413.

Ashraf, M., Hasnain, S., Berge, O., Mahmood, T., 2004. Inoculating wheat seedlings with exopolysaccharide-producing bacteria restricts sodium uptake and stimulates plant growth under salt stress. Biol. Fertil. Soils 40 (3), 157–162.

Atouei, M.T., Pourbabaee, A.A., Shorafa, M., 2019. Alleviation of salinity stress on some growth parameters of wheat by exopolysaccharide-producing bacteria. Iran. J. Sci. Technol. Trans. A Sci. 43 (5), 2725–2733.

Behera, B., Singdevsachan, S.K., Mishra, R., Dutta, S., Thatoi, H., 2014. Diversity, mechanism and biotechnology of phosphate solubilising microorganism in mangrove—a review. Biocatal. Agric. Biotechnol. 3 (2), 97–110.

Bernstein, N., 2019. Plants and salt: plant response and adaptations to salinity. In: Rampelotto, P., Gordon, R., Seckbach, J. (Eds.), (Eds.), Model Ecosystems in Extreme Environments. Elsevier, pp. 101–112.

Bhattacharyya, P.N., Jha, D.K., 2012. Plant growth-promoting rhizobacteria (PGPR): emergence in agriculture. World J. Microbiol. Biotechnol. 28 (4), 1327–1350.

Bhise, K.K., Dandge, P.B., 2019. Alleviation of salinity stress in rice plant by encapsulated salt tolerant plant growth promoting bacteria *Pantoea agglomerans* strain KL and its root colonization ability. Arch. Agron. Soil Sci. 65 (14), 1955–1968.

Bibi, F., Strobel, G.A., Naseer, M.I., Yasir, M., Khalaf Al-Ghamdi, A.A., Azhar, E.I., 2018. Halophytes-associated endophytic and rhizospheric bacteria: diversity, antagonism and metabolite production. Biocontrol Sci. Technol. 28 (2), 192–213.

Chaiharn, M., Lumyong, S., 2011. Screening and optimization of indole-3-acetic acid production and phosphate solubilization from rhizobacteria aimed at improving plant growth. Curr. Microbiol. 62 (1), 173–181.

Chandra, D., Sharma, A.K., 2020. Field evaluation of consortium of bacterial inoculants producing ACC deaminase on growth, nutrients and yield components of rice and wheat. J. Crop. Sci. Biotechnol. 24 (3), 293–305.

Chandra, D., Srivastava, R., Glick, B.R., Sharma, A.K., 2018a. Drought-tolerant *Pseudomonas* spp. improve the growth performance of finger millet (*Eleusine coracana* (L.) Gaertn.) under non-stressed and drought-stressed conditions. Pedosphere 28, 227–240.

Chandra, D., Srivastava, R., Sharma, A., 2018b. Influence of IAA and ACC deaminase producing fluorescent pseudomonads in alleviating drought stress in wheat (*Triticum aestivum*). Agric. Res. 7, 290–299.

Chandra, D., Srivastava, R., Gupta, V.V., Franco, C.M., Sharma, A.K., 2019a. Evaluation of ACC-deaminase-producing rhizobacteria to alleviate water-stress impacts in wheat (*Triticum aestivum* L.) plants. Can. J. Microbiol. 65 (5), 387–403.

Chandra, D., Srivastava, R., Gupta, V.V., Franco, C.M., Paasricha, N., Saifi, S.K., Tuteja, N., Sharma, A.K., 2019b. Field performance of bacterial inoculants to alleviate water stress effects in wheat (*Triticum aestivum* L.). Plant Soil 441 (1), 261–281.

Chandra, D., Srivastava, R., Glick, B.R., Sharma, A.K., 2020. Rhizobacteria producing ACC deaminase mitigate water-stress response in finger millet (*Eleusine coracana* (L.) Gaertn.). 3 Biotech 10, 65.

Cohen, A.C., Bottini, R., Pontin, M., Berli, F.J., Moreno, D., Boccanlandro, H., Travaglia, C.N., Piccoli, P.N., 2015. *Azospirillum brasilense* ameliorates the response of *Arabidopsis thaliana* to drought mainly via enhancement of ABA levels. Physiol. Plant. 153 (1), 79–90.

Čolo, J., Hajnal-Jafari, T.I., Durić, S., Stamenov, D., Hamidović, S., 2014. Plant growth promotion rhizobacteria in onion production. Pol. J. Microbiol. 63, 83–88.

Corral, P., Amoozegar, M.A., Ventosa, A., 2020. Halophiles and their biomolecules: recent advances and future applications in biomedicine. Mar. Drugs 18 (1), 33.

Egamberdieva, D., Wirth, S.J., Alqarawi, A.A., AbdAllah, E.F., Hashem, A., 2017. Phytohormones and beneficial microbes: essential components for plants to balance stress and fitness. Front. Microbiol. 8, 2104.

Egamberdieva, D., Hua, M., Reckling, M., Wirth, S., Bellingrath-Kimura, S.D., 2018. Potential effects of biochar-based microbial inoculants in agriculture. Environ. Sustain. 1 (1), 19–24.

Etesami, H., 2019. Plant growth promotion and suppression of fungal pathogens in rice (*Oryza sativa* L.) by plant growth-promoting Bacteria. In: Maheshwari, D.K., Dheeman, S. (Eds.), Field Crops: Sustainable Management by PGPR. Springer, pp. 351–383.

Etesami, H., Alikhani, H.A., 2019. Halotolerant plant growth-promoting fungi and bacteria as an alternative strategy for improving nutrient availability to salinity-stressed crop plants. In: Hassan Etesami, H., Kumar, M., Kumar, V. (Eds.), Saline soil-based agriculture by halotolerant microorganisms. Springer, pp. 103–146.

Etesami, H., Beattie, G.A., 2018. Mining halophytes for plant growth-promoting halotolerant bacteria to enhance the salinity tolerance of non-halophytic crops. Front. Microbiol. 9, 148.

Etesami, H., Glick, B.R., 2020. Halotolerant plant growth–promoting bacteria: prospects for alleviating salinity stress in plants. Environ. Exp. Bot. 178, 104124.

FAO, 2005. Global Network on Integrated Soil Management for Sustainable Use of Salt-Affected Soils. FAO, Rome.

Fatima, T., Mishra, I., Verma, R., Arora, N.K., 2020. Mechanisms of halotolerant plant growth promoting Alcaligenes sp. involved in salt tolerance and enhancement of the growth of rice under salinity stress. 3 Biotech 10 (8), 1–12.

Franche, C., Lindström, K., Elmerich, C., 2009. Nitrogen-fixing bacteria associated with leguminous and non-leguminous plants. Plant Soil 321 (1), 35–59.

Galinski, E.A., 1995. Osmoadaptation in bacteria. Adv. Microb. Physiol. 37, 273–328.

Gamalero, E., Glick, B.R., 2015. Bacterial modulation of plant ethylene levels. Plant Physiol. 169 (1), 13–22.

Grover, M., Ali, S.Z., Sandhya, V., Rasul, A., Venkateswarlu, B., 2011. Role of microorganisms in adaptation of agriculture crops to abiotic stresses. World J. Microbiol. Biotechnol. 27 (5), 1231–1240.

Gunde-Cimerman, N., Plemenitaš, A., Oren, A., 2018. Strategies of adaptation of microorganisms of the three domains of life to high salt concentrations. FEMS Microbiol. Rev. 42 (3), 353–375.

Gupta, G., Srivastava, S., Khare, S., Prakash, V., 2014. Extremophiles: an overview of microorganism from extreme environment. Int. J. Agric. Environ. Biotechnol. 7 (2), 371.

Hameeda, B., Harini, G., Rupela, O., Wani, S., Reddy, G., 2008. Growth promotion of maize by phosphate-solubilizing bacteria isolated from composts and macrofauna. Microbiol. Res. 163 (2), 234–242.

Han, H., Lee, K., 2005. Physiological responses of soybean-inoculation of *Bradyrhizobium japonicum* with PGPR in saline soil conditions. Res. J. Agric. Biol. Sci. 1 (3), 216–221.

Jamil, A., Riaz, S., Ashraf, M., Foolad, M.R., 2011. Gene expression profiling of plants under salt stress. Crit. Rev. Plant Sci. 30, 435–458.

Janda, K., Hideg, É., Szalai, G., Kovács, L., Janda, T., 2012. Salicylic acid may indirectly influence the photosynthetic electron transport. J. Plant Physiol. 169 (10), 971–978.

Jayakannan, M., Bose, J., Babourina, O., Shabala, S., Massart, A., Poschenrieder, C., Rengel, Z., 2015. The NPR1-dependent salicylic acid signalling pathway is pivotal for enhanced salt and oxidative stress tolerance in *Arabidopsis*. J. Exp. Bot. 66 (7), 1865–1875.

Jebbar, M., Hickman-Lewis, K., Cavalazzi, B., Taubner, R.S., Rittmann, S.K.M.R., Antunes, A., 2020. Microbial diversity and biosignatures: an icy moons perspective. Space Sci. Rev. 216 (1), 1–47.

Jha, B., Gontia, I., Hartmann, A., 2012. The roots of the halophyte *Salicornia brachiata* are a source of new halotolerant diazotrophic bacteria with plant growth-promoting potential. Plant Soil 356 (1), 265–277.

Kang, S.M., Khan, A.L., You, Y.H., Kim, J.G., Kamran, M., Lee, I.J., 2014. Gibberellin production by newly isolated strain *Leifsonia* soli SE134 and its potential to promote plant growth. J. Microbiol. Biotechnol. 24 (1), 106–112.

Karimzadeh, J., Alikhani, H.A., Etesami, H., Pourbabaei, A.A., 2021. Improved phosphorus uptake by wheat plant (*Triticum aestivum* L.) with rhizosphere fluorescent *Pseudomonads* strains under water-deficit stress. J. Plant Growth Regul. 40 (1), 162–178.

Kasotia, A., Varma, A., Tuteja, N., Choudhary, D.K., 2016. Amelioration of soybean plant from saline-induced condition by exopolysaccharide producing *Pseudomonas*-mediated expression of high affinity K^+-transporter (HKT1) gene. Curr. Sci., 1961–1967.

Kearl, J., McNary, C., Lowman, J.S., Mei, C., Aanderud, Z.T., Smith, S.T., West, J., Colton, E., Hamson, M., Nielsen, B.L., 2019. Salt-tolerant halophyte rhizosphere bacteria stimulate growth of alfalfa in salty soil. Front. Microbiol. 10, 1849.

Khadri, M., Tejera, N.A., Lluch, C., 2006. Alleviation of salt stress in common bean (*Phaseolus vulgaris*) by exogenous abscisic acid supply. J. Plant Growth Regul. 25 (2), 110–119.

Khalid, A., Aftab, F., 2020. Effect of exogenous application of IAA and GA_3 on growth, protein content, and antioxidant enzymes of *Solanum tuberosum* L. grown in vitro under salt stress. In Vitro Cell Dev. Biol. Plant 56, 377–389.

Khan, M.I.R., Asgher, M., Khan, N.A., 2014. Alleviation of salt-induced photosynthesis and growth inhibition by salicylic acid involves glycinebetaine and ethylene in mungbean (*Vigna radiata* L.). Plant Physiol. Biochem. 80, 67–74.

Khan, A.L., Waqas, M., Asaf, S., Kamran, M., Shahzad, R., Bilal, S., Khan, A.M., Kang, S.M., Kim, Y.H., Yun, B.W., 2017. Plant growth-promoting endophyte *Sphingomonas* sp. LK11 alleviates salinity stress in *Solanum pimpinellifolium*. Environ. Exp. Bot. 133, 58–69.

Kilic¸, C.C., Kukul, Y.S., Anac¸, D., 2008. Performance of purslane (*Portulaca oleracea* L.) as a salt-removing crop. Agric. Water Manag. 95 (7), 854–858.

Kim, K., Jang, Y.J., Lee, S.M., Oh, B.T., Chae, J.C., Lee, K.J., 2014. Alleviation of salt stress by Enterobacter sp. EJ01 in tomato and *Arabidopsis* is accompanied by up-regulation of conserved salinity responsive factors in plants. Mol. Cells 37 (2), 109.

Kobayashi, T., Nishizawa, N.K., 2012. Iron uptake, translocation, and regulation in higher plants. Annu. Rev. Plant Biol. 63, 131–152.

Komaresofla, B.R., Alikhani, H.A., Etesami, H., Khoshkholgh-Sima, N.A., 2019. Improved growth and salinity tolerance of the halophyte *Salicornia* sp. by co–inoculation with endophytic and rhizosphere bacteria. Appl. Soil Ecol. 138, 160–170.

Kour, D., Rana, K., Verma, P., Yadav, A., Kumar, V., Singh, D.H., 2017. Biofertilizers: eco-friendly technologies and bioresources for sustainable agriculture. In: Proceeding of International Conference on Innovative Research in Engineering Science and Technology. IREST/PP/014.

Kour, D., Rana, K.L., Kaur, T., Singh, B., Chauhan, V.S., Kumar, A., Gupta, V.K., 2019. Extremophiles for hydrolytic enzymes productions: biodiversity and potential biotechnological applications. Bioprocess. Biomol. Prod., 321–372.

Kudoyarova, G.R., Melentiev, A.I., Martynenko, E.V., Timergalina, L.N., Arkhipova, T.N., Shendel, G.V., Kuzmina, L.Y., Dodd, I.C., Veselov, S.Y., 2014. Cytokinin producing bacteria stimulate amino acid deposition by wheat roots. Plant Physiol. Biochem. 83, 285–291.

Kumar, S., Khare, S., 2012. Purification and characterization of maltooligosaccharide-forming α-amylase from moderately halophilic *Marinobacter* sp. EMB8. Bioresour. Technol. 116, 247–251.

Kumar, M., Etesami, H., Kumar, V., 2019. Saline Soil-Based Agriculture by Halotolerant Microorganisms. Springer.

Kumar, A., Singh, S., Mukherjee, A., Rastogi, R.P., Verma, J.P., 2021. Salt-tolerant plant growth-promoting *Bacillus pumilus* strain JPVS11 to enhance plant growth attributes of rice and improve soil health under salinity stress. Microbiol. Res. 242, 126616.

Ljung, K., 2013. Auxin metabolism and homeostasis during plant development. Development 140 (5), 943–950.

Misra, S., Chauhan, P.S., 2020. ACC deaminase-producing rhizosphere competent *Bacillus* spp. mitigate salt stress and promote *Zea mays* growth by modulating ethylene metabolism. 3 Biotech 10 (3), 1–14.

Mukherjee, P., Mitra, A., Roy, M., 2019. *Halomonas* rhizobacteria of *Avicennia marina* of Indian Sundarbans promote rice growth under saline and heavy metal stresses through exopolysaccharide production. Front. Microbiol. 10, 1207.

Mukhtar, S., Zareen, M., Khaliq, Z., Mehnaz, S., Malik, K.A., 2020. Phylogenetic analysis of halophyte-associated rhizobacteria and effect of halotolerant and halophilic phosphate-solubilizing biofertilizers on maize growth under salinity stress conditions. J. Appl. Microbiol. 128 (2), 556–573.

Munns, R., 2005. Genes and salt tolerance: bringing them together. New Phytol. 167 (3), 645–663.

Navarro-Torre, S., Barcia-Piedras, J.M., Mateos-Naranjo, E., Redondo-Gómez, S., Camacho, M., Caviedes, A.M., Pajuelo, E., Rodríguez-Llorente, I.D., 2017. Assessing the role of endophytic bacteria in the halophyte *Arthrocnemum macrostachyum* salt tolerance. Plant Biol. 19 (2), 249–256.

Olanrewaju, O.S., Glick, B.R., Babalola, O.O., 2017. Mechanisms of action of plant growth promoting bacteria. World J. Microbiol. Biotechnol. 33 (11), 1–16.

Oren, A., 1999. Bioenergetic aspects of halophilism. Microbiol. Mol. Biol. Rev. 63 (2), 334–348.

Oren, A., 2002. Diversity of halophilic microorganisms: environments, phylogeny, physiology, and applications. J. Ind. Microbiol. Biotechnol. 28 (1), 56–63.

Oren, A., 2008. Microbial life at high salt concentrations: phylogenetic and metabolic diversity. Saline Syst. 4, 2.

Orhan, F., 2016. Alleviation of salt stress by halotolerant and halophilic plant growth-promoting bacteria in wheat (*Triticum aestivum*). Braz. J. Microbiol. 47 (3), 621–627.

Orhan, F., Demirci, A., 2020. Salt stress mitigating potential of halotolerant/halophilic plant growth promoting. Geomicrobiol J. 37 (7), 663–669.

Othman, Y., Al-Karaki, G., Al-Tawaha, A.R., Al-Horani, A., 2006. Variation in germination and ion uptake in barley genotypes under salinity conditions. World J. Agric. Sci. 2, 11–15.

Ozawa, T., Wu, J., Fujii, S., 2007. Effect of inoculation with a strain of *Pseudomonas pseudoalcaligenes* isolated from the endorhizosphere of Salicornia europea on salt tolerance of the glasswort. Soil Sci. Plant Nutr. 53 (1), 12–16.

Panta, S., Flowers, T.J., Lane, P., Doyle, R., Haros, G., Shabala, S., 2014. Halophyte agriculture: success stories. Environ. Exp. Bot. 107, 71–83.

Patle, P., Navnage, N., Ramteke, P., 2018. Endophytes in plant system: roles in growth promotion, mechanism and their potentiality in achieving agriculture sustainability. Int. J. Chem. Stud. 6 (1), 270–274.

Pereira, S., Monteiro, C., Vega, A., Castro, P.M., 2016. Endophytic culturable bacteria colonizing *Lavandula dentata* L. plants: isolation, characterization and evaluation of their plant growth-promoting activities. Ecol. Eng. 87, 91–97.

Qadir, M., Oster, J., Schubert, S., Noble, A., Sahrawat, K., 2007. Phytoremediation of sodic and saline-sodic soils. Adv. Agron. 96, 197–247.

Rabhi, M., Barhoumi, Z., Ksouri, R., Abdelly, C., Gharsalli, M., 2007. Interactive effects of salinity and iron deficiency in *Medicago ciliaris*. C. R. Biol. 330 (11), 779–788.

Raheem, A., Ali, B., 2015. Halotolerant rhizobacteria: beneficial plant metabolites and growth enhancement of *Triticum aestivum* L. in salt-amended soils. Arch. Agron. Soil Sci. 61 (12), 1691–1705.

Rajkumar, M., Ae, N., Prasad, M.N.V., Freitas, H., 2010. Potential of siderophore-producing bacteria for improving heavy metal phytoextraction. Trends Biotechnol. 28 (3), 142–149.

Ryu, H., Cho, Y.G., 2015. Plant hormones in salt stress tolerance. J. Plant Biol. 58 (3), 147–155.

Safdar, H., Amin, A., Shafiq, Y., Ali, A., Yasin, R., Shoukat, A., Hussan, M.U., Sarwar, M.I., 2019. A review: impact of salinity on plant growth. Nat. Sci. 17 (1), 34–40.

Sah, S.K., Reddy, K.R., Li, J., 2016. Abscisic acid and abiotic stress tolerance in crop plants. Front. Plant Sci. 7, 571.

Sarwar, M.K., Azam, I., Iqbal, T., 2015. Biology and applications of halophilic bacteria and archaea: A. Electron. J. Biol. 11 (3), 98–103.

Sgroy, V., Cassán, F., Masciarelli, O., Del Papa, M.F., Lagares, A., Luna, V., 2009. Isolation and characterization of endophytic plant growth-promoting (PGPB) or stress homeostasis-regulating (PSHB) bacteria associated to the halophyte *Prosopis strombulifera*. Appl. Microbiol. Biotechnol. 85 (2), 371–381.

Shahzad, R., Waqas, M., Khan, A.L., Asaf, S., Khan, M.A., Kang, S.M., Yun, B.W., Lee, I.J., 2016. Seed-borne endophytic *Bacillus amyloliquefaciens* RWL-1 produces gibberellins and regulates endogenous phytohormones of *Oryza sativa*. Plant P hysiol. Biochem. 106, 236–243.

Shahzad, R., Khan, A.L., Bilal, S., Waqas, M., Kang, S.M., Lee, I.J., 2017. Inoculation of abscisic acid-producing endophytic bacteria enhances salinity stress tolerance in *Oryza sativa*. Environ. Exp. Bot. 136, 68–77.

Sharma, S.B., Sayyed, R.Z., Trivedi, M.H., Gobi, T.A., 2013. Phosphate solubilizing microbes: sustainable approach for managing phosphorus deficiency in agricultural soils. Springerplus 2 (1), 1–14.

Sharma, S., Kulkarni, J., Jha, B., 2016. Halotolerant rhizobacteria promote growth and enhance salinity tolerance in peanut. Front. Microbiol. 7, 1600.

Shaygan, M., Mulligan, D., Baumgartl, T., 2018. The potential of three halophytes (Tecticornia pergranulata, Sclerolaena longicuspis, and Frankenia serpyllifolia) for the rehabilitation of brine-affected soils. Land Degrad. Dev. 29 (6), 2002–2014.

Shrivastava, P., Kumar, R., 2015. Soil salinity: a serious environmental issue and plant growth promoting bacteria as one of the tools for its alleviation. Saudi J. Biol. Sci. 22 (2), 123–131.

Shukla, P.S., Agarwal, P.K., Jha, B., 2012. Improved salinity tolerance of *Arachis hypogaea* (L.) by the interaction of halotolerant plant-growth-promoting rhizobacteria. J. Plant Growth Regul. 31 (2), 195–206.

Siddikee, M.A., Chauhan, P.S., Anandham, R., Han, G.H., Sa, T.M., 2010. Isolation, characterization, and use for plant growth promotion under salt stress, of ACC deaminase-producing halotolerant bacteria derived from coastal soil. J. Picrobiol. Biotechnol. 20 (11), 1577–1584.

Silva, A.C.D., Suassuna, J.F., Melo, A.S.D., Costa, R.R., Andrade, W.L.D., Silva, D.C.D., 2017. Salicylic acid as attenuator of drought stress on germination and initial development of sesame. Rev. Bras. Eng. Agricola Ambient. 21 (3), 156–162.

Slama, H.B., Cherif-Silini, H., Chenari Bouket, A., Qader, M., Silini, A., Yahiaoui, B., Alenezi, F.N., Luptakova, L., Triki, M.A., Vallat, A., Oszako, T., Rateb, M.E., Belbahri, L., 2019. Screening for *Fusarium* antagonistic bacteria from contrasting niches designated the endophyte *Bacillus halotolerans* as plant warden against *Fusarium*. Front. Microbiol. 9, 3236.

Sorty, A.M., Meena, K.K., Choudhary, K., Bitla, U.M., Minhas, P., Krishnani, K., 2016. Effect of plant growth promoting bacteria associated with halophytic weed (*Psoralea corylifolia* L) on germination and seedling growth of wheat under saline conditions. Appl. Biochem. Biotechnol. 180 (5), 872–882.

Suman, A., Yadav, A.N., Verma, P., 2016. Endophytic microbes in crops: diversity and beneficial impact for sustainable agriculture. In: Singh, D.P., Singh, H.B., Prabha, R. (Eds.), Microbial Inoculants in Sustainable Agricultural Productivity. Springer, pp. 117–143.

Tao, J.J., Chen, H.W., Ma, B., Zhang, W.K., Chen, S.Y., Zhang, J.S., 2015. The role of ethylene in plants under salinity stress. Front. Plant Sci. 6, 1059.

Tripathi, A.K., Nagarajan, T., Verma, S.C., Le Rudulier, D., 2002. Inhibition of biosynthesis and activity of nitrogenase in *Azospirillum brasilense* Sp7 under salinity stress. Curr. Microbiol. 44 (5), 363–367.

Ullah, S., Bano, A., 2015. Isolation of plant-growth-promoting rhizobacteria from rhizospheric soil of halophytes and their impact on maize (*Zea mays* L.) under induced soil salinity. Can. J. Microbiol. 61 (4), 307–313.

Ullrich, W.R., 2002. Salinity and nitrogen nutrition. In: Läuchli, A., Lüttge, U. (Eds.), Salinity: Environment-Plants-Molecules. Springer, pp. 229–248.

Upadhyay, S., Singh, D., 2015. Effect of salt-tolerant plant growth-promoting rhizobacteria on wheat plants and soil health in a saline environment. Plant Biol. 17 (1), 288–293.

Upadhyay, S.K., Singh, D.P., Saikia, R., 2009. Genetic diversity of plant growth promoting rhizobacteria isolated from rhizospheric soil of wheat under saline condition. Curr. Microbiol. 59 (5), 489–496.

Walia, A., Mehta, P., Chauhan, A., Shirkot, C., 2014. Effect of *Bacillus subtilis* strain CKT1 as inoculum on growth of tomato seedlings under net house conditions. Proc. Natl. Acad. Sci. India B Biol. Sci. 84 (1), 145–155.

Wood, N.T., 2001. Nodulation by numbers: the role of ethylene in symbiotic nitrogen fixation. Trends Plant Sci. 6, 501–502.

Xiao, C., Chi, R., Li, X., Xia, M., Xia, Z., 2011. Biosolubilization of rock phosphate by three stress-tolerant fungal strains. Appl. Biochem. Biotechnol. 165 (2), 719–727.

Yadav, A.N., 2017. Agriculturally important microbiomes: biodiversity and multifarious PGP attributes for amelioration of diverse abiotic stresses in crops for sustainable agriculture. Biomed. J. Sci. Tech. Res. 1 (4), 861–864.

Yadav, A.N., Singh, S., Mishra, S., Gupta, A., 2019. Recent Advancement in White Biotechnology Through Fungi: Volume 2: Perspective for Value-Added Products and Environments. Springer.

Yadav, A.N., Kaur, T., Kour, D., Rana, K.L., Yadav, N., Rastegari, A.A., Kumar, M., Paul, D., Sachan, S.G., Saxena, A.K., 2020. Saline microbiome: biodiversity, ecological significance, and potential role in amelioration of salt stress. In: Gupta, V. (Ed.), New and Future Developments in Microbial Biotechnology and Bioengineering. Elsevier, pp. 283–309.

Yan, N., Marschner, P., Cao, W., Zuo, C., Qin, W., 2015. Influence of salinity and water content on soil microorganisms. Int. Soil Water Conserv. Res. 3 (4), 316–323.

Zhao, S., Zhou, N., Zhao, Z.Y., Zhang, K., Wu, G.H., Tian, C.Y., 2016. Isolation of endophytic plant growth-promoting bacteria associated with the halophyte *Salicornia europaea* and evaluation of their promoting activity under salt stress. Curr. Microbiol. 73 (4), 574–581.

Zhou, N., Zhao, S., Tian, C.Y., 2017. Effect of halotolerant rhizobacteria isolated from halophytes on the growth of sugar beet (*Beta vulgaris* L.) under salt stress. FEMS Microbiol. Lett. 364 (11), fnx091.

Zhu, J.K., 2001. Plant salt tolerance. Trends Plant Sci. 6 (2), 66–71.

CHAPTER 8

Inflection of the root microbiome by plants: Plant growth promotion and disease management

Aakansha Verma, Sudha Bind, and Jyoti Bajeli
Department of Biological Sciences, CBSH, GBPUA&T, Pantnagar, Uttarakhand, India

Chapter outline

Introduction

Plant microbiomes are crucial for deciphering plant health, relative abundance, and productivity. Plants have always experienced abiotic and biotic stressors in their local surroundings due to their sessile nature (De Zelicourt et al., 2013). Plants nurture a plenitude of microbial communities, particularly for each accessible plant organ, enclosing a range of microbes from ectophytes to endophytes, including a plethora of fungi, bacteria, protists, actinomycetes, nematodes, and viruses (collectively known as the plant microbiota).

Unravelling Plant-Microbe Synergy. https://doi.org/10.1016/B978-0-323-99896-3.00004-7

There occurs an intricate networking between plants and the associated microbiota. Out of these, as the most extensive habitat on Earth, the rhizosphere is the most predisposed habitat due to its immense potential for nutrient uptake, enhancing nutrient absorption, modulating plant hormone signaling, and increasing plant resilience to abiotic and biotic stresses and various ailments (Chandra et al., 2020; Chandra and Sharma, 2020; el Zahar Haichar et al., 2021). Lately, it has been depicted that some members of the microbiota are able to influence the disease resistance abilities of their hosts in a holistic manner (Matsumoto et al., 2021). Moreover, a horde of abiotic factors such as soil quality, nutritional status, and climatic circumstances help balance the morphological and functional diversity of plant-associated microbiomes (Berg and Smalla, 2009). Pathogenic microbes and host plants are always at odds with each other, and the outcomes determine whether plants develop resistance or disease. Microbes colonize plants in a deliberate manner rather arbitrarily, which is marked by the existence of a specific microbial matrix and concomitance patterns. The use and manipulation of plant microbiomes has opened up new avenues for sustainable agriculture (Berg et al., 2020; Chandra et al., 2018c).

Interactions between plants and the microbiota and associated soil

Plants and their environment are in a constant state of interaction, with plants keeping an eye on their surroundings and communicating changes. A variegated plethora of microorganisms comprising 10,000–50,000 species of microbes can be found in 1 g of soil (Schloss and Handelsman, 2006), and their concentration varies as per the physiochemical nature of the soil and its texture. Various evidences suggest that soil attributes such as nitrogen content (Frey et al., 2004), phosphorus content (Faoro et al., 2010; Chandra et al., 2015), and soil pH (Fierer and Jackson, 2006; Lauber et al., 2008; Rousk et al., 2010) play a significant role in the governance of microbial communities, and, out of these attributes, soil pH may be the key deciding factor in the establishment of a bacterial community in the soil. Bacterial growth in a relatively narrow pH range and its tolerance is assumed to be one of the reasons for the reposition of bacterial communities (Rousk et al., 2010). On the contrary, phosphorus concentration, altitude, and the ratio of cations in soils (Ca^{2+}, Mg^{2+}, and Al^{3+}) are considered as a driving force for regulating microbial diversity, demolishing the pH concept hypothesis (Faoro et al., 2010).

The profuseness and composition of microbial dwellers in soils are determined by the type of plant, resulting in plant–soil interactions (van der Putten et al., 2007). Roots are the key elements of a plant, providing mechanical support, water, and nutritional supplements and aiding in interacting with its surroundings (Bais et al., 2006). Plants detect and concede the chemical signals generated by microorganisms inhabiting the soil and then respond to these signals through the release of complex chemical consolidations in the form of root exudates. A diverse amalgam of compounds such as sugars, proteins, amino acids, flavonoids, aliphatic acids, and fatty acids are all released by plants in the form of root exudates, thus providing a unique microenvironment in the rhizosphere (Badri et al., 2009). This concoction of root exudates captivates a plethora of nonpathogenic (beneficial), pathogenic, and neutral microbiota within the bounds of the rhizosphere (Bais et al., 2006). There are some cost benefits associated with the intricate network existing between plants and the associated microbiota. The composition and concentration of root exudates change in response to the signals received from the environment and the associated microbiota, type of soil (Rovira, 1969), age of the plant (De-la-Peña et al., 2010), and biotic and abiotic variables (Flores et al., 1999). Soil microorganisms employ root exudates as a growth substrate (Vandenkoornhuyse et al., 2007). Therefore, microbes that live in the rhizosphere change when the composition and concentration of exudates change (Micallef et al., 2009), indicating the intricate relationship between the microbes and exudates. The root exudates of plants are released by numerous mechanisms that are primarily involved in diffusion, ion channels, and vesicle transport, and these processes of root exudation are mostly passive in nature (Neumann and Romheld, 2000; Bertin et al., 2003).

Ecology of plant microbiomes

Microbe–microbe interactions within a microbiome play a vital, although often neglected, role in regulating the overall microbiome assembly. To demonstrate this, Niu et al. (2017) used a simulated maize root community with seven species to investigate the role that bacterial interspecies cooperation plays in microbiome construction. They discovered that removing just one species, *Enterobacter cloacae*, resulted in *Curtobacterium pusillum* dominance and a considerable reduction in the other five members of the community, implying that *E. cloacae* is a prominent factor in the assemblage of a community. As a group, seven of the microbiome members protected maize roots from infection caused by the fungus *Fusarium verticillioides* in a better manner than did the members individually, supporting the findings that diversity in microbial communities and their interactions influence pathogenic microorganism invasion. Plant root secretions mediate the communication between the soil microbes and the roots of the plants, whose continued existence serves as a key factor in determining the health and growth of the plants (Tückmantel et al., 2017). Root exudates discharge ions, oxygen, and water and, in particular, incorporate carbon-containing compounds. Some root exudates work as anti-agents against microorganisms, whereas others work as attractants, depending on the physiological status, types of plants, and nature of microbes (Kang et al., 2010). In contrast to other plant compartments, the phyllosphere has less diversified but more differentiated networks due to the tremendous vacillation in abiotic and biotic environments (Knief et al., 2012).

There are certain advantageous characteristics that individuals in plant-associated microbial communities possess, and the manifestation of those traits in a community is an emergent property that cannot be predicted from individual members. The ability of the *Pseudomonas* species to suppress plant pathogens is partly due to antibiosis and partly due to competition, but, overall, disease suppression depends on several factors including pathogen population dynamics, genetic backgrounds, and biotic and abiotic conditions (Mendes et al., 2011; Gouda et al., 2018) and the framework and diversity of the plant microbiota. This chapter discusses the endophytes, epiphytes, and rhizobiomes that inhabit the endosphere, phyllosphere, and rhizosphere of a plant, respectively.

Endophytes

Endophytes are a group of microbes that colonize plant tissues in a manner similar to phytopathogens without causing harm to the plants. It has been reported that these endophytes exist virtually in all plants (Fadiji and Babalola, 2020). The diversity and structure of an endophytic community is much less diverse and structured in contrast to the variegation and composition of the microbial community found outside of the plant. Their association with plants can be obligative or facultative, as they are omnipresent and harmless to plants (Nair and Padmavathy, 2014). Some of the endophytes possess cell wall-degrading enzymes in low concentrations and the type III secretion system and can enter host tissues, whereas others recruit some active and passive processes to enter the host such as entry through root cracks, emergence points of lateral roots, lenticels, trichomes, stomatal openings, vegetative or reproductive parts, or some damage to the tissues as a passage route (Monteiro et al., 2012). Endophytes are considered to be either harmful or nonpathogenic based on their ecological niche rather than just the function they perform in the host (Brader et al., 2017). Once the endophytes invade the plant tissues, they start fabricating enzymes such as endoglucanases and endopolygalacturonases and respond to plant cues, in order to facilitate them to enter the endophytic life stage and spread to other tissues of the root cortex and beyond (Reinhold-Hurek et al., 2006).

A pronounced involvement of endophytes has been observed in the active biocontrol of plant-associated pathogens, in the augmentation of plant growth, and in the emanation of secondary metabolites such as terpenoids, flavonoids, alkaloids, and steroids (Sharma et al., 2017; Bind et al., 2021). For example, an endophyte *Metarhizium anisopliae*, facilitated by plant *Taxus chinensis* was discovered to be the fount of alkaloid taxol (Liu et al., 2009). Endophytes can exhibit strict obligative behavior as in *Xylella fastidiosa*, which is extremely adapted to live within plant tissues, or can show facultative behavior as in *Ralstonia solanacearum biovar*, which can survive in water systems as well as in the tissues of tomato plants in an avirulent form (Hardoim et al., 2008). In most instances, it has been discovered that endophytes maintain a symbiotic relationship with their host plant. The host plant provides endophytes with nutrients and habitat requirements, and, they, in turn, provide exceptional services to the growth, health, and development of the host plant, such as protecting the host from abiotic and biotic challenges and enhancing nutrient absorption (Omomowo and Babalola, 2019).

Apart from these extraordinary contributions, several endophytic microorganisms have been found to progressively find their way into the plant, and, as this association strengthens, they create novel ways for inhabiting, evolving, establishing, and improving their relationship with the host for their survival (Goyal et al., 2016). A detailed facet of the new-fangled genes and functions performed by these endophytes and associated community networking can reinforce plant development and health, and it can be mediated by extensive comprehension of the organization and function of endophytic microbes.

Epiphytes

Epiphytes include a group of pathogenic as well as nonpathogenic microbes that inhabit the external surface of plants and multiply nonparasitically there for procuring support, water, and nutritional benefits from their host. A diverse array of microbial groups such as microbial algae, bacteria, protozoa, viruses, archaea, fungi, nematodes, and plants manifest epiphytism (Lindow and Brandl, 2003), and mycorrhizae, rhizobium bacteria, plant growth-promoting rhizobacteria (PGPR), and biocontrol microorganisms are the most intensively investigated beneficial rhizosphere species. They develop adaptive traits and intimate associations with plants by encountering biotic and abiotic stressors and triggering basal plant defense by priming their host for enhanced resistance to pathogens and influencing plant growth and fitness (Vorholt, 2012; Helfrich et al., 2018). Other factors such as cell signaling, quorum-sensing system, biofilm formation, secondary metabolite production, competition, and cooperation among the microbiota may be the major drivers of bacterial colonization in the associated host.

The communities of microorganisms colonizing the phyllosphere and rhizosphere are believed to enhance plant growth and disease resilience by virtue of the possession of some combination of traits such as enhancement of plant growth through nitrogen fixation, production of plant hormones, protection against pathogens, regulation of the biogeochemical cycle in the environment, removal of harmful contaminants, and production of secondary metabolites (Andreote et al., 2014; Berg et al., 2014). Epiphytes can hold out against uttermost environmental conditions, set apart by sparse nutrients, inconsistent humidity, pH, UV radiation, and temperature (Andreote et al., 2014). Differences in these environmental conditions, as well as the genotype and functional metabolism of the plant, determine the diversity, composition, and distribution pattern of epiphytes in individual plant species (Berg et al., 2014).

Rhizobiomes

The microbial communities dwelling in the rhizosphere are designated the communal term ‘rhizobiome’ (Sasse et al., 2018; Olanrewaju et al., 2019). The rhizosphere is the hotspot of advantageous, neutral, and harmful microbial activities, which exist in a dynamic state (Kuzyakov and Blagodatskaya, 2015). These microbial communities existing in the soil are impacted by biotic as well as abiotic factors, which incorporate edaphic and environmental factors, root morphology, including root density, root length, root volume, root biomass, and surface area as well as root exudates, which play an important role in the rhizobiome's abundance and diversity by creating different ecological functional areas for them (Bach et al., 2018). In comparison to root secretions, rhizodeposits such as mucilages, lysates, collapsed root cells, and volatiles have been found to play a substantial role in swaying and influencing microbial populationsin the rhizosphere (Dennis et al., 2010).

The frequency of bacteria is higher in the rhizobiome as compared to the total biomass of fungi and protozoa (Zhou et al., 2016). Aside from the luxuriant bacterial population, the rhizosphere comprises splendid fungal inhabitants, including the *Aspergillus*, *Fusarium*, *Trichoderma*, and *Penicillium* species (Hossain et al., 2017). Rhizospheric microorganisms dispense plant growth and function by enhancing plant pathogen resistance and utilization, accumulating more nutrients, retaining more water, and increasing their growth, and, in response, plants manifest energy and food sources consumed by these microbes in the form of carbon metabolites. A maze of interactions and signaling pathways occur between plants and microbes, microbes and microbes, and plants and plants, which further regulate and induce responses in the rhizosphere (Pausch and Kuzyakov, 2011). The rhizosphere harbors a heterogeneous composition of microbiota in its different zones, viz., the endosphere, rhizosphere, and rhizoplane, and the root tips and mature roots are the hub of these microbes as they produce generous amounts of these root exudates (Pausch and Kuzyakov, 2011). It is evident that the interactions between the rhizobiome, secretions by roots, and various other components play an indispensable role in enhancing plant growth and development. However, the procedures for a thorough analysis of the dynamic process in progress have not yet been adequately defined (Olanrewaju et al., 2019).

Plant microbiome function and interaction

The diversity of the microbiota associated with plants is tremendous, existing in a dynamic state, and their interactions can be enduring or ephemeral (King et al., 2013; Huang et al., 2020). The association of microbiomes with plants is a sequential, multiphased procedure that is regulated by the interactions between the environment and the associated host. Soil comprises manifolds of microbes, ranging from actinomycetes, fungi, viruses, bacteria, and algae to nematodes and protozoans (Geisen et al., 2019). Plant growth progression, disease symptoms, and resilience to biotic and abiotic stress factors are some of the factors associated with plant-associated microorganisms (Pérez-Jaramillo et al., 2018). These can promptly affect some functional aspects of plants (Berg et al., 2014) through a sophisticated networking strategy involving production of siderophores and phytohormones, mineral phosphate mobilization, immobilization of heavy metals through bioaccumulation, production of antibiotics, and geochemical cycling of minerals.

This interaction varies from species to species and even within the same plant's roots. For example, grapevine roots were found to be impregnated with members of the phyla *Verrucomicrobia*, *Acidobacteria*, *Actinobacteria*, *Bacteroidetes*, *Proteobacteria*, *Firmicutes*, *Planctomycetes*, *Chloroflexi*, *Planctomycetes*, and *Gemmatimonadetes*; the most abundant families often found in the case of the roots of rice were *Bradyrhizobiaceae*, *Comamonadaceae*, *Streptomycetaceae*, and *Rhizobiaceae*; and, in the case of maize roots, *Bacteroidetes*, *Firmicutes*, and *Proteobacteria* were reported to be the preponderant phyla (Correa-Galeote et al., 2018). Rhizosphere networks are more labyrinthine than those in bulk soils, which signify that the rhizosphere has immense potential for sharing its fundamental niche and interconnection and thus entertain a plethora of microbiota (Shi et al., 2016).

Nutrient acquisition and growth promotion

Microbiomes associated with plants play a crucial role in the acquisition of the nutrient pool and enhance the nutritional status of the plants. A diverse group of bacteria such as *Paraburkholderia*, *Pantoea*, and *Pseudomonas* are predominantly associated with the roots of wheat and maize, which possess some plant growth-promoting characteristics such as indole acetic acid production, nitrogen fixation, solubilization of phosphates, potassium solubilization, iron uptake, and 1-aminocyclopropane1-carboxylate (ACC) deaminase production, all of which are involved in augmenting plant growth including acquisition of nutrients and tolerance to stress (Rascovan et al., 2016). Endogenous hormone levels in related plants are modulated by microbe-produced plant growth-promoting hormones. These hormones are effective even at extremely minute concentrations and regulate various physiological processes. Additionally, the microbiota improve nutrient uptake by plants through some biochemical processes such as phosphorus solubilization and nitrogen fixation (Rascovan et al., 2016). An enzyme called ACC deaminase is secreted by some strains of *Pseudomonas* spp., *Bacillus* spp., and *Arthrobacter*, which diminishes the concentration of ethylene (stress hormone) in plants and indirectly alleviates plant growth. Plant-derived secondary metabolites—coumarins—are known to shape the root bacterial community in *A. thaliana*, which facilitates iron mobilization in plants and production of reactive oxygen species (ROS) that reduce the growth of the *Pseudomonas* species that are relatively widespread and competes with plants for iron acquisition (Voges et al., 2019).

Disease suppression

In nature, plants are targeted by a range of pathogenic microorganisms and insect pests that undermine their health. Pathogen suppression is a significant feature of a plant root microbiome that acts as a protective shield against soil-borne pathogens (Weller et al., 2002). Direct and indirect interactions with pathogens involve stimulation of the plant immune system, facilitating pathogen suppression (Lugtenberg and Kamilova, 2009).

This leads to an intricate networking between the plant microbiome and associated pathogens and counteraction of pathogen assault by the fabrication of innumerable antibiotics and pathogen-suppressing volatile compounds, regulation of phytohormone levels, assemblage of lytic enzymes, siderophore production, and induction of plant systemic resistance (Hopkins et al., 2017; Berg and Koskella, 2018; De Vrieze et al., 2018). In the tomato plant, *Fusarium oxysporum* f. sp. *lycopersici*, an attack is encountered by a heterogeneous plethora of the present microbiota, which triggers the metabolism of phenylpropanoid, a key secondary metabolite involved in the synthesis of salicylic acid, leading to cell fortification, which serves as a barrier to pathogen attacks. The pronounced role of the arbuscular mycorrhiza fungus *Rhizophagus irregularis* was observed in enhancing resistance to *Xanthomonas campestris* and a tremendous resistance against *Erysiphe pisi* in the case of *Medicago truncatula* (Smigielski et al., 2019). Particularly, genera like *Burkholderia, Paraburkholderia, Bacillus, Paenibacillus, Streptomyces, Enterobacter, Pantoea* and *Pseudomonas* are renowned for their significant contribution to pathogen assault. A significant increase in resistance to the fungal pathogen *Gaeumannomyces graminis* was observed in barley inoculated with *Pseudomonas* spp. (Schlatter et al., 2017).

Various symbiotic microbes or some chemical morphogens prompt induced systemic resistance (ISR) besides pathogen-induced systemic acquired resistance (SAR) (Chester, 1933), which dispense resistance to the host tissues against the attack of pathogenic microorganisms. The genesis of various antimicrobial compounds including coumarins and benzoxazinoids is mediated by ISR-inducing, root-associated bacteria and fungi that further reinforce the deployment of ISR-inducing strains in the associated rhizosphere (Hu et al., 2018). Metabolites secreted by the roots drive the plant–soil interpretation of development and defense by framing the microbiota of the rhizosphere (Chialva et al., 2018). In sweet pepper, the herbivory pronounced by aphid infestation prompted a hike in the population of the nonpathogenic rhizobacterium *Bacillus subtilis*. In addition to this, the insect herbivory is also affected by induction of defenses regulated by the bacterial and fungal communities associated with the roots (Yuan et al., 2018).

Stress tolerance

Droughts, heat, excess of metals, salinity, herbivory, and infections are all believed to trigger plants to sense and respond to stress signals, which are augmented by episodes of climate change (Fedoroff et al., 2010). The physiological, biochemical, and molecular characteristics of plants are all altered when they are distressed. During periods of biotic or abiotic stresses, plants generally harbor a cornucopia of stress resistance-promoting microbiomes (Fitzpatrick et al., 2018). Endosymbionts are best known to confer resilience to stress. The pronounced role of plant growth-promoting bacteria (PGPB) is comprehended when plants are subjected to innumerable abiotic or biotic stressors (Fahad et al., 2015; Chandra et al., 2019a, b), and these can encounter multiple stresses at a time. In the case of durum wheat in salt-overblown soils and arid field conditions, a remarkable increase in the growth and yield was observed by application of *Azospirillum brasilense* (Nabti et al., 2010), whereas *Paraburkholderia phytofirmans* enhanced cold tolerance by prompting the strengthening of cell walls and increasing photosynthetic pigments, leading to improved cold tolerance in *A. thaliana* (Su et al., 2015). Additionally, an increased host resistance against innumerable bacterial and fungal pathogens was shown by *P. phytofirmans* (Miotto-Vilanova et al., 2016; Timmermann et al., 2017).

The recycling of agricultural wastes is an eminent feature of microorganisms (Singh and Nain, 2014). Actinomycetes are filamentous Gram-positive bacteria that possess the characteristics of both bacteria and fungi and obstruct the survival of various pathogens in the rhizosphere. They recycle the assemblage of the polymers present in dead plants, animals, and fungal materials, ensuring the production of various extracellular enzymes that are conducive to the production of biomass of plants. Besides, they possess some antibiotic properties and release metabolites that enhance plant development and ameliorate the availability of mineral nutrition to plants (Srinivasan et al., 1991; Jeffrey et al., 2007). Actinomycetes have been reported to accelerate the pace of mycorrhizal development that results in increased plant growth (Mohammadi and Lahdenpera, 1992). Another group of photosynthetically active organisms found in soils are algae, reported to enhance the binding of soil particles and increase the water retention capacity of soils, oxygenation of the soil environment, and organic matter content of soils upon their death (Crouzet et al., 2019). Algae present in soils comprise four main phyla, namely, *Cyanophyta*, *Chlorophyta*, *Xanthophyta*, and *Bacillariophyta*. Protozoans are organisms that have antimicrobial properties and play an important role in nutrient cycling. The constitution of plant secondary metabolites and development of various metabotypes are also impacted by plant microbiomes as seen in medicinal plants, where these microbiomes alter the production of bioactive compounds (Köberl et al., 2013; Schmidt et al., 2014). Besides these attributes, root microbiomes are also known to impact the phenology of plants. They maintain an ecological balance in the soils by keeping a check on the population of bacteria by their predation (Gaines et al., 2019). Additionally, they play a critical role in enhancing decomposition rates, soil aggregation, and mineralizing nutrients, ensuring their accessibility by various soil microorganisms and other plants.

Besides the above-listed organisms, there are innumerable viruses and nematodes harbored by soils, which are greatly implicated in plant diseases. Aside from their disease-engendering ability, nematodes are known to confer some benefits on the soil such as mineralization of organic nutrients, nutrient conversion, and contribution of dissolved N to the soil, which further aid in plant development and health (Gebremikael et al., 2016). Similarly, millions of viruses harbored by soils are involved in influencing various pathogenic and non-pathogenic microorganisms in soils and swaying plant health. Viruses play a crucial role in the assemblage and turnover of bacterial communities; this turnover in the soil is critically monitored by viruses, but a detailed account of their function in environments associated with plants is yet to be completely understood (Pratama and Van Elsas, 2018). However, fresh research approaches and a thorough understanding of the composition, richness, probable relationships, variety, and co-occurrence of microbial assemblages in their natural habitats are yet to be developed (Qi et al., 2019). A chasmic analysis of the reciprocal nature of the interactions between the host and the microbiota can aid in elucidating the organization of associated microbial groups (Layeghifard et al., 2017), besides apprehending the dynamic composition and structure of the concerned microbiota (Ma et al., 2016).

Factors affecting plant microbiomes

The amalgam of the microbial horde around the roots is altered by rhizodeposits and root exudates, which include carbohydrates, organic acids, amino acids, fatty acids, phenolic compounds, plant growth regulators, nucleotides, sugars, sterols, mucilages, and vitamins (Carvalhais et al., 2011). These exudates released from the roots are predominantly photosynthetic products that are transported to the rhizosphere (Grayston et al., 1997; Walker et al., 2003). However, it is still an enigma how the immune system of plants evolves and distinguishes between pathogenic and nonpathogenic microbes and inhibits colonization by the former and entertains habitation by the latter. *Streptomyces*, *Serratia*, *Arthrobacter*, *Pseudomonas*, *Azospirillum*, *Bacillus*, *Azotobacter*, *Mesorhizobium*, *Burkholderia*, *Rhodococcus*, *Burkholderia*, *Micrococcus*, *Agrobacterium*, *Alcaligenes*, *Enterobacter*, *Cellulomonas*, *Bradyrhizobium*, *Rhizobia*, and *Klebsiella* are some of the bacteria that dominate the rhizosphere and alter its physiochemical panorama (Prashar et al., 2014).

An intricate networking occurs among the colossal diversity of bacteria in an antagonistic, competitive, and mutualistic manner, which involves interactions via competition and exchange for nutrition and interdependence where the exchange of metabolites among microbes expedites the exemplary growth of various microbial species (Peterson et al., 2006). Bacterial–fungal interactions are also sustained in the same manner as the potential of a plant to develop symbiosis with arbuscular mycorrhiza fungi (AMF), or nitrogen-fixing *Rhizobia*, strongly affects the encompassing microbial communities (Pii et al., 2016; Zgadzaj et al., 2016; Zgadzaj et al., 2019). Thus, the composition of microbiomes and their end results on the host can be impacted by direct cooperative or competitive interactions among the community members, which is a paramount factor for the outcome of the plant–microbiota relationship. The composition of root exudates plays a key role in framing the prosperity of rhizospheric microbial communities corresponding with the respective plant (Zhang et al., 2014). Alterations in the microbial communities colonizing the roots of the plants are driven by transformation in the framework of exudates released by roots. Liljeroth et al. (1990) and Zhu et al. (2016) characterized the root exudates from maize, cultivated in a nitrogen-enriched environment, and concluded that the abundance of root microbiomes was affected by root secretions, such as mucilages and sugars, phenolic compounds, and sugar alcohols, which remarkably disposed the level of the fertilizers.

Moreover, wheat plants associated rhizospheric bacteria, deprived of inorganic nitrogen fertilizers, resulted in the reduction of genes associated with the anabolism and catabolism of amino acids and carbohydrates and enhanced the associated terpenoid metabolism pathways, further facilitating members of the phylum *Bacteroidetes* and other rhizospheric communities associated with it (Kavamura et al., 2018). Terpenoids are known to regulate loss of nitrogen by nitrification in nitrogen-deficient environments, denominated as inhibitors of nitrification (Coskun et al., 2017; Hartman and Tringe, 2019). However, it remains an enigma whether the terpenoids exuded by plants are for nitrifying the growth environment or they have some nitrification-related functions that need to be tracked down (Coskun et al., 2017). The composition of plant microbiomes is also influenced by innumerable rhizodeposits (Pascale et al., 2020). The shielding mechanism adopted by plants includes synthesis of various indolic and aliphatic glucosinolates (Xu et al., 2017). (*Avena barbata*, at different growth stages, has revealed that sucrose concentration was relatively high during the early stages of development, whereas at a later stage, amino acids and defense molecules were highly fabricated (Zhalnina et al., 2018).

The pronounced function of plant genotypes is also reported to ascend the relative prosperity of the microbiota associated with the rhizosphere of plants (Haney et al., 2015). It is evident that profusion of the microbiota and their role in conferring health benefits on plants subjected to different abiotic stresses are regulated by the genotypes of plants (Soussi et al., 2016; Chandra et al., 2018a, b). Host plants as well as the nature of the soil transform the diversity of bacteria, with a pronounced role of the latter (Latour et al., 1996). Evidences for the same are validated in a study conducted by Salles et al. (2004), in which plants such as barley, maize, grass, and oat were cultivated in a greenhouse with soils having varying land use applications, and the results indicated a prominence in the composition of the associated bacterial communities of *Burkholderia* and diversity of *Pseudomonas*. An investigation of microbial heterogeneity in the rhizosphere of *Reaumuria negevensis* reported that *Actinobacteria* were more prepotent in all the major seasons, except in winter during which *Acidobacteria* proliferated to their maximum. On the contrary, the phyla *Chloroflexi* and *Bacteroidetes* flourished during summer with remarkable decline during autumn dominated by the phylum *Gemmatimonadetes* (Saul-Tcherkas and Steinberger, 2011). This holds true for the fact that microbial communities associated with respective plants are also influenced by seasonal variations (Soussi et al., 2016).

Apprenticing and modulating plant microbiomes

For more than a decade, scientists have been researching the impact of microbial inoculants on plant microbiomes. Investigations of microbiomes have changed our perception of plant microbiome function while also opening up new avenues for sustainable agricultural applications (Berg et al., 2020). A comprehension of their mechanisms of action is critical for the development of prospective applications such as microbial inoculants for long-term agriculture (Köhl et al., 2019). The use of microbial consortia is a novel method for overcoming field challenges, and this method could be justified by a mixture of microorganisms with diverse properties, either complimenting one another or combining distinct mechanisms required for different outcomes, such as plant growth promotion and disease control (Samad et al., 2017). Plant–microbe as well as pathogen–microbe interactions are well-researched in vitro (Whipps et al., 1988; Weller et al., 2002; Köhl et al., 2019), whereas the interconnection with microbial inoculants in situ is not yet fully comprehended. Transient microbiome shifts, microbial diversity stabilization, microbiome uniformity, reduction of a pathogen-induced shift, targeted shifts toward potentially advantageous phyla, and exhaustion of potential pathogens are the six main types of microbiome modulation documented by research (Sessitsch et al., 2019).

In various instances, microbial inoculants have had a more striking and uniform outcome than inoculation with a single strain in different plant growth mechanisms, such as nutrient supply, plant hormone production, pest control, nitrogen fixation, higher shoot and root biomass, and soil disease suppression (Toyota and Watanabe, 2013). In the case of maize, a pronounced resistance against drought was observed by application of *Azospirillum brasilense*, *Sphingomonas* sp., *Pseudomonas putida*, and *Acinetobacter* sp. (Molina-Romero et al., 2017), and application of *Microbacterium* sp., *Stenotrophomonas* sp., and *Xanthomonas* sp. enhanced resistance against fungal pathogens in *A. thaliana* (Berendsen et al., 2018). Studies have revealed that these co-colonizers can drive plant health in a neutral or beneficial manner or may exert detrimental effects on growth and yield by engrossing carbon sinks of the plant as seen in the colonization of *Lotus japonicus* by *Rhizobia*, which facilitated other endophytes to colonize the nodule by thumbing along the infection thread, serving as a route for the *Rhizobia* to enter the nodule (Zgadzaj et al., 2015). In addition to this, a cocktail of these microbiota conferred more advantages on plants corresponding to a single inoculum, as in the case of tomato, where co-inoculation of the arbuscular mycorrhiza fungus *Glomus mosseae* with a plant growth-promoting rhizobacterium (either *Bacillus* sp. or *Bacillus polymyxa*) provided a luxuriant growth with any of the microorganisms alone.

Microbial fingerprinting techniques have already revealed significant alterations among microbial communities, such as those generated by *Stenotrophomonas* or *Pseudomonas* treatments, and the use of next-generation sequencing techniques has allowed researchers to unravel these alterations to a greater depth (Götz et al., 2006; Adesina et al., 2009). This indicates that plant immune responses can be modulated in order to attract a plethora of microorganisms that provide long-term plant protection (Köhl et al., 2019). Engineering of microbiomes accompanying plants to manage plant diseases will continue to necessitate a greater knowledge of plant–microbiome–environment interactions in a variety of crops. It is envisaged that a better tool chest of technologies will eliminate bias caused by individual technologies in the future, resulting in a more comprehensive perspective of the biological system as a whole (Berg et al., 2020).

Plant–microbiome operation

Plant-microbiome interactions are dynamic, involving a wide spectrum of microbes and often producing collaborative effects (Mendes et al., 2011). High microbial diversity and evenness within plant-associated indigenous microbial populations describe healthy plant microbiomes (Berg et al., 2017; Trivedi et al., 2020).

Endophytes and mutualistic symbionts are notable examples, which just like pathogens, are potent enough to invade the host tissue but remain there in harmony with the host, imparting various advantages to it. These findings indicate that the pathogenic and beneficial lifestyle of microbiomes can be found to be associated with the host. Microbial relatives with a contrasting nature are found within the same genus, e.g., among the *Paenibacillus*: *P. azotofixans* and *P. amylolyticus* (Grady et al., 2016), among the *Bacillus*: *B. velezensis* and *B. cereus* (Radhakrishnan et al., 2017), among the *Pseudomonas: P. simiae* and *P. syringae* (Anderson et al., 2018), and even within the same species, e.g., *Pseudomonas aeruginosa* (Steindler et al., 2009; Ndeddy Aka and Babalola, 2016). A striking example is seen in the case of *Streptomyces* (Viaene et al., 2016), where *S. lividans* can shield plants against fungal pathogens (Meschke and Schrempf, 2010), and, on the contrary, *S. scabiei* causes rot on tubers and the roots of carrots, beets, and potatoes (Hiltunen et al., 2009).

Positive interactions

A plethora of beneficial microbiota present in the rhizobiome employ various mechanisms to repress pathogens and supplement plant growth and development (Olanrewaju et al., 2017). Biofertilization, bioremediation, and biocontrol are some mechanisms deployed by these microbiota indulging in disparate activities such as nitrogen fixation, solubilization of phosphates, fabrication of cytokinin, ACC deaminase, auxin, and gibberellin, siderophore synthesis, modulation of ethylene levels, induced systemic resistance, acquired systemic resistance, and supplementing plant growth (Babalola, 2010; Martínez-Viveros et al., 2010; Olanrewaju et al., 2017).

Siderophores are iron-chelating compounds produced by a great variety of microorganisms that thrive in inadequate iron conditions (Kaszubiak, 1998). They scavenge trace elements like iron from soils, are primarily present in an insoluble form, are inaccessible to plants (Aznar and Dellagi, 2015), and hinder the proliferation of pathogens by depriving the access of these elements to pathogens. Various beneficial microbial activities, such as quorum-sensing interference, antibiosis, and competition for space and nutrients as well as induction of systemic resistance, chelation of iron, and metabolite formation, are used to block pathogenic activities, but they also impair or eliminate the beneficial microbiota (Babalola, 2010; Raaijmakers and Mazzola, 2012; Olanrewaju et al., 2017). In terms of siderophore production, *Actinobacteria* is one of the most important groups.

Antibiotics can impede the growth of bacterial biofilms and protozoa by blocking enzymes involved in the formation of cell walls, anabolism and catabolism of nucleic acid and repair, disruption of protein synthesis, disruption of membrane structure, or acting as mediators of cellular signals, depending on their dose (Raaijmakers and Mazzola, 2012). Studies have also found that low-abundance taxa are crucial for preventing undesired microbial incursions into soil communities. Some rhizospheric bacterial species can also activate the plant immune system, which is largely controlled by jasmonic acid and ethylene signaling (Berendsen et al., 2012; Nambara, 2013). The jasmonic acid pathway may be engaged during plant interactions with beneficial microorganisms, resulting in induced systemic resistance (ISR). Advantageous bacteria actuate the jasmonic acid–ethylene pathway during ISR, which improves the responses of plants to pathogen infections (Zamioudis and Pieterse, 2012).

Plants have a substantial impact on the metabolic activities of beneficial microorganisms, as these are helpful to plants. Sugars, amino acids, organic acids, and other innumerable primary and secondary metabolites are synthesized by plants. These are consumed by bacteria as a source of food, allowing them to grow and perform other metabolic functions (Geetanjali and Jain, 2016). The nature of microorganisms present in the rhizosphere is determined by exudates generated by plant roots, and the types of bacteria present there, in turn, modify the root exudates produced (Prashar et al., 2014). As a result, microorganisms are believed to be plant-specific. Furthermore, the capacity to anticipate plant performance based on microbiome composition will go a long way toward increasing the soil microbiome's role in agriculture.

Negative interactions

Various pathogenic microbiota thriving in the rhizobiome of plants are known to confer several diseases in associated plants; pathogenic fungi, several nematodes and actinomycetes, virulent bacteria, and some viruses belong to this category. They primarily encompass *Pectobacterium carotovorum*, *Rhizoctonia solanacearum*, *Pectobacterium atrosepticum*, *Dickeya dadantii*, *Dickeya solani*, *Agrobacterium tumefaciens*, and various other pathogenic organisms (Mansfield et al., 2012). Various plant-associated microorganisms such as *Pseudomonas syringae*, *Erwinia amylovora*, *Xanthomonas* spp., and *Xylella fastidiosa* produce proteinaceous toxic chemicals that cause diseases in numerous crops such as bananas, potatoes, green beans, and tomatoes (Mansfield et al., 2012). Nematodes serve as carriers of some pathogenic viruses and fungi into the rhizosphere of plants (Rochon, 2009) as well as exhaust nutrients and confer various diseases on the associated host (Rasmann et al., 2012). The release of some phytotoxins such as ethylene and hydrogen cyanide has been reported to hinder plant root development and serve as indirect means of negative plant–microbe interactions (Martínez-Viveros et al., 2010).

Many human pathogenic bacteria including *Pseudomonas aeruginosa*, *Burkholderia cepacia*, and *Salmonella enterica* release metabolites that allow them to procure nourishment, colonize, and totally dominate other helpful resident microbes on plant surfaces (Berg et al., 2005). A number of factors influence the severity of plant diseases, including pathogen population size, host sensitivity, a favorable environment, and abiotic and biotic factors such as the plant microbiome, all of which collaborate to influence the outcomes of the plant–pathogen interaction (Schwessinger and Ronald, 2012). Beneficial and pathogenic bacteria share physiological and evolutionary similarities to the point where the appearance of a pathogenic phenotype can be influenced by minor variations in the microbes and, in certain cases, the host. Plants, on the other hand, have evolved sophisticated ways to discern between a possibly beneficial bacterium and a closely related potentially lethal infection.

Conclusion and future perspectives

The repercussions of plant–microbe interactions and their variegated activities have piqued the interest of agronomists, evolutionary biologists, plant biologists, and ecologists in the current scenario. The divergent plethora of microbiota can function as phytopathogens as well as biostimulants, biofertilizers, biopesticides, and plant growth promoters and can drive the health and performance of associated plants by a subset of intricate networking. In the context of climate change and future precision agriculture, these microbial functions have attracted growing interest, where dispersion of selected beneficial microbial cocktails on a targeted basis may result in a long-term resolution to counteract biotic and abiotic stress conditions and ensure crop yield reliability. Various corroborations have revealed the excellent capability of the microbiota and the complicated networking that exists between them, but there is still a void that must be addressed by the requisition of strong networking models that are required to analyze these interactions. The establishment of large collections of microorganisms and development in meta-omics have dramatically increased our knowledge of the structure and function of plant microbiome composition and dynamicity. Further research will likely include the development of vigorous network modeling to gain insights into the structure and synergistic effects of these microbiomes, which will aid in the development of new management approaches.

References

Adesina, M.F., Grosch, R., Lembke, A., Vatchev, T.D., Smalla, K., 2009. In vitro antagonists of *Rhizoctonia solani* tested on lettuce: rhizosphere competence, biocontrol efficiency and rhizosphere microbial community response. FEMS Microbiol. Ecol. 69 (1), 62–74.

Anderson, J.A., Staley, J., Challender, M., Heuton, J., 2018. Safety of *Pseudomonas chlororaphis* as a gene source for genetically modified crops. Transgenic Res. 27 (1), 103–113.

Andreote, F.D., Gumiere, T., Durrer, A., 2014. Exploring interactions of plant microbiomes. Sci. Agric. 71 (6), 528–539.

Aznar, A., Dellagi, A., 2015. New insights into the role of siderophores as triggers of plant immunity: what can we learn from animals? J. Exp. Bot. 66 (11), 3001–3010.

Babalola, O.O., 2010. Beneficial bacteria of agricultural importance. Biotechnol. Lett. 32 (11), 1559–1570.

Bach, E.M., Williams, R.J., Hargreaves, S.K., Yang, F., Hofmockel, K.S., 2018. Greatest soil microbial diversity found in micro-habitats. Soil Biol. Biochem. 118, 217–226.

Badri, D.V., Weir, T.L., Van der Lelie, D., Vivanco, J.M., 2009. Rhizosphere chemical dialogues: plant–microbe interactions. Curr. Opin. Biotechnol. 20 (6), 642–650.

Bais, H.P., Weir, T.L., Perry, L.G., Gilroy, S., Vivanco, J.M., 2006. The role of root exudates in rhizosphere interactions with plants and other organisms. Annu. Rev. Plant Biol. 57, 233–266.

Berendsen, R.L., Pieterse, C.M., Bakker, P.A., 2012. The rhizosphere microbiome and plant health. Trends Plant Sci. 17 (8), 478–486.

Berendsen, R.L., Vismans, G., Yu, K., Song, Y., de Jonge, R., Burgman, W.P., Burmølle, M., Herschend, J., Bakker, P.A., Pieterse, C.M., 2018. Disease-induced assemblage of a plant-beneficial bacterial consortium. ISME J. 12 (6), 1496–1507.

Berg, M., Koskella, B., 2018. Nutrient-and dose-dependent microbiome-mediated protection against a plant pathogen. Curr. Biol. 28 (15), 2487–2492.

Berg, G., Smalla, K., 2009. Plant species and soil type cooperatively shape the structure and function of microbial communities in the rhizosphere. FEMS Microbiol. Ecol. 68 (1), 1–13.

Berg, G., Eberl, L., Hartmann, A., 2005. The rhizosphere as a reservoir for opportunistic human pathogenic bacteria. Environ. Microbiol. 7 (11), 1673–1685.

Berg, G., Grube, M., Schloter, M., Smalla, K., 2014. Unraveling the plant microbiome: looking back and future perspectives. Front. Microbiol. 5, 148.

Berg, G., Köberl, M., Rybakova, D., Müller, H., Grosch, R., Smalla, K., 2017. Plant microbial diversity is suggested as the key to future biocontrol and health trends. FEMS Microbiol. Ecol. 93 (5).

Berg, G., Rybakova, D., Fischer, D., Cernava, T., Vergès, M.C.C., Charles, T., Chen, X., Cocolin, L., Eversole, K., Corral, G.H., Kazou, M., Schloter, M., 2020. Microbiome definition re-visited: old concepts and new challenges. Microbiome 8 (1), 1–22.

Bertin, C., Yang, X., Weston, L.A., 2003. The role of root exudates and allelochemicals in the rhizosphere. Plant Soil 256 (1), 67–83.

Bind, S., Bind, S., Chandra, D., 2021. An endophytic bacterial approach: a key regulator of drought stress tolerance in plants. In: Bhatt, P., Gangola, S., Udayanga, D., Kumar, G. (Eds.), Microbial Technology for Sustainable Environment. Springer, Singapore, pp. 551–569.

Brader, G., Compant, S., Vescio, K., Mitter, B., Trognitz, F., Ma, L.J., Sessitsch, A., 2017. Ecology and genomic insights into plant-pathogenic and plant-nonpathogenic endophytes. Annu. Rev. Phytopathol. 55, 61–83.

Carvalhais, L.C., Dennis, P.G., Fedoseyenko, D., Hajirezaei, M.R., Borriss, R., von Wirén, N., 2011. Root exudation of sugars, amino acids, and organic acids by maize as affected by nitrogen, phosphorus, potassium, and iron deficiency. J. Plant Nutr. Soil Sci. 174 (1), 3–11.

Chandra, D., Sharma, A.K., 2020. Field evaluation of consortium of bacterial inoculants producing ACC deaminase on growth, nutrients and yield components of rice and wheat. J. Crop. Sci. Biotechnol. 24 (3), 293–305.

Chandra, D., Srivastava, R., Sharma, A.K., 2015. Environment-friendly phosphorus biofertilizer as an alternative to chemical fertilizers. In: Pati, B.R., Mandal, S.M. (Eds.), Recent Trends in Biofertilizers. IK International Publishing House, New Delhi, pp. 43–71.

Chandra, D., Srivastava, R., Glick, B.R., Sharma, A.K., 2018a. Drought-tolerant *Pseudomonas* spp. improve the growth performance of finger millet (*Eleusine coracana* (L.) Gaertn.) under non-stressed and drought-stressed conditions. Pedosphere 28, 227–240.

Chandra, D., Srivastava, R., Sharma, A., 2018b. Influence of IAA and ACC deaminase producing fluorescent pseudomonads in alleviating drought stress in wheat (*Triticum aestivum*). Agric. Res. 7, 290–299.

Chandra, D., Barh, A., Sharma, I.P., 2018c. Plant growth promoting bacteria: a gateway to sustainable agriculture. In: Bhatt, P., Sharma, A. (Eds.), Microbial Biotechnology in Environmental Monitoring and Cleanup. IGI Global, pp. 318–338.

Chandra, D., Srivastava, R., Gupta, V.V., Franco, C.M., Sharma, A.K., 2019a. Evaluation of ACC-deaminase-producing rhizobacteria to alleviate water-stress impacts in wheat (*Triticum aestivum* L.) plants. Can. J. Microbiol. 65 (5), 387–403.

Chandra, D., Srivastava, R., Gupta, V.V., Franco, C.M., Paasricha, N., Saifi, S.K., Tuteja, N., Sharma, A.K., 2019b. Field performance of bacterial inoculants to alleviate water stress effects in wheat (*Triticum aestivum* L.). Plant Soil 441 (1), 261–281.

Chandra, D., Srivastava, R., Glick, B.R., Sharma, A.K., 2020. Rhizobacteria producing ACC deaminase mitigate water-stress response in finger millet (*Eleusine coracana* (L.) Gaertn.). 3. Biotech 10, 65.

Chester, K.S., 1933. The problem of acquired physiological immunity in plants. Q. Rev. Biol. 8 (3), 275–324.

Chialva, M., Salvioli di Fossalunga, A., Daghino, S., Ghignone, S., Bagnaresi, P., Chiapello, M., Novero, M., Spadaro, D., Perotto, S., Bonfante, P., 2018. Native soils with their microbiotas elicit a state of alert in tomato plants. New Phytol. 220 (4), 1296–1308.

Correa-Galeote, D., Bedmar, E.J., Arone, G.J., 2018. Maize endophytic bacterial diversity as affected by soil cultivation history. Front. Microbiol. 9, 484.

Coskun, D., Britto, D.T., Shi, W., Kronzucker, H.J., 2017. Nitrogen transformations in modern agriculture and the role of biological nitrification inhibition. Nat. Plants 3 (6), 1–10.

Crouzet, O., Consentino, L., Pétraud, J.P., Marrauld, C., Aguer, J.P., Bureau, S., Le Bourvellec, C., Touloumet, L., Berard, A., 2019. Soil photosynthetic microbial communities mediate aggregate stability: influence of cropping systems and herbicide use in an agricultural soil. Front. Microbiol. 10, 1319.

De Vrieze, M., Germanier, F., Vuille, N., Weisskopf, L., 2018. Combining different potato-associated *Pseudomonas* strains for improved biocontrol of Phytophthora infestans. Front. Microbiol. 9, 2573.

De Zelicourt, A., Al-Yousif, M., Hirt, H., 2013. Rhizosphere microbes as essential partners for plant stress tolerance. Mol. Plant 6 (2), 242–245.

De-la-Peña, C., Badri, D.V., Lei, Z., Watson, B.S., Brandão, M.M., Silva-Filho, M.C., Sumner, L.W., Vivanco, J.M., 2010. Root secretion of defense-related proteins is development-dependent and correlated with flowering time. J. Biol. Chem. 285 (40), 30654–30665.

Dennis, P.G., Miller, A.J., Hirsch, P.R., 2010. Are root exudates more important than other sources of rhizodeposits in structuring rhizosphere bacterial communities? FEMS Microbiol. Ecol. 72 (3), 313–327.

el Zahar Haichar, F., Cernava, T., Liu, J., Timm, C.M., 2021. Novel insights into the response of the plant microbiome to abiotic factors. Front. Plant Sci. 12.

Fadiji, A.E., Babalola, O.O., 2020. Elucidating mechanisms of endophytes used in plant protection and other bioactivities with multifunctional prospects. Front. Bioeng. Biotechnol. 8, 467.

Fahad, S., Hussain, S., Bano, A., Saud, S., Hassan, S., Shan, D., Khan, F.A., Khan, F., Chen, Y., Wu, C., Tabassum, M.A., Huang, J., 2015. Potential role of phytohormones and plant growth-promoting rhizobacteria in abiotic stresses: consequences for changing environment. Environ. Sci. Pollut. Res. 22 (7), 4907–4921.

Faoro, H., Alves, A.C., Souza, E.M., Rigo, L.U., Cruz, L.M., Al-Janabi, S.M., Monteiro, R.A., Baura, V.A., Pedrosa, F.O., 2010. Influence of soil characteristics on the diversity of bacteria in the Southern Brazilian Atlantic forest. Appl. Environ. Microbiol. 76 (14), 4744–4749.

Fedoroff, N.V., Battisti, D.S., Beachy, R.N., Cooper, P.J., Fischhoff, D.A., Hodges, C.N., Knauf, V.C., Lobell, D., Mazur, B.J., Molden, D., Reynolds, M.P., Zhu, J.K., 2010. Radically rethinking agriculture for the 21st century. Science 327 (5967), 833–834.

Fierer, N., Jackson, R.B., 2006. The diversity and biogeography of soil bacterial communities. Proc. Natl. Acad. Sci. U. S. A. 103 (3), 626–631.

Fitzpatrick, C.R., Copeland, J., Wang, P.W., Guttman, D.S., Kotanen, P.M., Johnson, M.T., 2018. Assembly and ecological function of the root microbiome across angiosperm plant species. Proc. Natl. Acad. Sci. U. S. A. 115 (6), E1157–E1165.

Flores, H.E., Vivanco, J.M., Loyola-Vargas, V.M., 1999. Radicle' biochemistry: the biology of root-specific metabolism. Trends Plant Sci. 4 (6), 220–226.

Frey, S.D., Knorr, M., Parrent, J.L., Simpson, R.T., 2004. Chronic nitrogen enrichment affects the structure and function of the soil microbial community in temperate hardwood and pine forests. For. Ecol. Manag. 196 (1), 159–171.

Gaines, A., Ludovice, M., Xu, J., Zanghi, M., Meinersmann, R.J., Berrang, M., Daley, W., Britton, D., 2019. The dialogue between protozoa and bacteria in a microfluidic device. PLoS One 14 (10), e0222484.

Gebremikael, M.T., Steel, H., Buchan, D., Bert, W., De Neve, S., 2016. Nematodes enhance plant growth and nutrient uptake under C and N-rich conditions. Sci. Rep. 6 (1), 1–10.

Geetanjali, Jain, P., 2016. Antibiotic production by rhizospheric soil microflora—a review. Int. J. Pharm. Sci. Res. 7 (11), 4304–4314.

Geisen, S., Briones, M.J., Gan, H., Behan-Pelletier, V.M., Friman, V.P., de Groot, G.A., Hannula, S.E., Lindo, Z., Philippot, L., Tiunov, A.V., Wall, D.H., 2019. A methodological framework to embrace soil biodiversity. Soil Biol. Biochem. 136, 107536.

Götz, M., Gomes, N.C., Dratwinski, A., Costa, R., Berg, G., Peixoto, R., Mendonça-Hagler, L., Smalla, K., 2006. Survival of gfp-tagged antagonistic bacteria in the rhizosphere of tomato plants and their effects on the indigenous bacterial community. FEMS Microbiol. Ecol. 56 (2), 207–218.

Gouda, S., Kerry, R.G., Das, G., Paramithiotis, S., Shin, H.S., Patra, J.K., 2018. Revitalization of plant growth promoting rhizobacteria for sustainable development in agriculture. Microbiol. Res. 206, 131–140.

Goyal, S., Ramawat, K.G., Mérillon, J.M., 2016. Different shades of fungal metabolites: an overview. Fungal Metab. 1, 29.

Grady, E.N., MacDonald, J., Liu, L., Richman, A., Yuan, Z.C., 2016. Current knowledge and perspectives of Paenibacillus: a review. Microb. Cell Factories 15 (1), 1–18.

Grayston, S.J., Vaughan, D., Jones, D., 1997. Rhizosphere carbon flow in trees, in comparison with annual plants: the importance of root exudation and its impact on microbial activity and nutrient availability. Appl. Soil Ecol. 5 (1), 29–56.

Haney, C.H., Samuel, B.S., Bush, J., Ausubel, F.M., 2015. Associations with rhizosphere bacteria can confer an adaptive advantage to plants. Nat. Plants 1 (6), 1–9.

Hardoim, P.R., Van Overbeek, L.S., Van Elsas, J.D., 2008. Properties of bacterial endophytes and their proposed role in plant growth. Trends Microbiol. 16, 463–471.

Hartman, K., Tringe, S.G., 2019. Interactions between plants and soil shaping the root microbiome under abiotic stress. Biochem. J. 476 (19), 2705–2724.

Helfrich, E.J., Vogel, C.M., Ueoka, R., Schäfer, M., Ryffel, F., Müller, D.B., Probst, S., Kreuzer, M., Piel, J., Vorholt, J.A., 2018. Bipartite interactions, antibiotic production and biosynthetic potential of the Arabidopsis leaf microbiome. Nat. Microbiol. 3 (8), 909–919.

Hiltunen, L.H., Ojanperä, T., Kortemaa, H., Richter, E., Lehtonen, M.J., Valkonen, J.P.T., 2009. Interactions and biocontrol of pathogenic Streptomyces strains co-occurring in potato scab lesions. J. Appl. Microbiol. 106 (1), 199–212.

Hopkins, S.R., Wojdak, J.M., Belden, L.K., 2017. Defensive symbionts mediate host–parasite interactions at multiple scales. Trends Parasitol. 33 (1), 53–64.

Hossain, M.M., Sultana, F., Islam, S., 2017. Plant growth-promoting fungi (PGPF): phytostimulation and induced systemic resistance. In: Singh, D.P., Singh, H.B., Prabha, R. (Eds.), Plant-Microbe Interactions in Agro-Ecological Perspectives. Springer, Singapore, pp. 135–191.

Hu, L., Robert, C.A., Cadot, S., Zhang, X., Ye, M., Li, B., Manzo, D., Chervet, N., Steinger, T., Heijden, M.G.A., Schlaeppi, K., Erb, M., 2018. Root exudate metabolites drive plant-soil feedbacks on growth and defense by shaping the rhizosphere microbiota. Nat. Commun. 9 (1), 1–13.

Huang, R., Zeng, J., Zhao, D., Cook, K.V., Hambright, K.D., Yu, Z., 2020. Sediment microbiomes associated with the rhizosphere of emergent macrophytes in a shallow, subtropical lake. Limnol. Oceanogr. 65, S38–S48.

Jeffrey, L.S.H., Sahilah, A.M., Son, R., Tosiah, S., 2007. Isolation and screening of actinomycetes from Malaysian soil for their enzymatic and antimicrobial activities. J. Trop. Agric. Food Sci. 35 (1), 159.

Kang, B.G., Kim, W.T., Yun, H.S., Chang, S.C., 2010. Use of plant growth-promoting rhizobacteria to control stress responses of plant roots. Plant Biotechnol. Rep. 4 (3), 179–183.

Kaszubiak, H., 1998. Mutual antagonism between fluorescent pseudomonads and soil actinomycetes. Pol. J. Environ. Stud. 7, 207–212.

Kavamura, V.N., Hayat, R., Clark, I.M., Rossmann, M., Mendes, R., Hirsch, P.R., Mauchline, T.H., 2018. Inorganic nitrogen application affects both taxonomical and predicted functional structure of wheat rhizosphere bacterial communities. Front. Microbiol. 9, 1074.

King, G.M., Smith, C., Tolar, B., Hollibaugh, J.T., 2013. Analysis of composition and structure of coastal to mesopelagic bacterioplankton communities in the northern gulf of Mexico. Front. Microbiol. 3, 438.

Knief, C., Delmotte, N., Chaffron, S., Stark, M., Innerebner, G., Wassmann, R., Von Mering, C., Vorholt, J.A., 2012. Metaproteogenomic analysis of microbial communities in the phyllosphere and rhizosphere of rice. ISME J. 6 (7), 1378–1390.

Köberl, M., Schmidt, R., Ramadan, E.M., Bauer, R., Berg, G., 2013. The microbiome of medicinal plants: diversity and importance for plant growth, quality and health. Front. Microbiol. 4, 400.

Köhl, J., Kolnaar, R., Ravensberg, W.J., 2019. Mode of action of microbial biological control agents against plant diseases: relevance beyond efficacy. Front. Plant Sci. 10, 845.

Kuzyakov, Y., Blagodatskaya, E., 2015. Microbial hotspots and hot moments in soil: concept & review. Soil Biol. Biochem. 83, 184–199.

Latour, X., Corberand, T., Laguerre, G., Allard, F., Lemanceau, P., 1996. The composition of fluorescent pseudomonad populations associated with roots is influenced by plant and soil type. Appl. Environ. Microbiol. 62 (7), 2449–2456.

Lauber, C.L., Strickland, M.S., Bradford, M.A., Fierer, N., 2008. The influence of soil properties on the structure of bacterial and fungal communities across land-use types. Soil Biol. Biochem. 40 (9), 2407–2415.

Layeghifard, M., Hwang, D.M., Guttman, D.S., 2017. Disentangling interactions in the microbiome: a network perspective. Trends Microbiol. 25 (3), 217–228.

Liljeroth, E., Bååth, E., Mathiasson, I., Lundborg, T., 1990. Root exudation and rhizoplane bacterial abundance of barley (Hordeum vulgare L.) in relation to nitrogen fertilization and root growth. Plant Soil 127 (1), 81–89.

Lindow, S.E., Brandl, M.T., 2003. Microbiology of the phyllosphere. Appl. Environ. Microbiol. 69 (4), 1875–1883.

Liu, K., Ding, X., Deng, B., Chen, W., 2009. Isolation and characterization of endophytic taxol-producing fungi from Taxus chinensis. J. Ind. Microbiol. Biotechnol. 36 (9), 1171.

Lugtenberg, B., Kamilova, F., 2009. Plant-growth-promoting rhizobacteria. Annu. Rev. Microbiol. 63, 541–556.

Ma, B., Wang, H., Dsouza, M., Lou, J., He, Y., Dai, Z., Brookes, P.C., Xu, J., Gilbert, J.A., 2016. Geographic patterns of co-occurrence network topological features for soil microbiota at continental scale in eastern China. ISME J. 10 (8), 1891–1901.

Mansfield, J., Genin, S., Magori, S., Citovsky, V., Sriariyanum, M., Ronald, P., Dow, M.A.X., Verdier, V., Beer, S.V., Machado, M.A., Toth, I.A.N., Foster, G.D., 2012. Top 10 plant pathogenic bacteria in molecular plant pathology. Mol. Plant Pathol. 13 (6), 614–629.

Martínez-Viveros, O., Jorquera, M.A., Crowley, D.E., Gajardo, G.M.L.M., Mora, M.L., 2010. Mechanisms and practical considerations involved in plant growth promotion by rhizobacteria. J. Soil Sci. Plant Nutr. 10 (3), 293–319.

Matsumoto, H., Fan, X., Wang, Y., Kusstatscher, P., Duan, J., Wu, S., Chen, S., Qiao, K., Wang, Y., Ma, B., Zhu, G., Wang, M., 2021. Bacterial seed endophyte shapes disease resistance in rice. Nat. Plants 7 (1), 60–72.

Mendes, R., Kruijt, M., De Bruijn, I., Dekkers, E., van der Voort, M., Schneider, J.H., Piceno, Y.M., DeSantis, T.Z., Andersen, G.L., Bakker, P.A., Raaijmakers, J.M., 2011. Deciphering the rhizosphere microbiome for disease-suppressive bacteria. Science 332 (6033), 1097–1100.

Meschke, H., Schrempf, H., 2010. Streptomyces lividans inhibits the proliferation of the fungus *Verticillium dahliae* on seeds and roots of *Arabidopsis thaliana*. Microb. Biotechnol. 3 (4), 428–443.

Micallef, S.A., Shiaris, M.P., Colón-Carmona, A., 2009. Influence of *Arabidopsis thaliana* accessions on rhizobacterial communities and natural variation in root exudates. J. Exp. Bot. 60 (6), 1729–1742.

Miotto-Vilanova, L., Jacquard, C., Courteaux, B., Wortham, L., Michel, J., Clément, C., Barka, E.A., Sanchez, L., 2016. *Burkholderia phytofirmans* PsJN confers grapevine resistance against *Botrytis cinerea* via a direct antimicrobial effect combined with a better resource mobilization. Front. Plant Sci. 7, 1236.

Mohammadi, O., Lahdenpera, M.L., 1992. Mycostop biofungicide in practice. In: Modern fungicides and antifungal compounds. German Phytomedical Society Series, vol. 4. Springer, pp. 289–295.

Molina-Romero, D., Baez, A., Quintero-Hernández, V., Castañeda-Lucio, M., Fuentes-Ramírez, L.E., Bustillos-Cristales, M.D.R., Rodríguez-Andrade, O., Morales-García, Y.E., Munive, A., Muñoz-Rojas, J., 2017. Compatible bacterial mixture, tolerant to desiccation, improves maize plant growth. PLoS One 12 (11), e0187913.

Monteiro, R.A., Balsanelli, E., Wassem, R., Marin, A.M., Brusamarello-Santos, L.C., Schmidt, M.A., Tadra-Sfeir, M.Z., Pankievicz, V.C., Cruz, L.M., Chubatsu, L.S., Pedrosa, F.O., Souza, E.M., 2012. *Herbaspirillum*-plant interactions: microscopical, histological and molecular aspects. Plant Soil 356 (1), 175–196.

Nabti, E., Sahnoune, M., Ghoul, M., Fischer, D., Hofmann, A., Rothballer, M., Schmid, M., Hartmann, A., 2010. Restoration of growth of durum wheat (*Triticum durum* var. waha) under saline conditions due to inoculation with the rhizosphere bacterium *Azospirillum brasilense* NH and extracts of the marine alga *Ulva lactuca*. J. Plant Growth Regul. 29 (1), 6–22.

Nair, D.N., Padmavathy, S., 2014. Impact of endophytic microorganisms on plants, environment and humans. Sci. World J. 2014.

Nambara, E., 2013. Plant Hormones. Brenner's Encyclopedia of Genetics, second ed. Academic Press, San Diego, pp. 346–348.

Ndeddy Aka, R.J., Babalola, O.O., 2016. Effect of bacterial inoculation of strains of *Pseudomonas aeruginosa*, *Alcaligenes feacalis* and *Bacillus subtilis* on germination, growth and heavy metal (Cd, Cr, and Ni) uptake of *Brassica juncea*. Int. J. Phytoremediation 18 (2), 200–209.

Neumann, G., Romheld, V., 2000. The release of root exudates as affected by the plant's physiological status. In: Pinton, R., Varanini, Z., Nannipieri, P. (Eds.), The Rhizosphere. CRC Press, pp. 57–110.

Niu, B., Paulson, J.N., Zheng, X., Kolter, R., 2017. Simplified and representative bacterial community of maize roots. Proc. Natl. Acad. Sci. U. S. A. 114 (12), E2450–E2459.

Olanrewaju, O.S., Glick, B.R., Babalola, O.O., 2017. Mechanisms of action of plant growth promoting bacteria. World J. Microbiol. Biotechnol. 33 (11), 1–16.

Olanrewaju, O.S., Ayangbenro, A.S., Glick, B.R., Babalola, O.O., 2019. Plant health: feedback effect of root exudates-rhizobiome interactions. Appl. Microbiol. Biotechnol. 103 (3), 1155–1166.

Omomowo, O.I., Babalola, O.O., 2019. Bacterial and fungal endophytes: tiny giants with immense beneficial potential for plant growth and sustainable agricultural productivity. Microorganisms 7 (11), 481.

Pascale, A., Proietti, S., Pantelides, I.S., Stringlis, I.A., 2020. Modulation of the root microbiome by plant molecules: the basis for targeted disease suppression and plant growth promotion. Front. Plant Sci. 10, 1741.

Pausch, J., Kuzyakov, Y., 2011. Photoassimilate allocation and dynamics of hotspots in roots visualized by ^{14}C phosphor imaging. J. Plant Nutr. Soil Sci. 174 (1), 12–19.

Pérez-Jaramillo, J.E., Carrión, V.J., de Hollander, M., Raaijmakers, J.M., 2018. The wild side of plant microbiomes. Microbiome 6 (1), 1–6.

Peterson, S.B., Dunn, A.K., Klimowicz, A.K., Handelsman, J., 2006. Peptidoglycan from *Bacillus cereus* mediates commensalism with rhizosphere bacteria from the Cytophaga-Flavobacterium group. Appl. Environ. Microbiol. 72 (8), 5421–5427.

Pii, Y., Borruso, L., Brusetti, L., Crecchio, C., Cesco, S., Mimmo, T., 2016. The interaction between iron nutrition, plant species and soil type shapes the rhizosphere microbiome. Plant Physiol. Biochem. 99, 39–48.

Prashar, P., Kapoor, N., Sachdeva, S., 2014. Rhizosphere: its structure, bacterial diversity and significance. Rev. Environ. Sci. Biotechnol. 13 (1), 63–77.

Pratama, A.A., Van Elsas, J.D., 2018. The 'neglected soil virome–potential role and impact. Trends Microbiol 26 (8), 649–662.

Qi, G., Ma, G., Chen, S., Lin, C., Zhao, X., 2019. Microbial network and soil properties are changed in bacterial wilt-susceptible soil. Appl. Environ. Microbiol. 85 (13), e00162-19.

Raaijmakers, J.M., Mazzola, M., 2012. Diversity and natural functions of antibiotics produced by beneficial and plant pathogenic bacteria. Annu. Rev. Phytopathol. 50, 403–424.

Radhakrishnan, R., Hashem, A., Abd Allah, E.F., 2017. Bacillus: a biological tool for crop improvement through bio-molecular changes in adverse environments. Front. Physiol. 8, 667.

Rascovan, N., Carbonetto, B., Perrig, D., Díaz, M., Canciani, W., Abalo, M., Alloati, J., González-Anta, G., Vazquez, M.P., 2016. Integrated analysis of root microbiomes of soybean and wheat from agricultural fields. Sci. Rep. 6 (1), 1–12.

Rasmann, S., Ali, J.G., Helder, J., van der Putten, W.H., 2012. Ecology and evolution of soil nematode chemotaxis. J. Chem. Ecol. 38 (6), 615–628.

Reinhold-Hurek, B., Maes, T., Gemmer, S., Van Montagu, M., Hurek, T., 2006. An endoglucanase is involved in infection of rice roots by the not-cellulose-metabolizing endophyte Azoarcus sp. strain BH72. Mol. Plant-Microbe Interact. 19 (2), 181–188.

Rochon, D.A., 2009. Fungal transmission of plant viruses. Curr. Protoc. Microbiol. 12 (1), 16B–4.

Rousk, J., Brookes, P.C., Bååth, E., 2010. Investigating the mechanisms for the opposing pH relationships of fungal and bacterial growth in soil. Soil Biol. Biochem. 42 (6), 926–934.

Rovira, A.D., 1969. Plant root exudates. Bot. Rev. 35 (1), 35–57.

Salles, J.F., Van Veen, J.A., Van Elsas, J.D., 2004. Multivariate analyses of Burkholderia species in soil: effect of crop and land use history. Appl. Environ. Microbiol. 70 (7), 4012–4020.

Samad, A., Antonielli, L., Sessitsch, A., Compant, S., Trognitz, F., 2017. Comparative genome analysis of the vineyard weed endophyte *Pseudomonas viridiflava* CDRTc14 showing selective herbicidal activity. Sci. Rep. 7 (1), 1–15.

Sasse, J., Martinoia, E., Northen, T., 2018. Feed your friends: do plant exudates shape the root microbiome? Trends Plant Sci. 23 (1), 25–41.

Saul-Tcherkas, V., Steinberger, Y., 2011. Soil microbial diversity in the vicinity of a Negev Desert shrub—*Reaumuria negevensis*. Microb. Ecol. 61 (1), 64–81.

Schlatter, D., Kinkel, L., Thomashow, L., Weller, D., Paulitz, T., 2017. Disease suppressive soils: new insights from the soil microbiome. Phytopathology 107 (11), 1284–1297.

Schloss, P.D., Handelsman, J., 2006. Toward a census of bacteria in soil. PLoS Comput. Biol. 2 (7), e92.

Schmidt, R., Köberl, M., Mostafa, A., Ramadan, E.M., Monschein, M., Jensen, K.B., Bauer, R., Berg, G., 2014. Effects of bacterial inoculants on the indigenous microbiome and secondary metabolites of chamomile plants. Front. Microbiol. 5, 64.

Schwessinger, B., Ronald, P.C., 2012. Plant innate immunity: perception of conserved microbial signatures. Annu. Rev. Plant Biol. 63, 451–482.

Sessitsch, A., Pfaffenbichler, N., Mitter, B., 2019. Microbiome applications from lab to field: facing complexity. Trends Plant Sci. 24 (3), 194–198.

Sharma, I.P., Chandra, S., Kumar, N., Chandra, D., 2017. PGPR: heart of soil and their role in soil fertility. In: Meena, V., Mishra, P., Bisht, J., Pattanayak, A. (Eds.), Agriculturally Important Microbes for Sustainable Agriculture. Springer, Singapore, pp. 51–67.

Shi, S., Nuccio, E.E., Shi, Z.J., He, Z., Zhou, J., Firestone, M.K., 2016. The interconnected rhizosphere: high network complexity dominates rhizosphere assemblages. Ecol. Lett. 19 (8), 926–936.

Singh, S., Nain, L., 2014. Microorganisms in the conversion of agricultural wastes to compost. Proc. Indian Natl. Sci. Acad. 80 (2), 473–481.

Smigielski, L., Laubach, E.M., Pesch, L., Glock, J.M.L., Albrecht, F., Slusarenko, A., Panstruga, R., Kuhn, H., 2019. Nodulation induces systemic resistance of *Medicago truncatula* and *Pisum sativum* against *Erysiphe pisi* and primes for powdery mildew-triggered salicylic acid accumulation. Mol. Plant-Microbe Interact. 32 (9), 1243–1255.

Soussi, A., Ferjani, R., Marasco, R., Guesmi, A., Cherif, H., Rolli, E., Mapelli, F., Ouzari, H.I., Daffonchio, D., Cherif, A., 2016. Plant-associated microbiomes in arid lands: diversity, ecology and biotechnological potential. Plant Soil 405 (1), 357–370.

Srinivasan, M.C., Laxman, R.S., Deshpande, M.V., 1991. Physiology and nutritional aspects of actinomycetes: an overview. World J. Microbiol. Biotechnol. 7 (2), 171–184.

Steindler, L., Bertani, I., De Sordi, L., Schwager, S., Eberl, L., Venturi, V., 2009. LasI/R and RhlI/R quorum sensing in a strain of *Pseudomonas aeruginosa* beneficial to plants. Appl. Environ. Microbiol. 75 (15), 5131–5140.

Su, F., Jacquard, C., Villaume, S., Michel, J., Rabenoelina, F., Clément, C., Barka, E.A., Dhondt-Cordelier, S., Vaillant-Gaveau, N., 2015. *Burkholderia phytofirmans* PsJN reduces impact of freezing temperatures on photosynthesis in *Arabidopsis thaliana*. Front. Plant Sci. 6, 810.

Timmermann, T., Armijo, G., Donoso, R., Seguel, A., Holuigue, L., González, B., 2017. *Paraburkholderia phytofirmans* PsJN protects *Arabidopsis thaliana* against a virulent strain of *Pseudomonas syringae* through the activation of induced resistance. Mol. Plant-Microbe Interact. 30 (3), 215–230.

Toyota, K., Watanabe, T., 2013. Recent trends in microbial inoculants in agriculture. Microbes Environ. 28 (4), 403–404.

Trivedi, P., Leach, J.E., Tringe, S.G., Sa, T., Singh, B.K., 2020. Plant–microbiome interactions: from community assembly to plant health. Nat. Rev. Microbiol. 18 (11), 607–621.

Tückmantel, T., Leuschner, C., Preusser, S., Kandeler, E., Angst, G., Mueller, C.W., Meier, I.C., 2017. Root exudation patterns in a beech forest: dependence on soil depth, root morphology, and environment. Soil Biol. Biochem. 107, 188–197.

Van der Putten, W.H., Klironomos, J.N., Wardle, D.A., 2007. Microbial ecology of biological invasions. ISME J. 1 (1), 28–37.

Vandenkoornhuyse, P., Mahé, S., Ineson, P., Staddon, P., Ostle, N., Cliquet, J.B., Francez, A.J., Fitter, A.H., Young, J.P.W., 2007. Active root-inhabiting microbes identified by rapid incorporation of plant-derived carbon into RNA. Proc. Natl. Acad. Sci. U. S. A. 104 (43), 16970–16975.

Viaene, T., Langendries, S., Beirinckx, S., Maes, M., Goormachtig, S., 2016. Streptomyces as a plant's best friend? FEMS Microbiol. Ecol. 92 (8).

Voges, M.J., Bai, Y., Schulze-Lefert, P., Sattely, E.S., 2019. Plant-derived coumarins shape the composition of an *Arabidopsis* synthetic root microbiome. Proc. Natl. Acad. Sci. U. S. A. 116 (25), 12558–12565.

Vorholt, J.A., 2012. Microbial life in the phyllosphere. Nat. Rev. Microbiol. 10 (12), 828–840.

Walker, T.S., Bais, H.P., Grotewold, E., Vivanco, J.M., 2003. Root exudation and rhizosphere biology. Plant Physiol. 132 (1), 44–51.

Weller, D.M., Raaijmakers, J.M., Gardener, B.B.M., Thomashow, L.S., 2002. Microbial populations responsible for specific soil suppressiveness to plant pathogens. Annu. Rev. Phytopathol. 40 (1), 309–348.

Whipps, J.M., Lewis, K., Cooke, R.C., 1988. Mycoparasitism and plant disease control. In: Fungi in Biological Control Systems. Elsevier, pp. 161–187.

Xu, D., Hanschen, F.S., Witzel, K., Nintemann, S.J., Nour-Eldin, H.H., Schreiner, M., Halkier, B.A., 2017. Rhizosecretion of stele-synthesized glucosinolates and their catabolites requires GTR-mediated import in *Arabidopsis*. J. Exp. Bot. 68 (12), 3205–3214.

Yuan, J., Zhao, J., Wen, T., Zhao, M., Li, R., Goossens, P., Huang, Q., Bai, Y., Vivanco, J.M., Kowalchuk, G.A., Berendsen, R.L., Shen, Q., 2018. Root exudates drive the soil-borne legacy of aboveground pathogen infection. Microbiome 6 (1), 1–12.

Zamioudis, C., Pieterse, C.M., 2012. Modulation of host immunity by beneficial microbes. Mol. Plant-Microbe Interact. 25 (2), 139–150.

Zgadzaj, R., James, E.K., Kelly, S., Kawaharada, Y., de Jonge, N., Jensen, D.B., Madsen, L.H., Radutoiu, S., 2015. A legume genetic framework controls infection of nodules by symbiotic and endophytic bacteria. PLoS Genet. 11 (6), e1005280.

Zgadzaj, R., Garrido-Oter, R., Jensen, D.B., Koprivova, A., Schulze-Lefert, P., Radutoiu, S., 2016. Root nodule symbiosis in *Lotus japonicus* drives the establishment of distinctive rhizosphere, root, and nodule bacterial communities. Proc. Natl. Acad. Sci. U. S. A. 113 (49), E7996–E8005.

Zgadzaj, R., Thiergart, T., Bozsóki, Z., Garrido-Oter, R., Radutoiu, S., Schulze-Lefert, P., 2019. *Lotus japonicus* symbiosis signaling genes and their role in the establishment of root-associated bacterial and fungal communities. BioRxiv, 547687.

Zhalnina, K., Louie, K.B., Hao, Z., Mansoori, N., da Rocha, U.N., Shi, S., Cho, H., Karaoz, U., Loqué, D., Bowen, B.P., Firestone, M.K., Brodie, E.L., 2018. Dynamic root exudate chemistry and microbial substrate preferences drive patterns in rhizosphere microbial community assembly. Nat. Microbiol. 3 (4), 470–480.

Zhang, M., Pereira e Silva, M.D.C., De Mares Maryam, C., van Elsas, J.D., 2014. The mycosphere constitutes an arena for horizontal gene transfer with strong evolutionary implications for bacterial-fungal interactions. FEMS Microbiol. Ecol. 89 (3), 516–526.

Zhou, D., Huang, X.F., Chaparro, J.M., Badri, D.V., Manter, D.K., Vivanco, J.M., Guo, J., 2016. Root and bacterial secretions regulate the interaction between plants and PGPR leading to distinct plant growth promotion effects. Plant Soil 401 (1–2), 259–272.

Zhu, S., Vivanco, J.M., Manter, D.K., 2016. Nitrogen fertilizer rate affects root exudation, the rhizosphere microbiome and nitrogen-use-efficiency of maize. Appl. Soil Ecol. 107, 324–333.

CHAPTER 9

The use of microbes as a combative strategy for alleviation of abiotic and biotic stresses

N.S. Raja Gopalan*, P.T. Nikhil*, Raunak Sharma*, and Sridev Mohapatra
Department of Biological Sciences, Birla Institute of Technology and Science (Pilani), Hyderabad Campus, Secunderabad, Telangana, India

Chapter outline

* Contributed equally.

Unravelling Plant-Microbe Synergy. https://doi.org/10.1016/B978-0-323-99896-3.00010-2

Introduction

Plant growth-promoting microbes (PGPMs) are a diverse group of beneficial soil microorganisms that colonize plant roots or the rhizosphere (soil adhering to the roots), contributing positively to plant growth and environmental stress amelioration (Kohler et al., 2008; Timmusk et al., 2014; Ngumbi and Kloepper, 2016; Chandra et al., 2018c, 2020; Chandra and Sharma, 2020). These microbes thus constitute a widely diverse soil microbiome that has co-evolved with plants for mutual benefit (Chandra et al., 2018c). There are broadly two types of PGPMs: (a) plant growth-promoting bacteria/rhizobacteria (PGPB/R) and plant growth-promoting fungi (PGPF). Some soil microbes endocolonize the roots by penetrating the root cells (intracellular or i-PGPMs), whereas others are free-living in the rhizosphere (extracellular or e-PGPMs). i-PGPMs are obligate symbionts, and examples include root nodulation, nitrogen-fixing PGPR (*Allorhizobium*, *Azorhizobium*, *Rhizobium* spp.), and arbuscular vesicular mycorrhizae (AVM). e-PGPMs are free-living and associate with plants through facultative symbiosis. Examples of e-PGPR include *Azotobacter*, *Azospirillum*, *Bacillus*, *Burkholderia*, *Pseudomonas*, etc. (Bhattacharyya and Jha, 2012; Ahemad and Kibret, 2014) and those of e-PGPF include several species of the genus *Trichoderma* (Harman et al., 1991). An important attribute of these soil microbes is their ability to help plants mitigate biotic and abiotic stresses, which is the predominant focus of this chapter (the mechanisms are summarized in Fig. 9.1). However, before understanding how these PGPMs help in stress tolerance/resistance in plants, it is essential to understand the different types of environmental stresses that plants encounter and how they deal with them.

Abiotic and biotic stresses encountered by plants and how they inherently cope with them

Plants encounter a wide array of environmental stresses, both biotic and abiotic. Since plants are sessile, they cannot 'escape' their 'enemies' but must 'endure' them. Therefore, nature has equipped plants with several endogenous mechanisms to survive these stresses.

Abiotic stresses

Plants experience a variety of abiotic stresses, predominantly caused by adverse environmental conditions either in the atmosphere or in the soil. Fluctuations in optimum atmospheric temperatures lead to either thermal (higher than optimal temperatures), chilling (low temperatures), or freezing stress (extremely low/freezing temperatures).

Similarly, the lower-than-optimal soil water potential causes drought/dehydration stress, whereas excess water can cause stress due to waterlogging. Contamination of soil water with excess salts can cause salinity and heavy metal stresses. Besides, lower- or higher-than-optimal oxygen levels, either in the atmosphere or in the soil, can cause hypoxia or hyperoxia,

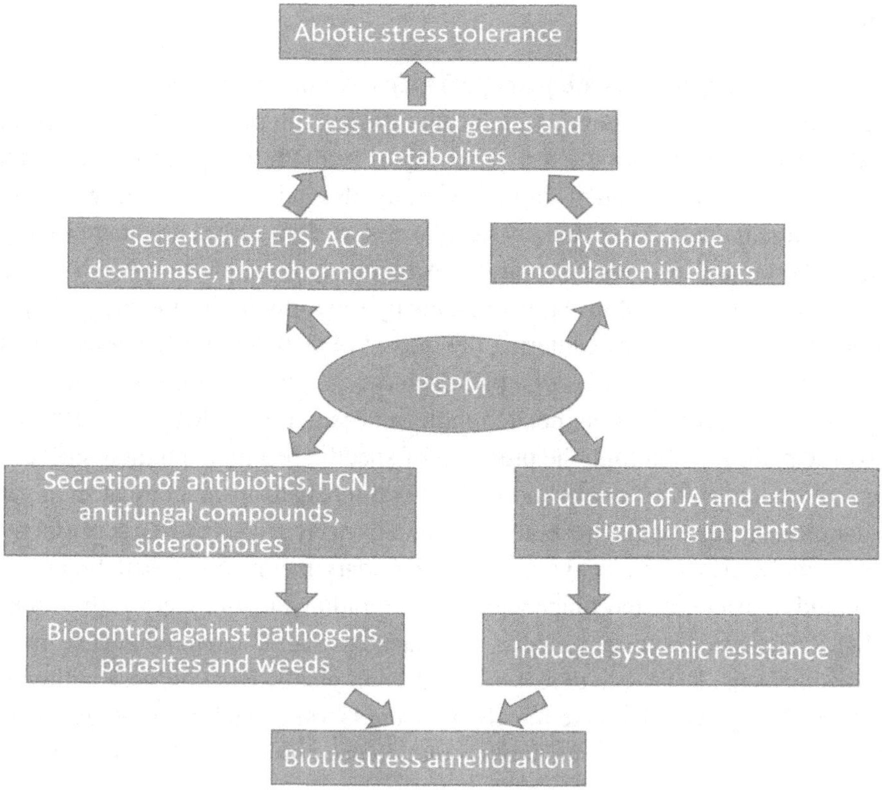

Fig. 9.1
An overview of the important attributes and mechanisms involved in PGPM-mediated stress alleviation in plants.

respectively. As with biotic stresses, plants have also evolved mechanisms to ameliorate abiotic stresses. Dehydration in the soil is perceived by the roots that transmit signals to the aerial parts, which respond through several means such as stomatal closure and accumulation of compatible osmolytes (proline, glycine betaine, sugar alcohols, etc.). Besides, the roots themselves change their morphology and architecture. All these changes help plants prevent water loss while extending their roots in search of water (Taiz et al., 2015). As far as salinity tolerance is concerned, plant cells have evolved specific mechanisms to prevent the uptake of Na^+ from their surroundings and facilitate the export of excess Na^+ outside of the cells. These responses involve a wide variety of genes and, thus, signaling pathways, some of which are ABA-dependent, whereas others follow ABA-independent signals (Taiz et al., 2015). Responses to low-temperature stress have significant overlaps with dehydration tolerance. Most abiotic stresses induce changes in the oxidative status of plant cells. On several occasions, toxic levels of reactive oxygen species are generated, which the plants attempt to scavenge through the upregulation of ROS scavenging enzymes (Taiz et al., 2015).

Biotic stresses

The predominant biotic stressors are plant pathogens and pests. Pathogens are viruses and microorganisms (bacteria and fungi), which cause diseases in plants, whereas, typically, insects and insect larvae are classified as pests. Apart from these organisms, plants may also encounter weeds and parasitic plants in their vicinity, thus adversely affecting their survival. The mode of stress induction in plants by these stressors varies from organism to organism. Pathogenic microorganisms either inhabit the rhizosphere or the phyllosphere (surfaces of leaves). They cause a variety of diseases in plants by releasing toxins that degrade plant cells. Plants respond to pathogens by countering the impact of such toxins at various stages of a pathogen attack. In the initial stages of a pathogen attack, plants prepare for the impending danger by recognizing pathogen-associated molecular patterns (PAMPS/MAMPS) or elicitors released by the pathogens through the presence of specific receptors on their surfaces. Following this, plants employ one or more of several mechanisms to combat such pathogenic invasions, such as hypersensitive response (via reactive oxygen species), recruitment of antimicrobial molecules such as phytoalexins (secondary metabolites), pathogenesis-related proteins, including bacterial and fungal cell wall-degrading enzymes, etc. (Ab Rahman et al., 2018). The phytohormones salicylic acid, jasmonic acid, and ethylene are involved in signaling mechanisms, leading to defense responses in plants. Besides these pathogens, plants are also attacked by pests, which are mostly in insect larval and adult forms. Jasmonic acid-based signaling pathways are also known to produce proteins that can become toxic to such pests (Taiz et al., 2015).

Mechanisms of PGPM-mediated stress tolerance

While there are specific mechanisms by which PGPMs help in the overall plant growth and development, such as nitrogen fixation, phosphate solubilization, etc., thereby acting as biofertilizers, there are also specific mechanisms by which they impart abiotic and biotic stress tolerance to plants whose rhizospheres they inhabit (Chandra et al., 2015; Chandra et al., 2018a, b). Some examples of such mechanisms are given below. Each of the topics—abiotic and biotic stress—is discussed with respect to (a) microbial secretions/exudates that help in conferring tolerance/resistance and (b) changes in signaling and metabolic pathways in plants in response to PGPM inoculation.

Microbial secretions that help in abiotic stress tolerance

Exopolysaccharide production

Several strains of PGPR produce exopolysaccharides (EPSs) which are dense, long-chain polymers of carbohydrate and protein residues. The secreted exopolysaccharides aid in the formation of biofilms (that act as protectants against dehydration) and improve the ability of

PGPR to colonize the roots (Sandhya et al., 2009; Sandhya et al., 2010a, b). During drought conditions, several PGPR strains belonging to *Azospirillum*, *Pseudomonas*, and *Bacillus* spp. produce EPSs, which cause changes in the soil structure, the ratio of root-associated soil (soil/root), and its macro properties, which allows plants to expand their root system better, thereby improving the uptake and availability of water and nutrients (Ahn et al., 2007; Sandhya et al., 2009, 2010a, b; Vardharajula et al., 2011). Some strains of the *Pseudomonas* spp. are known to prevent the hardening of roots by producing EPSs that help form hydrophilic biofilms (Rolli et al., 2015). The chemical properties of EPSs (particularly lipopolysaccharides) make them ideal emulsifying agents that can quench free radicals and protect plants against the adverse effects of ROS (Dimitrova et al., 2013). The changes in the composition of EPSs under water-stressed conditions to contain high amounts of specific sugars like glucose, rhamnose, mannose, and trehalose enhance their water retention properties (Tewari and Arora, 2014). Some fungi are also known to produce EPSs, including members of higher-order basidiomycetes, lower-order filamentous fungi, and yeasts. Some saprophytic fungi that produce EPSs include *Penicillium simplicissimum*, *Penicillium radicum*, *Aspergillus fumigatus*, *Aspergillus niger*, etc. The EPSs produced from these fungal strains are known to solubilize phosphates from Ca and Al salts, thus acting as biofertilizers during abiotic stress conditions (Osińska-Jaroszuk et al., 2015).

Production of phytohormones

Plant growth-promoting microbes can produce important phytohormones such as auxins, gibberellins, cytokinins, and zeatins, which act as crucial growth regulators in plants (Ghosh et al., 2019b). Since phytohormones play a significant role in plant growth, development, and stress tolerance, it can be hypothesized that PGPR secreted phytohormones are the key players in plant–PGPM interactions. It is hypothesized that microbe-released phytohormones can be directly taken up and utilized by plants. Phytohormone-producing bacteria alleviate the impacts of abiotic stresses in plants and improve their growth and development. It is reported that IAA-producing PGPR enhance the uptake of water and nutrients in plants by increasing the root surface area (Mantelin and Touraine, 2003). Such a specific benefit was seen in case of wheat inoculated with auxin-producing *Azospirillum* (Arzanesh et al., 2011). Similarly, GA-producing PGPR strains such as *Pseudomonas putida* H-2-3 and *Azospirillum lipoferum* increase plant growth and alleviate the effects of drought stress in soybean, wheat, and maize plants (Creus et al., 2004; Cohen et al., 2009; Kang et al., 2014b). Inoculation of plants with cytokinin-producing PGPR strains such as *Micrococcus luteus* chp37 and *B. subtilis* has been shown to improve the overall physiology of plants such as shoot and root biomass under water stress conditions (Arkhipova et al., 2007; Raza and Faisal, 2013). In a study by Salomon et al. (2014), inoculation of ABA-producing *Bacillus licheniformis* and *Pseudomonas fluorescens* improved the growth of grapevine under drought conditions. An auxin-overproducing strain *P. putida* GR12–2 promoted root growth in canola plants.

In contrast, an IAA biosynthetic mutant did not show any improvement, suggesting the importance of IAA in plant growth promotion (Xie et al., 1996; Patten and Glick, 2002).

A drought-tolerant PGPR strain *P. putida* GAP P45 produces phytohormones like IAA, GA, and ABA and is shown to confer drought tolerance on maize and sunflower (*Arabidopsis thaliana*) (Sandhya et al., 2009, 2010a, b; Ghosh et al., 2017, 2018, 2019a). Several strains of fungal species such as *Aspergillus fumigatus*, *Penicillium citrinum*, etc. have been known to secrete GA in a culture medium (Waqas et al., 2012). The cytokinins produced by fungi are involved in drought and salinity stress through cross talk with ABA (Ansari et al., 2013). Other examples, such as inoculation with the *Penicillium* species during salt stress, promoted shoot growth due to the enhanced secretion of active GAs in cucumber plants (Waqas et al., 2012; Gupta et al., 2021).

Secretion of 1-aminocyclopropane-1 carboxylate (ACC) deaminase

Ethylene is an important plant hormone mainly responsible for senescence and promoting leaf abscission. The level of ethylene increases drastically during severe stress conditions, leading to leaf abscission, chlorosis, and senescence. It may also lead to inhibition of growth and development in plants (Jalili et al., 2009; Ahemad and Kibret, 2014; Kumari et al., 2016). Some PGPMs can secrete ACC deaminase, an important enzyme that degrades the ethylene precursor ACC, leading to the production of α-ketobutyrate and ammonia (NH_3), which can be used as a nitrogen source by microbes (Glick, 2005; Saharan and Nehra, 2011b; Chandra et al., 2020; Chandra and Sharma, 2020). When inoculated with ACC deaminase-producing PGPR under water-limited conditions, the development and survival of plants improved from enhanced root growth (Belimov et al., 2009; Glick, 2014; Chandra et al., 2019a, b). Similarly, the ACC deaminase-producing strains *Achromobacter piechaudii* and *Variovorax paradoxus* confer salinity and drought tolerance on tomato and pea plants (Mayak et al., 2004; Belimov et al., 2009). Several strains of beneficial fungi such as *Penicillium citrinum*, *Trichoderma atroviride*, and *Trichoderma* spp. are known to exhibit ACC deaminase activity, which improves plant growth and confers drought and salinity tolerance on plants and by reducing ethylene levels (Viterbo et al., 2010; Ansari et al., 2013; Begum et al., 2019). For example, the growth-promoting activity of *Trichoderma atroviride*-inoculated tomato plants was shown to be associated with lowering of ethylene levels resulting from fungal ACC deaminase activity (Viterbo et al., 2010).

Production of compatible osmolytes and other metabolites

Similar to the responses of plants to osmotic stress conditions, drought-tolerant rhizobacteria accumulate compatible osmolytes in the cytosol and positively modulate cytosolic pH, cellular redox status, etc. under dehydrating conditions (Vurukonda et al., 2016; Anjum et al., 2017). Examples of such compatible osmolytes include the amino acid proline (Pro), sugar alcohols sorbitol, mannitol, etc. However, whether or not these compatible osmolytes of bacterial origin are excreted into the soil and whether or not the plants take them up is unclear. It is known, though, that PGPR inoculation does enhance the biosynthesis and accumulation of compatible osmolytes within plant tissues under abiotic stress conditions (Sandhya et al., 2010a, b; Grover et al., 2011; Chandra et al., 2018a, b). Beneficial fungi can also maintain plant growth by producing enzymes such as urease, catalase, siderophores, etc., which modify plant biochemical mechanisms to improve water and nutrient uptake for maintenance of the structure and function of plant cells under stress conditions (Ripa et al., 2019).

Production of volatile organic compounds

Many PGPR are known to release a wide range of volatile organic compounds (VOCs) into their surroundings, which can suppress or promote plant growth (Chandra et al., 2017). The VOCs produced by the PGPR strain *Bacillus amyloliquefaciens* confers salt stress tolerance on *Codonopsis pilosula* by regulating stomatal activity and photosynthesis (Han et al., 2017). The VOCs released by the same strain also cause drought tolerance in *Arabidopsis thaliana* (Cho et al., 2008) and modulate sulfur metabolism in *Arabidopsis thaliana* (Aziz et al., 2016). The *Bacillus* sp. strain B55 produces sulfur-containing volatiles that combat the adverse effects of sulfur deprivation in *Nicotiana attenuate* (Meldau et al., 2013). Bacterial volatiles modulate the function of AtHKT1 sodium transporters in plants to reduce the levels of sodium inside the plants (Vaishnav et al., 2015). VOCs produced by PGPR increase metallic iron uptake in *Arabidopsis* and combat iron deficiency (Zhang et al., 2009).

Plant responses to PGPM inoculation under abiotic stress

Expression of stress-inducible genes

Plant growth-promoting microbes regulate the gene expression levels of host plants to improve their growth and development under different environmental conditions. The expression pattern changes in a multitude of host genes can be attributed to the various molecular and signaling mechanisms that PGPR modulate. Inoculation of *Arabidopsis* plants with the PGPR strain *P. putida* (MTCC5279) improved plant growth by the upregulation of crucial auxin-responsive genes and calcium-dependent protein kinases involved in signaling during salinity stress responses (Srivastava et al., 2012). *Arthrobacter nitroguajacolicus* improves nutrient uptake and confers salinity tolerance on wheat by upregulation of ATP-binding cassettes, ion transporters, P450 genes, ascorbate peroxidase genes, and oligopeptide transporters (Safdarian et al., 2019). *Pseudomonas putida* NBRIRA regulates the expression of miRNAs and their downstream genes under salinity and drought stress to confer tolerance on chickpea plants (Jatan et al., 2019). A previous research from our laboratory has shown that the drought-mitigating plant growth-promoting rhizobacterium *Pseudomonas putida* GAP P45 maintains osmotic balance under drought in *Arabidopsis* seedlings by upregulating several essential genes involved in drought tolerance such as proline and polyamine (PA) metabolism and catabolism genes (Ghosh et al., 2017; Sen et al., 2018; Sen and Mohapatra, 2022). Similarly, there are some reports of PGPF-mediated changes in gene expression in plants under abiotic stress. For example, it was shown that in rice plants, AMF (arbuscular mycorrhizal fungi) association leads to the modulation of the expression of ion transporter genes (*OsSOS1*, *OsNHX3*, *OsHKT2;1*, and *OsHKT1*) to maintain ion homeostasis during salinity stress (Evelin et al., 2019). The fungus *Piriformospora indica* promotes the growth of Chinese cabbage under drought by expressing drought-related genes (Ansari et al., 2013).

Modulation in the levels of stress-induced metabolites

It has been shown that several PGPM strains upregulate the biosynthesis of compatible osmolytes (proline, glycine betaine, sorbitol, etc.), inside plant tissues, especially in roots, creating a concentration gradient between the soil and roots and facilitating the movement of water inside the plants under abiotic stress (Ait Barka et al., 2006; Kohler et al., 2008; Sandhya et al., 2010a, b; Grover et al., 2011). Several strains of PGPR are known to upregulate proline biosynthetic pathways in plants under abiotic stresses, thus helping improve the overall water status in plants (Sandhya et al., 2010a, b; Szabados and Savouré, 2010; Vardharajula et al., 2011; Pii et al., 2015; Ghosh et al., 2017; Ghosh et al., 2018). Sen et al. (2018, 2021) have shown that drought-mitigating *Pseudomonas putida* GAP-P45 modulated the metabolism of the three major polyamines, namely, putrescine, spermidine, and spermine, in *Arabidopsis* under water stress.

Several studies have reported the role of endophytic fungi in the accumulation of soluble sugars under salinity stress. Colonization by *Phomopsis liquidambari* was shown to increase soluble sugar content in peanut leaves due to the tripartite symbiotic association between peanut and the *Rhizobia* (Zhang et al., 2016). Similarly, inoculation of cucumber seedlings with *Trichoderma harzianum* alleviated salt stress by enhancing sugar and proline accumulation (Zhang et al., 2019). Fungal endophytes have also been reported to enhance iron acquisition by host plants upon colonization by endophytic fungi under salt stress through secretion of organic acids that can chelate and solubilize iron in soil (Gupta et al., 2021). Endophytic fungi can also reduce putrescine content in plants under stress conditions. Under salinity stress, colonization by *Epichloe bromicola* resulted in the enhanced conversion of putrescine to spermidine and spermine. Moreover, this fungus could convert the free forms and soluble conjugated polyamines to insoluble bound forms of polyamines (Gupta et al., 2021). Combined drought and salt stress in tomato plants was ameliorated by *Scolecobasidium constrictum*. This strain improved plant biomass, leaf water content, and stomatal conductance through enhanced nutrient uptake, osmoprotectant accumulation, and antioxidant induction activity (Begum et al., 2019).

Regulation of phytohormone signaling

Susceptibility or tolerance of plants to different environmental stress factors can be directly attributed to the ability of plants to modulate the levels of key phytohormones such as auxins, cytokinins, gibberellins, etc. and relocate them to different tissues in response to specific stress conditions such as drought, heat, salinity, etc. (Dobra et al., 2010; Sreenivasulu et al., 2012; Fahad et al., 2015). Many IAA-producing PGPR like *P. putida* strains, *Bacillus megaterium* strains, *Burkholderia cepacia* SE4, *Promicromonospora* spp. SE188, *Bacillus subtilis* LK14, and *Acinetobacter calcoaceticus* SE370 can increase biomass and chlorophyll content, modify the endogenous levels of phytohormones, and change the root structure architecture in clove, tomato, and cucumber plants (Mantelin and Touraine, 2003; Marulanda et al., 2009; Kang et al., 2014a; Khan et al., 2016), thus helping the plants in accelerating soil water under drought conditions. Inoculation of *Azospirillum* spp. changes the xylem architecture and enhances root growth and lateral root formation by upregulating the indole-3-pyruvate decarboxylase genes and IAA production in wheat (Arzanesh et al., 2011; Pereyra et al., 2012). Inoculation of *Phyllobacterium* reduces the transpiration rate in the leaves of *Arabidopsis* by increasing the levels of ABA inside plants under osmotic stress conditions. An ABA-mediated decrease in leaf transpiration was reported in *A. thaliana* under osmotic stress when inoculated with *Phyllobacterium* (Bresson et al., 2013). Modifications of the root architecture and endogenous hormonal signaling by *Phyllobacterium brassicacearum* STM196 promote overall plant growth in *Arabidopsis* seedlings (Contesto et al., 2010; Galland et al., 2012; Bresson et al., 2013). Cytokinin producers *Citricoccus zhacaiensis* and *B. amyloliquefaciens* increase the relative water content, the rate of photosynthesis, transpiration, and yield under water-stressed conditions in tomato plants (Selvakumar et al., 2018). It is hypothesized that the release of phytohormones near the roots by PGPR can lead to overall changes/modulation in the levels of phytohormones inside the plants (Dodd et al., 2010; Barnawal et al., 2017). However, the influence of the bacteria-released phytohormones on endogenous phytohormone modulation in plants remains somewhat speculative. Ghosh et al. (2019b) observed PGPR-mediated fluctuations in the endogenous content of major phytohormones in *Arabidopsis* plants, inoculated with *Pseudomonas putida* GAP–P45, under water stress.

While ABA levels were downregulated, IAA and cytokinin accumulation increased in shoots and roots. This bacterium caused an increase in endogenous GA content in *Arabidopsis* shoots, while downregulating the same in roots under water stress. A beneficial species of the fungus *Trichoderma virens* caused alterations in the root structure architecture of *Arabidopsis thaliana*, by modulating IAA concentration (Contreras-Cornejo et al., 2009). The beneficial fungus *Piriformospora indica* has been reported to alleviate drought and salinity stress in several plants by secretion of the phytohormone cytokinin and its cross talk with plant ABA levels (Ansari et al., 2013).

Mechanisms of PGPM-mediated biotic stress tolerance

The ability of PGPMs to restrict the growth of pathogenic organisms in the rhizosphere has earned them the title of 'biocontrol agents' (i.e., they help 'control' the growth of pathogenic microorganisms). The mechanisms by which PGPMs achieve this are called 'indirect mechanisms' of plant growth and development since they indirectly help in plant growth by suppressing pathogens' growth. The following are the important biocontrol mechanisms in PGPMs.

Production of antibiotics, lytic enzymes, and hydrogen cyanide (HCN)

Beneficial soil microbes secrete a wide variety of antibiotics such as 2,4-diacetyl phloroglucinol and phenazine, which prevent the growth of pathogenic soil microbes (Saharan and Nehra, 2011a, b). Some important strains of PGPR reported to prevent pathogen infections are *Pseudomonas fluorescens* TDK1, *Arthrobacter* sp., *Pseudomonas putida* UW4, *Bacillus subtilis, Pseudomonas fluorescens* MSP-393, *Pseudomonas aeruginosa* Sha8, *Bacillus luciferensis* KJ2C12, and *Pseudomonas fluorescens* CHA0 and AMF (Mayak et al., 2004; Haas and Défago, 2005; Barriuso et al., 2008; Saharan and Nehra, 2011a, b). The secretion of cell wall-degrading enzymes such as chitinases, cellulases, etc. leads to the destruction of pathogenic fungi (Beneduzi et al., 2012; Bhattacharyya and Jha, 2012; Ahemad and Kibret, 2014). Gram-negative *Pseudomonas fluorescens, Pseudomonas aeruginosa, Chromobacterium violaceum*, and a wide variety of *Bacillus* strains produce hydrogen cyanide as a secondary metabolite, which is volatile in nature. It suppresses the growth of various microorganisms and weeds and shows negative effects on the growth and development of plants (Jalili et al., 2009; Saharan and Nehra, 2011a, b; Raza and Faisal, 2013). Most of the time, HCN concentrations are too low to affect the growth and development of the host plant. Apart from these, other antifungal chemicals secreted by PGPR include phenazines, pyrrolnitrin, 2,4-diacetylphloroglucinol (DAPG), and pyoluteorin (Ab Rahman et al., 2018). Similarly, many fungi like *Aspergillus niger*, *Aspergillus oryzae*, and *Trichoderma reesei* secrete enzymes like amylases, proteases, cellulases, etc., which are antimicrobial (Wiebe, 2003).

Production of siderophores

Iron is an essential element for cellular growth and metabolism in all organisms. It is present in the ferric ion form in soil that plants and microorganisms cannot assimilate (Ahemad and Kibret, 2014). Siderophores are Fe-chelating molecules produced and utilized by bacteria for Fe absorption. Siderophores bind ferric ions with higher affinity than any other pathogenic bacteria or fungi, depriving them of the essential growth requirement of iron (Saharan and Nehra, 2011a, b). A ferric–siderophore complex is recognized by specific membrane receptors of bacteria (Beneduzi et al., 2012; Bhattacharyya and Jha, 2012). Siderophores play an essential role in plant development by facilitating the uptake of iron by plants (Saharan and Nehra, 2011a, b; Numan et al., 2018). Siderophore-producing bacteria belong to the genera *Rhizobium*, *Bradyrhizobium*, *Pseudomonas*, *Serratia*, and *Streptomyces* (Beneduzi et al., 2012; Sandhya et al., 2010a, b; Numan et al., 2018; Chandra and Sharma, 2020). Many fungi are also known to secrete siderophores in response to iron stress. Several species of *Aspergillus* produce the hydroxymate kind of siderophore (Neilands, 1984). Apart from this, other fungi like *Aspergillus ochraceous*, *Penicillium chrysogenum*, and *Penicillium citrinum* are known to secrete siderophores (Altaf et al., 2018). Several PGPF are also known to release antibiotics and enzymes that keep the host pathogens in check (Imran et al., 2021).

Plant responses to PGPM inoculation under biotic stress

Plant responses to PGPMs with respect to biotic stress tolerance are predominantly manifested through the phenomenon of induced systemic resistance (ISR). Inoculation of plant roots with PGPMs has been reported to activate signaling cascades that lead to the host plant's acquisition of resistances to numerous pathogenic organisms. This phenomenon of 'priming' plants to combat pathogen invasions is known as ISR (Dimkpa et al., 2009; Hayat et al., 2010; Saharan and Nehra, 2011a, b; Ab Rahman et al., 2018). The immune system of plants responds in two different ways: (1) response to the virulence factors imposed by infections of pathogenic organisms (called systemic acquired resistance or SAR) and (2) memory-based recognition and response to the elicitor molecules secreted by nonpathogenic PGPR that ultimately confer resistance to other pathogens or abiotic stresses that constitute ISR. Typically, ISR involves jasmonic acid and ethylene signaling, stimulating host defense responses to a range of pathogens (Saharan and Nehra, 2011a, b; Ahemad and Kibret, 2014; Ab Rahman et al., 2018). These signal transduction processes lead to the accumulation of phytochemicals toxic to the invading pathogens, such as phytoalexins and pathogenesis-related proteins. A list of PGPMs and the plants to which they impart ISR against specific pathogens is provided in Table 9.1. Numerous PGPR strains such as *Pseudomonas fluorescens* ENPF1, *Pseudomonas chlororaphis* BCA, *Bacillus* sp., *Paenibacillus polymyxa*, *Bacillus subtilis* GBO3, *Bacillus amyloliquefaciens* IN937a, have been reported to confer ISR on a variety of host plant species (Timmusk and Wagner, 1999; Dimkpa et al., 2009; Hayat et al., 2010).

Use of PGPR as agents of abiotic and biotic stress tolerance for sustainable agriculture

As described in this chapter, a significant number of studies have been conducted to assess the role of PGPM as biological agents of stress mitigation in crops. While many of these studies have been conducted under laboratory conditions using growth rooms/chambers and greenhouses, several studies have been carried out under field conditions as well. There are

Table 9.1: Selected examples of PGPMs that help combat biotic stress.

Fungi	Host	Pathogen	References
Trichoderma harzianum T22	Maize	*Colletotrichum graminicola*	Harman et al. (1991)
Trichoderma virens or *Trichoderma atroviride*	*Arabidopsis*	*Botrytis cinerea*	Contreras-Cornejo et al. (2011)
Trichoderma asperellum T34	Cucumber	*Pseudomonas syringae* pv. *lachrymans*	Segarra et al. (2007)
Trichoderma harzianum	Melon	*Fusarium oxysporum*	Martínez-Medina et al. (2010)
Talaromyces sp. NBP-61, *Penicillium* sp. NBP-45, *Trichoderma* sp. NBP-67	Chili	*Colletotrichum capsici*	Naziya et al. (2020)
Trichoderma harzianum T39	Grapevine	*Plasmopara viticola*	Palmieri et al. (2012)
Penicillium sp. EU0013	Tomato, cabbage	*Fusarium oxysporum*	Alam et al. (2011)
Bacteria			
Multiple PGPR formulations	Musk melon	Root-knot nematode	Hayat et al. (2010)
Paenibacillus polymyxa	*Arabidopsis*	*Erwinia carotovora*	Timmusk and Wagner (1999)
Bacillus cereus AR156	*Arabidopsis*	*P. syringae* pv. Tomato	Niu et al. (2016)
Klebsiella sp. SBP-8	Sorghum	*Fusarium graminearum*	Singh and Jha (2017)

obvious gaps in transferring the techniques from the laboratory to the field, which is the major challenge in translating such research to practical, real-life conditions. Hence, despite a substantial body of work in the use of PGPM for stress tolerance/resistance, there are relatively fewer reports on the use of such soil microbes for sustained stress mitigation of crops in fields, resulting in enhanced crop survival under adverse environmental conditions. One important consideration toward field studies is that, in nature, plants do not usually encounter one environmental stress at a time but a combination of stresses, which can be termed a 'stress matrix.' Typically, high-temperature stress is also accompanied by drought and vice versa.

Similarly, severe rainfall and cold stress may together affect crop productivity. Plants can be exposed to any number of combinations of different biotic stressors and/or in conjunction with abiotic stressors. Hence, there needs to be more research on the simultaneous, combinatorial impact of multiple stresses and the role of PGPMs in alleviating them. Similarly, more studies are required on applying multiple strains of PGPMs together (a PGPM cocktail) on crops for enhanced and synergistic impacts on stress mitigation. Field studies are also somewhat limited by the availability of different types of 'formulations' for the ease and efficacy of inoculating crops with PGPMs. Therefore, continued efforts are required to enhance the productivity of crops in terms of yield and quality using PGPMs to ameliorate biotic and abiotic stresses.

References

Ab Rahman, S.F.S., Singh, E., Pieterse, C.M., Schenk, P.M., 2018. Emerging microbial biocontrol strategies for plant pathogens. Plant Sci. 267, 102–111. https://doi.org/10.1016/j.plantsci.2017.11.012.

Ahemad, M., Kibret, M., 2014. Mechanisms and applications of plant growth promoting rhizobacteria: current perspective. J. King Saud Univ. Sci. 26 (1), 1–20. https://doi.org/10.1016/j.jksus.2013.05.001.

Ahn, T.S., Ka, J.O., Lee, G.H., Song, H.G., 2007. Microcosm study for revegetation of barren land with wild plants by some plant growth-promoting rhizobacteria. J. Microbiol. Biotechnol. 17 (1), 52–57.

Ait Barka, E., Nowak, J., Clément, C., 2006. Enhancement of chilling resistance of inoculated grapevine plantlets with a plant growth-promoting rhizobacterium, *Burkholderia phytofirmans* strain PsJN. Appl. Environ. Microbiol. 72 (11), 7246–7252. https://doi.org/10.1128/AEM.01047-06.

Alam, S.S., Sakamoto, K., Inubushi, K., 2011. Biocontrol efficiency of *fusarium* wilt diseases by a root-colonizing fungus Penicillium sp. Soil Sci. Plant Nutr. 57 (2), 204–212. https://doi.org/10.1080/00380768.2011.564996.

Altaf, M.M., Imran, M., Abulreesh, H.H., Khan, M.S.A., Ahmad, I., 2018. Diversity and applications of penicillium spp. in plant-growth promotion. In: New and future developments in microbial biotechnology and bioengineering: Penicillium system properties and applications, pp. 261–276, https://doi.org/10.1016/B978-0-444-63501-3.00015-6.

Anjum, S.A., Ashraf, U., Tanveer, M., Khan, I., Hussain, S., Shahzad, B., Zohaib, A., Abbas, F., Saleem, M.F., Ali, I., Wang, L.C., 2017. Drought induced changes in growth, osmolyte accumulation and antioxidant metabolism of three maize hybrids. Front. Plant Sci. 08. https://doi.org/10.3389/fpls.2017.00069.

Ansari, M.W., Trivedi, D.K., Sahoo, R.K., Gill, S.S., Tuteja, N., 2013. A critical review on fungi mediated plant responses with special emphasis to *Piriformospora indica* on improved production and protection of crops. Plant Physiol. Biochem. 70, 403–410. https://doi.org/10.1016/J.PLAPHY.2013.06.005.

Arkhipova, T.N., Prinsen, E., Veselov, S.U., Martinenko, E.V., Melentiev, A.I., Kudoyarova, G.R., 2007. Cytokinin producing bacteria enhance plant growth in drying soil. Plant and Soil 292 (1–2), 305–315. https://doi.org/10.1007/s11104-007-9233-5.

Arzanesh, M.H., Alikhani, H.A., Khavazi, K., Rahimian, H.A., Miransari, M., 2011. Wheat (*Triticum aestivum* L.) growth enhancement by *Azospirillum* sp. under drought stress. World J. Microbiol. Biotechnol. 27 (2), 197–205. https://doi.org/10.1007/s11274-010-0444-1.

Aziz, M., Nadipalli, R.K., Xie, X., Sun, Y., Surowiec, K., Zhang, J.L., Paré, P.W., 2016. Augmenting sulfur metabolism and herbivore defense in *Arabidopsis* by bacterial volatile signaling. Front. Plant Sci. https://doi.org/10.3389/fpls.2016.00458.

Barnawal, D., Bharti, N., Pandey, S.S., Pandey, A., Chanotiya, C.S., Kalra, A., 2017. Plant growth-promoting rhizobacteria enhance wheat salt and drought stress tolerance by altering endogenous phytohormone levels and TaCTR1/TaDREB2 expression. Physiol. Plant. 161 (4), 502–514. https://doi.org/10.1111/ppl.12614.

Barriuso, J., Ramos Solano, B., Fray, R.G., Cámara, M., Hartmann, A., Gutiérrez Mañero, F.J., 2008. Transgenic tomato plants alter quorum sensing in plant growth-promoting rhizobacteria. Plant Biotechnol. J. 6 (5), 442–452. https://doi.org/10.1111/j.1467-7652.2008.00331.x.

Begum, N., Qin, C., Ahanger, M.A., Raza, S., Khan, M.I., Ashraf, M., Ahmed, N., Zhang, L., 2019. Role of arbuscular mycorrhizal fungi in plant growth regulation: implications in abiotic stress tolerance. Front. Plant Sci. 10. https://doi.org/10.3389/FPLS.2019.01068/FULL.

Belimov, A.A., Dodd, I.C., Hontzeas, N., Theobald, J.C., Safronova, V.I., Davies, W.J., 2009. Rhizosphere bacteria containing 1-aminocyclopropane-1-carboxylate deaminase increase yield of plants grown in drying soil via both local and systemic hormone signalling. New Phytol. 181 (2), 413–423. https://doi.org/10.1111/j.1469-8137.2008.02657.x.

Beneduzi, A., Ambrosini, A., Passaglia, L.M.P., 2012. Plant growth-promoting rhizobacteria (PGPR): their potential as antagonists and biocontrol agents. Genet. Mol. Biol. 35 (4 (suppl)), 1044–1051.

Bhattacharyya, P.N., Jha, D.K., 2012. Plant growth-promoting rhizobacteria (PGPR): emergence in agriculture. World J. Microbiol. Biotechnol. https://doi.org/10.1007/s11274-011-0979-9.

Bresson, J., Varoquaux, F., Bontpart, T., Touraine, B., Vile, D., 2013. The PGPR strain *Phyllobacterium brassicacearum* STM196 induces a reproductive delay and physiological changes that result in improved drought tolerance in Arabidopsis. New Phytol. 200 (2), 558–569. https://doi.org/10.1111/nph.12383.

Chandra, D., Sharma, A.K., 2020. Field evaluation of consortium of bacterial inoculants producing ACC deaminase on growth, nutrients and yield components of rice and wheat. J. Crop. Sci. Biotechnol. 24 (3), 293–305. https://doi.org/10.1007/s12892-020-00077-y.

Chandra, D., Srivastava, R., Sharma, A.K., 2015. Environment-friendly phosphorus biofertilizer as an alternative to chemical fertilizers. In: Pati, B.R., Mandal, S.M. (Eds.), Recent Trends in Biofertilizers. IK International Publishing House, New Delhi, pp. 43–71. ISBN NO. "978-93-84588-65-6".

Chandra, D., Pallavi, A., Sharma, A.K., 2017. Harnessing endophytic microbial volatile organic compound (MVOC) for sustainable agroecosystem. In: Maheshwari, D., Annapurna, K. (Eds.), Endophytes: Crop Productivity and Protection. Sustainable Development and Biodiversity. vol 16. Springer, Cham, https://doi.org/10.1007/978-3-319-66544-3_11.

Chandra, D., Srivastava, R., Glick, B.R., Sharma, A.K., 2018a. Drought-tolerant *Pseudomonas* spp. improve the growth performance of finger millet (*Eleusine coracana* (L.) Gaertn.) under non-stressed and drought-stressed conditions. Pedosphere 28, 227–240. https://doi.org/10.1016/S1002-0160(18)60013-X.

Chandra, D., Srivastava, R., Sharma, A., 2018b. Influence of IAA and ACC deaminase producing fluorescent pseudomonads in alleviating drought stress in wheat (*Triticum aestivum*). Agric. Res. 7, 290–299. https://doi.org/10.1007/s40003-018-0305-y.

Chandra, D., Barh, A., Sharma, I.P., 2018c. Plant growth promoting bacteria: a gateway to sustainable agriculture. In: Bhatt, P., Sharma, A. (Eds.), Microbial Biotechnology in Environmental Monitoring and Cleanup. IGI Global, pp. 318–338.

Chandra, D., Srivastava, R., Gupta, V.V., Franco, C.M., Sharma, A.K., 2019a. Evaluation of ACC-deaminase-producing rhizobacteria to alleviate water-stress impacts in wheat (*Triticum aestivum* L.) plants. Can. J. Microbiol. 65 (5), 387–403. https://doi.org/10.1139/cjm-2018-0636.

Chandra, D., Srivastava, R., Gupta, V.V., Franco, C.M., Paasricha, N., Saifi, S.K., Tuteja, N., Sharma, A.K., 2019b. Field performance of bacterial inoculants to alleviate water stress effects in wheat (*Triticum aestivum* L.). Plant and Soil 441 (1), 261–281. https://doi.org/10.1007/s11104-019-04115-9.

Chandra, D., Srivastava, R., Glick, B.R., Sharma, A.K., 2020. Rhizobacteria producing ACC deaminase mitigate water-stress response in finger millet (*Eleusine coracana* (L.) Gaertn.). 3 Biotech 10, 65. https://doi.org/10.1007/s13205-019-2046-4.

Cho, S.M., Kang, B.R., Han, S.H., Anderson, A.J., Park, J.-Y., Lee, Y.-H., Cho, B.H., Yang, K.-Y., Ryu, C.-M., Kim, Y.C., 2008. 2R,3R-butanediol, a bacterial volatile produced by *Pseudomonas chlororaphis* O6, is involved in induction of systemic tolerance to drought in Arabidopsis thaliana. Mol. Plant Microbe Interact. 21 (8), 1067–1075. https://doi.org/10.1094/MPMI-21-8-1067.

Cohen, A.C., Travaglia, C.N., Bottini, R., Piccoli, P.N., 2009. Participation of abscisic acid and gibberellins produced by endophytic *Azospirillum* in the alleviation of drought effects in maize. Botany 87 (5), 455–462. https://doi.org/10.1139/B09-023.

Contesto, C., Milesi, S., Mantelin, S., Zancarini, A., Desbrosses, G., Varoquaux, F., Bellini, C., Kowalczyk, M., Touraine, B., 2010. The auxin-signaling pathway is required for the lateral root response of Arabidopsis to the rhizobacterium *Phyllobacterium brassicacearum*. Planta 232 (6), 1455–1470. https://doi.org/10.1007/s00425-010-1264-0.

Contreras-Cornejo, H.A., Macías-Rodríguez, L., Cortés-Penagos, C., López-Bucio, J., 2009. *Trichoderma* virens, a plant beneficial fungus, enhances biomass production and promotes lateral root growth through an auxin-dependent mechanism in Arabidopsis. Plant Physiol. 149 (3), 1579–1592. https://doi.org/10.1104/pp.108.130369.

Contreras-Cornejo, H.A., Macías-Rodríguez, L., Beltrán-Peña, E., Herrera-Estrella, A., López-Bucio, J., 2011. *Trichoderma*-induced plant immunity likely involves both hormonal- and camalexin-dependent mechanisms in *Arabidopsis thaliana* and confers resistance against necrotrophicfungi *Botrytis cinerea*. Plant Signal. Behav. 6 (10), 1554–1563. https://doi.org/10.4161/psb.6.10.17443.

Creus, C.M., Sueldo, R.J., Barassi, C.A., 2004. Water relations and yield in *Azospirillum*- inoculated wheat exposed to drought in the field. Can. J. Bot. 82 (2), 273–281. https://doi.org/10.1139/b03-119.

Dimitrova, S., Pavlova, K., Lukanov, L., Korotkova, E., Petrova, E., Zagorchev, P., Kuncheva, M., 2013. Production of metabolites with antioxidant and emulsifying properties by antarctic strain *Sporobolomyces salmonicolo*r AL1. Appl. Biochem. Biotechnol. 169 (1), 301–311. https://doi.org/10.1007/s12010-012-9983-2.

Dimkpa, C., Weinand, T., Asch, F., 2009. Plant-rhizobacteria interactions alleviate abiotic stress conditions. Plant Cell Environ. 32 (12), 1682–1694. https://doi.org/10.1111/j.1365-3040.2009.02028.x.

Dobra, J., Motyka, V., Dobrev, P., Malbeck, J., Prasil, I.T., Haisel, D., Gaudinova, A., Havlova, M., Gubis, J., Vankova, R., 2010. Comparison of hormonal responses to heat, drought and combined stress in tobacco plants with elevated proline content. J. Plant Physiol. 167 (16), 1360–1370. https://doi.org/10.1016/J.JPLPH.2010.05.013.

Dodd, I.C., Zinovkina, N.Y., Safronova, V.I., Belimov, A.A., 2010. Rhizobacterial mediation of plant hormone status. Ann. Appl. Biol. 157 (3), 361–379. https://doi.org/10.1111/j.1744-7348.2010.00439.x.

Evelin, H., Devi, T.S., Gupta, S., Kapoor, R., 2019. Mitigation of salinity stress in plants by arbuscular mycorrhizal symbiosis: current understanding and new challenges. Front. Plant Sci. 10. https://doi.org/10.3389/FPLS.2019.00470.

Fahad, S., Hussain, S., Matloob, A., Khan, F.A., Khaliq, A., Saud, S., Hassan, S., Shan, D., Khan, F., Ullah, N., Faiq, M., Khan, M.R., Tareen, A.K., Khan, A., Ullah, A., Ullah, N., Huang, J., 2015. Phytohormones and plant responses to salinity stress: a review. Plant Growth Regul. 75 (2), 391–404. https://doi.org/10.1007/s10725-014-0013-y.

Galland, M., Gamet, L., Varoquaux, F., Touraine, B., Touraine, B., Desbrosses, G., 2012. The ethylene pathway contributes to root hair elongation induced by the beneficial bacteria *Phyllobacterium brassicacearum* STM196. Plant Sci. 190, 74–81. https://doi.org/10.1016/j.plantsci.2012.03.008.

Ghosh, D., Sen, S., Mohapatra, S., 2017. Modulation of proline metabolic gene expression in Arabidopsis thaliana under water-stressed conditions by a drought-mitigating *Pseudomonas putida* strain. Ann. Microbiol. 67 (10). https://doi.org/10.1007/s13213-017-1294-y.

Ghosh, D., Sen, S., Mohapatra, S., 2018. Drought-mitigating *Pseudomonas putida* GAP-P45 modulates proline turnover and oxidative status in *Arabidopsis thaliana* under water stress. Ann. Microbiol. https://doi.org/10.1007/s13213-018-1366-7.

Ghosh, D., Gupta, A., Mohapatra, S., 2019a. A comparative analysis of exopolysaccharide and phytohormone secretions by four drought-tolerant rhizobacterial strains and their impact on osmotic-stress mitigation in *Arabidopsis thaliana*. World J. Microbiol. Biotechnol. https://doi.org/10.1007/s11274-019-2659-0.

Ghosh, D., Gupta, A., Mohapatra, S., 2019b. Dynamics of endogenous hormone regulation in plants by phytohormone secreting rhizobacteria under water-stress. Symbiosis. https://doi.org/10.1007/s13199-018-00589-w.

Glick, B.R., 2005. Modulation of plant ethylene levels by the bacterial enzyme ACC deaminase. FEMS Microbiol. Lett. 251 (1), 1–7. https://doi.org/10.1016/j.femsle.2005.07.030.

Glick, B.R., 2014. Bacteria with ACC deaminase can promote plant growth and help to feed the world. Microbiol. Res. 169 (1), 30–39. https://doi.org/10.1016/J.MICRES.2013.09.009.

Grover, M., Ali, S.Z., Sandhya, V., Rasul, A., Venkateswarlu, B., 2011. Role of microorganisms in adaptation of agriculture crops to abiotic stresses. World J. Microbiol. Biotechnol. 27 (5), 1231–1240. https://doi.org/10.1007/s11274-010-0572-7.

Gupta, S., Schillaci, M., Walker, R., Smith, P.M.C., Watt, M., Roessner, U., 2021. Alleviation of salinity stress in plants by endophytic plant-fungal symbiosis: current knowledge, perspectives and future directions. Plant and Soil 461 (1–2), 219–244. https://doi.org/10.1007/S11104-020-04618-W.

Haas, D., Défago, G., 2005. Biological control of soil-borne pathogens by fluorescent pseudomonads. Nat. Rev. Microbiol. 3 (4), 307–319. https://doi.org/10.1038/nrmicro1129.

Han, Q.Q., Wu, Y.N., Gao, H.J., Xu, R., Paré, P.W., Shi, H., Zhao, Q., Li, H.R., Khan, S.A., Wang, Y.Q., Wang, S.M., Zhang, J.L., 2017. Improved salt tolerance of medicinal plant Codonopsis pilosula by *Bacillus amyloliquefaciens* GB03. Acta Physiol. Plant. https://doi.org/10.1007/s11738-016-2325-1.

Harman, G.E., Jin, X., Stasz, T.E., Peruzzotti, G., Leopold, A.C., Taylor, A.G., 1991. Production of conidial biomass of *Trichoderma harzianu*m for biological control. Biol. Control 1 (1), 23–28. https://doi.org/10.1016/1049-9644(91)90097-J.

Hayat, R., Ali, S., Amara, U., Khalid, R., Ahmed, I., 2010. Soil beneficial bacteria and their role in plant growth promotion: a review. Ann. Microbiol. 60 (4), 579–598. https://doi.org/10.1007/s13213-010-0117-1.

Imran, M., Abulreesh, H.H., Monjed, M.K., Elbanna, K., Samreen, Ahmad, I., 2021. Multifarious functional traits of free-living rhizospheric fungi, with special reference to Aspergillus spp. isolated from North Indian soil, and their inoculation effect on plant growth. Ann. Microbiol. 71 (1), 31. https://doi.org/10.1186/s13213-021-01643-7.

Jalili, F., Khavazi, K., Pazira, E., Nejati, A., Rahmani, H.A., Sadaghiani, H.R., Miransari, M., 2009. Isolation and characterization of ACC deaminase-producing fluorescent pseudomonads, to alleviate salinity stress on canola (*Brassica napu*s L.) growth. J. Plant Physiol. 166 (6), 667–674. https://doi.org/10.1016/j.jplph.2008.08.004.

Jatan, R., Chauhan, P.S., Lata, C., 2019. Pseudomonas putida modulates the expression of miRNAs and their target genes in response to drought and salt stresses in chickpea (*Cicer arietinum* L.). Genomics 111 (4), 509–519. https://doi.org/10.1016/j.ygeno.2018.01.007.

Kang, S.M., Khan, A.L., Waqas, M., You, Y.H., Kim, J.H., Kim, J.G., Hamayun, M., Lee, I.J., 2014a. Plant growth-promoting rhizobacteria reduce adverse effects of salinity and osmotic stress by regulating phytohormones and antioxidants in *Cucumis sativu*s. J. Plant Interact. 9 (1), 673–682. https://doi.org/10.1080/17429145.2014.894587.

Kang, S.M., Radhakrishnan, R., Khan, A.L., Kim, M.J., Park, J.M., Kim, B.R., Shin, D.H., Lee, I.J., 2014b. Gibberellin secreting rhizobacterium, *Pseudomonas puti*da H-2-3 modulates the hormonal and stress physiology of soybean to improve the plant growth under saline and drought conditions. Plant Physiol. Biochem. 84, 115–124. https://doi.org/10.1016/j.plaphy.2014.09.001.

Khan, A.L., Halo, B.A., Elyassi, A., Ali, S., Al-Hosni, K., Hussain, J., Al-Harrasi, A., Lee, I.-J., 2016. Indole acetic acid and ACC deaminase from endophytic bacteria improves the growth of *Solanum lycopersicum*. Electron. J. Biotechnol. 21, 58–64. https://doi.org/10.1016/J.EJBT.2016.02.001.

Kohler, J., Hernández, J.A., Caravaca, F., Roldán, A., 2008. Plant-growth-promoting rhizobacteria and arbuscular mycorrhizal fungi modify alleviation biochemical mechanisms in water-stressed plants. Funct. Plant Biol. 35 (2), 141. https://doi.org/10.1071/FP07218.

Kumari, S., Varma, A., Tuteja, N., Choudhary, D.K., 2016. Bacterial ACC-Deaminase: An Eco-Friendly Strategy to Cope Abiotic Stresses for Sustainable Agriculture. pp. 165–185, https://doi.org/10.1007/978-981-10-2854-0_8.

Mantelin, S., Touraine, B., 2003. Plant growth-promoting bacteria and nitrate availability: impacts on root development and nitrate uptake. J. Exp. Bot. 55 (394), 27–34. https://doi.org/10.1093/jxb/erh010.

Martínez-Medina, A., Pascual, J.A., Pérez-Alfocea, F., Albacete, A., Roldán, A., 2010. *Trichoderma harzianum* and *Glomus intraradic*es modify the hormone disruption induced by *fusarium oxysporum* infection in melon plants. Phytopathology 100 (7). https://doi.org/10.1094/PHYTO-100-7-0682.

Marulanda, A., Barea, J.-M., Azcón, R., 2009. Stimulation of plant growth and drought tolerance by native microorganisms (AM fungi and bacteria) from dry environments: mechanisms related to bacterial effectiveness. J. Plant Growth Regul. 28 (2), 115–124. https://doi.org/10.1007/s00344-009-9079-6.

Mayak, S., Tirosh, T., Glick, B.R., 2004. Plant growth-promoting bacteria that confer resistance to water stress in tomatoes and peppers. Plant Sci. 166 (2), 525–530. https://doi.org/10.1016/j.plantsci.2003.10.025.

Meldau, D.G., Meldau, S., Hoang, L.H., Underberg, S., Wünsche, H., Baldwin, I.T., 2013. Dimethyl disulfide produced by the naturally associated bacterium *Bacillus* sp B55 promotes *Nicotiana attenuata* growth by enhancing sulfur nutrition. Plant Cell. https://doi.org/10.1105/tpc.113.114744.

Naziya, B., Murali, M., Amruthesh, K.N., 2020. Plant growth-promoting fungi (PGPF) instigate plant growth and induce disease resistance in *Capsicum annuum* L. upon infection with *Colletotrichum capsici* (syd.) butler & bisby. Biomolecules 10 (1). https://doi.org/10.3390/biom10010041.

Neilands, J.B., 1984. Methodology of siderophores. Siderophores from Microorganisms and Plants. Structure and Bonding. vol. 58 Springer, Berlin, Heidelberg.

Ngumbi, E., Kloepper, J., 2016. Bacterial-mediated drought tolerance: current and future prospects. Appl. Soil Ecol. 105, 109–125. https://www.sciencedirect.com/science/article/pii/S0929139316301111.

Niu, D., Wang, X., Wang, Y., Song, X., Wang, J., Guo, J., Zhao, H., 2016. *Bacillus cereus* AR156 activates PAMP-triggered immunity and induces a systemic acquired resistance through a NPR1-and SA-dependent signaling pathway. Biochem. Biophys. Res. Commun. 469 (1). https://doi.org/10.1016/j.bbrc.2015.11.081.

Numan, M., Bashir, S., Khan, Y., Mumtaz, R., Shinwari, Z.K., Khan, A.L., Khan, A., AL-Harrasi, A., 2018. Plant growth promoting bacteria as an alternative strategy for salt tolerance in plants: a review. Microbiol. Res. 209, 21–32. https://doi.org/10.1016/J.MICRES.2018.02.003.

Osińska-Jaroszuk, M., Jarosz-Wilkołazka, A., Jaroszuk-Ściseł, J., Szałapata, K., Nowak, A., Jaszek, M., Ozimek, E., Majewska, M., 2015. Extracellular polysaccharides from Ascomycota and Basidiomycota: production conditions, biochemical characteristics, and biological properties. World J. Microbiol. Biotechnol. 31 (12), 1823–1844. https://doi.org/10.1007/S11274-015-1937-8.

Palmieri, M.C., Perazzolli, M., Matafora, V., Moretto, M., Bachi, A., Pertot, I., 2012. Proteomic analysis of grapevine resistance induced by *Trichoderma harzianum* T39 reveals specific defence pathways activated against downy mildew. J. Exp. Bot. 63 (17). https://doi.org/10.1093/jxb/ers279.

Patten, C.L., Glick, B.R., 2002. Role of *Pseudomonas putida* indoleacetic acid in development of the host plant root system. Appl. Environ. Microbiol. https://doi.org/10.1128/AEM.68.8.3795-3801.2002.

Pereyra, M.A., García, P., Colabelli, M.N., Barassi, C.A., Creus, C.M., 2012. A better water status in wheat seedlings induced by *Azospirillum* under osmotic stress is related to morphological changes in xylem vessels of the coleoptile. Appl. Soil Ecol. 53, 94–97. https://doi.org/10.1016/J.APSOIL.2011.11.007.

Pii, Y., Mimmo, T., Tomasi, N., Terzano, R., Cesco, S., Crecchio, C., 2015. Microbial interactions in the rhizosphere: beneficial influences of plant growth-promoting rhizobacteria on nutrient acquisition process. A review. Biol. Fertil. Soils 51 (4), 403–415. https://doi.org/10.1007/s00374-015-0996-1.

Raza, F.A., Faisal, M., 2013. Growth promotion of maize by desiccation tolerant *Micrococcus luteus*-chp37 isolated from Cholistan desert, Pakistan. Aust. J. Crop. Sci. 7 (11).

Ripa, F.A., Cao, W.D., Tong, S., Sun, J.G., 2019. Assessment of plant growth promoting and abiotic stress tolerance properties of wheat endophytic fungi. Biomed. Res. Int. 2019. https://doi.org/10.1155/2019/6105865.

Rolli, E., Marasco, R., Vigani, G., Ettoumi, B., Mapelli, F., Deangelis, M.L., Gandolfi, C., Casati, E., Previtali, F., Gerbino, R., Pierotti Cei, F., Borin, S., Sorlini, C., Zocchi, G., Daffonchio, D., 2015. Improved plant resistance to drought is promoted by the root-associated microbiome as a water stress-dependent trait. Environ. Microbiol. 17 (2), 316–331. https://doi.org/10.1111/1462-2920.12439.

Safdarian, M., Askari, H., Shariati, J.V., Nematzadeh, G., 2019. Transcriptional responses of wheat roots inoculated with *Arthrobacter nitroguajacolicus* to salt stress. Sci. Rep. https://doi.org/10.1038/s41598-018-38398-2.

Saharan, B.S., Nehra, V., 2011a. Plant growth promoting rhizobacteria: a critical review. In: Life Sciences and Medicine Research. vol. 2011.

Saharan, B.S., Nehra, V., 2011b. Plant growth promoting Rhizobacteria: a critical review. In: Life Sciences and Medicine Research. vol. 21, pp. 1–30. http://astonjournals.com/manuscripts/Vol2011/LSMR-21_Vol2011.pdf.

Salomon, M.V., Bottini, R., de Souza Filho, G.A., Cohen, A.C., Moreno, D., Gil, M., Piccoli, P., 2014. Bacteria isolated from roots and rhizosphere of Vitis vinifera retard water losses, induce abscisic acid accumulation and synthesis of defense-related terpenes in in vitro cultured grapevine. Physiol. Plant. 151 (4), 359–374. https://doi.org/10.1111/ppl.12117.

Sandhya, V., Zulfikar, S.K., Grover, M., Reddy, G., Venkateswarlu, B., 2009. Alleviation of drought stress effects in sunflower seedlings by the exopolysaccharides producing *Pseudomonas putida* strain GAP-P45. Biol. Fertil. Soils 46 (1), 17–26. https://doi.org/10.1007/s00374-009-0401-z.

Sandhya, V., Ali, S.Z., Grover, M., Reddy, G., Venkateswarlu, B., 2010a. Effect of plant growth promoting *Pseudomonas* spp. on compatible solutes, antioxidant status and plant growth of maize under drought stress. Plant Growth Regul. 62 (1), 21–30. https://doi.org/10.1007/s10725-010-9479-4.

Sandhya, V., Ali, S.Z., Venkateswarlu, B., Reddy, G., Grover, M., 2010b. Effect of osmotic stress on plant growth promoting *Pseudomonas* spp. Arch. Microbiol. 192 (10), 867–876. https://doi.org/10.1007/s00203-010-0613-5.

Segarra, G., Casanova, E., Bellido, D., Odena, M.A., Oliveira, E., Trillas, I., 2007. Proteome, salicylic acid, and jasmonic acid changes in cucumber plants inoculated with *Trichoderma asperellum* strain T34. Proteomics 7 (21). https://doi.org/10.1002/pmic.200700173.

Selvakumar, G., Bindu, G.H., Bhatt, R.M., Upreti, K.K., Paul, A.M., Asha, A., Shweta, K., Sharma, M., 2018. Osmotolerant cytokinin producing microbes enhance tomato growth in deficit irrigation conditions. Proc. Natl. Acad. Sci. India B: Biol. Sci. 88 (2), 459–465. https://doi.org/10.1007/s40011-016-0766-3.

Sen, S., Ghosh, D., Mohapatra, S., 2018. Modulation of polyamine biosynthesis in Arabidopsis thaliana by a drought mitigating *Pseudomonas putida* strain. Plant Physiol. Biochem. https://doi.org/10.1016/j.plaphy.2018.05.034.

Sen, S., Mohapatra, S., 2022. Drought-mitigating *Pseudomonas putida* strain modulates polyamine catabolism in *Arabidopsis thaliana*. J. Plant Growth Regul. 41, 300–313. https://doi.org/10.1007/s00344-021-10297-3.

Singh, R.P., Jha, P.N., 2017. Analysis of fatty acid composition of PGPR *Klebsiella* sp. SBP-8 and its role in ameliorating salt stress in wheat. Symbiosis 73, 213–222.

Sreenivasulu, N., Harshavardhan, V.T., Govind, G., Seiler, C., Kohli, A., 2012. Contrapuntal role of ABA: does it mediate stress tolerance or plant growth retardation under long-term drought stress? Gene 506 (2), 265–273. https://doi.org/10.1016/J.GENE.2012.06.076.

Srivastava, S., Chaudhry, V., Mishra, A., Chauhan, P.S., Rehman, A., Yadav, A., Tuteja, N., Nautiyal, C.S., 2012. Gene expression profiling through microarray analysis in *Arabidopsis thaliana* colonized by *Pseudomonas putida* MTCC5279, a plant growth promoting rhizobacterium. Plant Signal. Behav. https://doi.org/10.4161/psb.18957.

Szabados, L., Savouré, A., 2010. Proline: a multifunctional amino acid. Trends Plant Sci. 15 (2), 89–97. https://doi.org/10.1016/j.tplants.2009.11.009.

Taiz, L., Zeiger, E., Møller, I.M., Murphy, A., 2015. Plant Physiology and Development, sixth ed. vol. 1 Sinauer Associates.

Tewari, S., Arora, N.K., 2014. Multifunctional exopolysaccharides from Pseudomonas aeruginosa PF23 involved in plant growth stimulation, biocontrol and stress amelioration in sunflower under saline conditions. Curr. Microbiol. 69 (4), 484–494. https://doi.org/10.1007/s00284-014-0612-x.

Timmusk, S., Wagner, E.G.H., 1999. The plant-growth-promoting Rhizobacterium Paenibacillus polymyxa induces changes in *Arabidopsis thaliana* gene expression: A possible connection between biotic and abiotic stress responses. Mol. Plant Microbe Interact. 12 (11), 951–959. https://doi.org/10.1094/MPMI.1999.12.11.951.

Timmusk, S., El-Daim, I.A., Copolovici, L., Tanilas, T., Kännaste, A., Behers, L., Nevo, E., Seisenbaeva, G., Stenström, E., N. Ü., 2014. Drought-tolerance of wheat improved by rhizosphere bacteria from harsh environments: enhanced biomass production and reduced emissions of stress volatiles. PLoS One 9 (5), e96086. https://journals.plos.org/plosone/article?id=10.1371/journal.pone.0096086.

Vaishnav, A., Kumari, S., Jain, S., Varma, A., Choudhary, D.K., 2015. Putative bacterial volatile-mediated growth in soybean (*Glycine max* L. Merrill) and expression of induced proteins under salt stress. J. Appl. Microbiol. https://doi.org/10.1111/jam.12866.

Vardharajula, S., Zulfikar Ali, S., Grover, M., Reddy, G., Bandi, V., 2011. Drought-tolerant plant growth promoting *Bacillus* spp.: effect on growth, osmolytes, and antioxidant status of maize under drought stress. J. Plant Interact. 6 (1), 1–14. https://doi.org/10.1080/17429145.2010.535178.

Viterbo, A., Landau, U., Kim, S., Chernin, L., Chet, I., 2010. Characterization of ACC deaminase from the biocontrol and plant growth-promoting agent *Trichoderma asperellum* T203. FEMS Microbiol. Lett. 305 (1), 42–48. https://doi.org/10.1111/j.1574-6968.2010.01910.x.

Vurukonda, S.S.K.P., Vardharajula, S., Shrivastava, M., SkZ, A., 2016. Enhancement of drought stress tolerance in crops by plant growth promoting rhizobacteria. Microbiol. Res. 184, 13–24. https://doi.org/10.1016/J.MICRES.2015.12.003.

Waqas, M., Khan, A., Kamran, M., Hamayun, M., 2012. Endophytic fungi produce gibberellins and indoleacetic acid and promotes host-plant growth during stress. Molecules 17, 10754–10773. https://doi.org/10.3390/molecules170910754.

Wiebe, M.G., 2003. Stable production of recombinant proteins in filamentous fungi—problems and improvements. Mycologist 17 (3), 140–144. https://doi.org/10.1017/S0269-915X(03)00303-3.

Xie, H., Pasternak, J.J., Glick, B.R., 1996. Isolation and characterization of mutants of the plant growth-promoting rhizobacterium *Pseudomonas putida* GR12-2 that overproduce indoleacetic acid. Curr. Microbiol. https://doi.org/10.1007/s002849900012.

Zhang, H., Sun, Y., Xie, X., Kim, M.-S., Dowd, S.E., Paré, P.W., 2009. A soil bacterium regulates plant acquisition of iron via deficiency-inducible mechanisms. Plant J. 58 (4), 568–577. https://doi.org/10.1111/j.1365-313X.2009.03803.x.

Zhang, W., Wang, H.W., Wang, X.X., Xie, X.G., Siddikee, M.A., Xu, R.S., Dai, C.C., 2016. Enhanced nodulation of peanut when co-inoculated with fungal endophyte *Phomopsis liquidambari* and *Bradyrhizobium*. Plant Physiol. Biochem. 98. https://doi.org/10.1016/j.plaphy.2015.11.002.

Zhang, F., Wang, Y., Liu, C., Chen, F., Ge, H., Tian, F., Yang, T., Ma, K., Zhang, Y., 2019. *Trichoderma harzianum* mitigates salt stress in cucumber via multiple responses. Ecotoxicol. Environ. Saf. 170. https://doi.org/10.1016/j.ecoenv.2018.11.084.

CHAPTER 10

Microbial nanotechnology: A green approach towards sustainable agriculture

Sudha Bind[a], Sandhya Bind[a], and Dinesh Chandra[a,b]
[a]Department of Biological Sciences, College of Basic Sciences and Humanities, Govind Ballabh Pant University of Agriculture and Technology, US Nagar, Uttarakhnad, India, [b]Govt. Inter Collge Chamtola, Almora, Uttarakhand, India

Chapter outline

Introduction

The current global population is 7 billion, and about 50% of this population is present in Asia. Agriculture is the pillar of various growing countries. According to Brock et al. (2011), approximately 60% of the world's population depends on agriculture for their survival. Increased food demands by the ever-growing population have decreased global

Unravelling Plant-Microbe Synergy. https://doi.org/10.1016/B978-0-323-99896-3.00008-4

agricultural land. According to the UNFAO (United Nations Food and Agriculture Organization) (2018), the agricultural land area is 4.82 billion hectares out of the world's total land area (13.6 billion hectares), which represents 36% of the land area. This area decreases by 0.4% every decade. At present, the other major issues that need to be addressed are soil health and fertility. The 2014 FAO World Soil Day forum reported that one-third of the world's topmost soil has already been degraded and an alarming fact is that the world's topmost soil might be washed out within the next 60 years. Therefore, there is an urgency for sustainable approaches to meet the food demands of the ever-growing population and to ensure global food security. Various new methods have been developed in agriculture for increasing crop production through utilization of efficient chemical fertilizers and pesticides. A major drawback of this technique is the discharge of undesirable and toxic substances into the environment. Extreme use of chemical fertilizers and pesticides leads to environmental pollution and, ultimately, causes various human diseases. Nowadays, nanotechnology is evolving as a promising approach to increasing food production to achieve sustainable agriculture. After biotechnology, nanotechnology is the fifth revolutionary technology of this century. Nanotechnology is an emerging area that includes the synthesis and application of nano-sized materials. Nanoparticles exhibit exceptional properties and behaviors. Nanotechnology has the potential to contribute to a new technology-based revolution (Shakiba et al., 2020). It plays a vital role in various fields, i.e., agriculture, medicine, biology, chemistry, physics, and materials science. Nanotechnology has wide applications in agriculture due to the eco-friendly nature of its applications. Microbial nanotechnology is a new and fast-growing area with broad applications in many fields. Nanotechnological tools such as nanoparticles, nanotubes, nanocapsules, and nanomolecules represent a sustainable approach to the modern-day agricultural system. It is intended that nanotechnology will transform agricultural production and agro-food systems in a sustainable manner. Utilization of nanotechnology-based nanofertilizers and nanopesticides increases the efficiency of nutrient uptake, improves fertility and soil health, and thus significantly enhances plant productivity and reduces the application of chemical fertilizers (Kour et al., 2020). Nanofertilizers are nutrient carriers with a dimension range of 30–40 nm, due to their large surface area and their ability to retain ample nutrient ions. Nanotechnology occupies a prominent position among technological innovations, in propelling agriculture toward enhanced food production. Nanotechnology has the potential to increase the efficiency of agricultural inputs, and, thus, it represents a significant method for sustainable agriculture. For this, nano plant growth promoters, nanofertilizers, nanopesticides, nanoherbicides, and agrochemical-encapsulated nanocarrier systems are being used. Nanomaterials can be manufactured through physical, chemical, and biological methods. Nanomaterial production through microorganisms is considered as a green approach to nanomaterial synthesis. Various microbes such as bacteria, fungi, and algae are used for nanoparticle synthesis. Microorganisms form nanoparticles by aggregating inorganic materials within or outside their cells. Nanomaterials have various features, i.e., they can be conjugated, encapsulated, and linked to other materials like drugs and other chemicals, and thus have wide biological applications (El-Sayed and Kamel, 2020). Nanoparticles exhibit different unique properties such as optical, mechanical, catalytical, magnetic, and thermal, thus making them more applicable in various fields. Each nanoparticle exhibits different physicochemical properties.

Nanomaterials

Nanomaterials are substances with dimensions in the nanoscale (the range of length is 1–100 nm). According to the Environmental Protection Agency, nanomaterials have exceptional properties than do their bulky counterparts or equivalent chemical compounds (United Nations, 2014). Nanoparticles have more reactivity and catalytic activity in comparison to that of the same element in the macroscale. Nanoparticles having greater chemical and physical properties play a key role in the preparation of novel materials with vast applications in various fields. Nowadays, there is large-scale production of nanomaterials owing to their enormous benefits. The application of nanotechnology has been proven to offer significant benefits to the agricultural field, such as reduction of the use of pesticides, fungicides, and weedicides that causes loss of nutrients during fertilization and enhanced crop yield through optimization of nutrient management. It is proven that nanomaterials play a key role in the detection and treatment of various plant diseases, increase the absorption of nutrients, and also deliver specific or active ingredients to a specific site.

Synthesis of nanoparticles

Nanoparticles are produced through physical, chemical, and biological methods. The physical methods of nanoparticle synthesis include pulsed laser desorption, ball milling, plasma arching, lithographic technique, sputter deposition, and diffusion flame (Kuppusamy et al., 2016). Nanoparticles prepared from physical methods get agglomerated, and a stabilizer is used to prevent their aggregation. Chemical-based methods of nanomaterial synthesis are mostly used. These include hydrolysis, electrodeposition, the sol–gel process, and the wet chemical method (Kuppusamy et al., 2016). Due to the involvement of toxic chemicals, these methods are not eco-friendly. Both chemical and physical methods of nanoparticle synthesis are cost-intensive and time-consuming. Therefore, there is immense need for eco-friendly approaches for nanoparticle synthesis. Synthesis of nanoparticles using biological methods is an eco-friendly and safe method.

Nanoparticle synthesis by microbes

In green nanotechnology, various microbes are used for the synthesis of nanoparticles. It is reported that various microbes are capable of aggregating inorganic materials outside or inside of cells for nanoparticle synthesis. Most of the microbes are capable of producing metallic nanoparticles. According to Narayanan (2010), various nanoparticles such as Ag, Au, Ag–Au alloy, Pt, Pd, Si, Se, Te, Ti, Fe_3O_4, UO_2, quantum dots (QDs), and Zr have been synthesized from bacteria, fungi, *Actinomycetes*, algae, and viruses. One of the most synthesized and used nanoparticles are Ag NPs, which are prepared using bacteria, fungi, algae, and *Actinomycetes*. During the synthesis process, microbes may interact with nanomaterials or may produce nanostructured compounds (Wang and Chen, 2009); this arises due to the efficiency of microbes in the selective absorption of metal ions, performance on a broad range of environmental conditions such as temperature, pH, and ionic strength, regeneration, and high biosorption capacity (Rai et al., 2009). Nanoparticles are precipitated inside microbial cells when incubated in a metal-containing medium. According to the study by Beveridge and Murray (1980), when bacteria are incubated with Au^3 +, gold nanoparticles are precipitated inside bacterial cells. Klaus-Joerger et al. (2001) synthesized silver nanoparticles using *Pseudomonas stutzeri* after incubating in silver nitrate solution. This bacterium was isolated from silver mines. These Ag NPs accumulated in the periplasmic space of bacteria. Magnetic nanoparticles have been synthesized within magnetosomes by *Magnetospirillum* (magnetotactic bacteria). Microbial preparations of nanoparticles are threatened by chances of contamination. Nanomaterials formed by microbes are categorized into various categories on the basis of their location of formation.

The mechanism of nanoparticle synthesis

The exact mechanism of nanoparticle synthesis by all microbes and plant materials still remains a mystery, and many research studies have been conducted in different ways to determine the actual mechanism (Meyer, 2008; Duran et al., 2005). Various methods exist, which may explain the metal reduction process in microbes. Fungi used for extracellular nanoparticle synthesis include three main mechanisms, namely, nitrate reductase action, electron quinone reduction, or both. In the nitrate reductase mechanism, nitrite reacts with 2,3-diaminonaphthalene (Kumar et al., 2007). Nitrate reductase is used for nanoparticle synthesis by various fungi including *Penicillium.* Other enzymes, like nitrate-dependent reductases, NADPH-dependent reductases, and electron shuttle quinones are used for Ag NP synthesis by *Fusarium oxysporum*. Various researchers support the nitrate reductase mechanism for extracellular nanoparticle synthesis (Deepa and Panda, 2014).

Nanoparticle synthesis by fungi

Fungi represent a group of nonphotosynthetic, eukaryotic microbes with rigid cell walls. Fungi require simple nutrition for their survival. They can grow on readily available and inexpensive substrates. The ability of fungi to secrete various kinds of metabolites and enzymes into their surroundings makes them as key factor in the production of various kinds of nanoparticles (Dhillon et al., 2012). These enzymes catalyze nanoparticle synthesis by reducing salts to different nanoparticles. Fungi are preferred over bacteria due to their better tolerance and metal accumulation properties. Ingle et al. (2008), for the first time, used the term 'mycosynthesis' for the synthesis of nanomaterials by fungi. According to Rai et al. (2009), filamentous fungi have more potential and offer more advantages over bacteria or other microbes for nanoparticle synthesis. Utilization of fungi in nanomaterial synthesis is potentially important because they are easy to handle in lab conditions and produce large a number of enzymes (Mohanpuria et al., 2007). Fungal strains, for example, *Aspergillus*, *Penicillium*, *Fusarium*, and *Verticillium*, have been proven to be a good source for nanomaterial synthesis. Various fungal strains are used for the production of nanoparticles, extracellularly or intracellularly. It is easy to purify nanoparticles since they are produced extracellularly (Gaikwad et al., 2013). Various studies have suggested that fungal filtrates can produce monodispersive nanoparticles (Mohammad Fayaz et al., 2009). To date, fungi used for silver nanoparticle synthesis include *Aspergillus terreus*, *Fusarium solani*, *Fusarium oxysporum*, *Fusarium culmorum*, *Penicillium* sp., *Penicillium chrysogenum*, *Phoma glomerata*, and *Trichoderma viride* (Khan et al., 2018). According to the study by Rai et al. (2021), *Colletotrichum* sp., *Fusarium solani*, and *Verticillium* sp. were utilized for the synthesis of gold nanoparticles. Riddin et al. (2006) synthesized platinum nanoparticles using *F. solani* and *F. oxysporum* sp. f. sp. *lycopersici*., and *Aspergillus flavus* was used for titanium nanoparticle synthesis. Ingle et al. (2009) synthesized barium titrate, zirconia (ZrO_2), and silica (SiO_2) using *Fusarium oxysporum*. Banu et al. (2011) synthesized Ag NPs using *Rhizopus stolonifer*, and Saravanan et al. (2018) synthesized Ag NPs using *Phanerochaete chrysosporium*. Some of the nanoparticles synthesized by fungi are listed in Table 10.1.

Nanoparticle synthesis by algae

Microalgae, for example, *Chaetoceros calcitrans*, *Chaetoceros salina*, *Isochrysis galbana*, and *Tetraselmis gracilis* 199, have been reported for Ag NP synthesis. Prasad et al. (2013) used the marine alga *Cystophora moniliformis* for Ag NP synthesis. Ag NPs were also formed using macroalgae *Sargassum wightii* Greville (Govindaraju et al., 2009), *Caulerpa racemose* (Kathiraven et al., 2015), *Colpomenia sinuosa*, *Pterocladia capillacea*, *Ulva fasciata*, *Gelidium amansii* (El-Rafie et al., 2013), and *Chaetomorpha linum* (Kannan et al., 2013). Salaam et al. (2018) synthesized silver nanoparticles exerting antibacterial effects using *Chlorella vulgaris*.

Nanoparticle synthesis by bacteria

Bacteria have been reported to show the potential for intracellular and extracellular nanoparticle synthesis. Nanoparticles get precipitated easily when incubated with respective metal ions in bacteria. According to the study by Rane et al. (2017), cyanobacteria were used for extracellular gold nanoparticle synthesis and *Bacillus subtilis* produced gold nanoparticles

Table 10.1: Nanoparticle synthesis using fungi.

Fungi	Intracellular/ extracellular	Nanoparticles	References
Aspergillus sp.	Extracellular	Zinc	Raliya and Tarafdar (2014)
Aspergillus alternata	Extracellular	Silver	Gajbhiye et al. (2009)
Aspergillus clavatus	Extracellular	Silver	Verma et al. (2010)
Aspergillus flavus	Extracellular	Silver	Vigneshwaran et al. (2006)
Aspergillus fumigatus	Extracellular	Silver	Bhainsa and D'souza (2006)
Aspergillus niger	Extracellular	Silver	Gajbhiye et al. (2009)
Aspergillus tamarii	Extracellular	Silver	Kumar et al. (2012)
Aspergillus terreus	Extracellular	Silver	Li et al. (2012)
Cladosporium cladosporioides	Extracellular	Silver	Balaji et al. (2009)
Coriolus versicolor	Extracellular	Cadmium sulfide	Sanghi and Verma (2009)
Fusarium acuminatum	Extracellular	Silver	Ingle et al. (2008)
Fusarium semitectum	Extracellular	Silver	Basavaraja et al. (2007)
Fusarium solani	Extracellular	Silver	Ingle et al. (2009)
Fusarium oxysporum	Extracellular	Silver	Ahmad et al., 2003b
Fusarium oxysporum	Extracellular	Zirconia, silica, and titania	Bansal et al. (2004)
Fusarium oxysporum	Extracellular	Magnetite	Bharde et al. (2006)
Fusarium oxysporum f. sp. *lycopersici*	Intracellular or extracellular	Platinum	Riddin et al. (2006)
Fusarium oxysporum f. sp. *vasinfectum*	Extracellular	Silver	Joshi et al. (2013)
Hormoconis resinae	Extracellular	Gold	Mishra et al. (2010)
Neurospora crassa	Extracellular	Cobalt, silver	Rashmi et al. (2004); Castro-Longoria et al. (2011)
Penicillium brevicompactum	Extracellular	Silver	Shaligram et al. (2009)
Penicillium fellutanum	Extracellular	Silver	Kathiresan et al. (2009)
Penicillium corylophilum	Extracellular	Selenium	Salem et al. (2020)
Penicillium purpurogenum NPMF	Extracellular	Silver	Nayak et al. (2010)
Phanerochaete chrysosporium	Extracellular	Silver	Vigneshwaran et al. (2006)
Phoma sp. 3.2883	Extracellular	Silver	Chen et al. (2003)
Phoma glomerata	Extracellular	Silver	Birla et al. (2009)
Pleurotus sajor-caju	Extracellular	Silver	Nithya and Ragunathan (2009)
Pythium oligandrum	Extracellular	Silver	Noshad et al. (2019)
Thermomonospora sp.	Extracellular	Gold	Ahmad et al. (2003a)
Trichothecium sp.	Intracellular/ extracellular	Gold	Ahmad et al. (2005)
Trichoderma sp.	Extracellular	Silver	Devi et al. (2013)
Trichoderma asperellum	Extracellular	Silver	Mukherjee et al. (2008)
Usnea longissima	Extracellular	Usnic acid	Shahi and Patra (2003)
Verticillium sp.	Intracellular	Silver and gold	Sastry et al. (2003)
Verticillium luteoalbum	Extracellular	Gold	Gericke and Pinches (2006)
Verticillium sp.	Extracellular	Magnetite	Bharde et al. (2006)
Volvariella volvacea	Extracellular	Silver	Philip (2009)
Yeast	Extracellular	CdTe	Bao et al. (2010)

intracellularly. Ag NPs were synthesized by *Escherichia coli*, *B. subtilis*, *Klebsiella pneumonia*, *Pseudomonas aeruginosa*, and *the* periplasmic space of *Pseudomonas stutzeri*. *Bacillus licheniformis*, *Brevibacterium casei*, *Rhodopseudomonas capsulate*, and *Morganella* sp. have been reported to produce various types of nanoparticles (Shende et al., 2020). Ag NPs are synthesized extracellularly using *B. subtilis* (Kannan et al., 2013).

Nanoparticle synthesis by Actinomycetes

Actinomycetes also have the ability to produce extracellular and intracellular nanoparticles. In intracellular nanoparticle synthesis, nanoparticles are produced on mycelial surfaces, which occur due to the binding of metal ions to the carboxylate groups (negatively charged) of enzymes existing on the mycelia (Manimaran and Kannabiran, 2017). According to Ahmad et al. (2003b), an alkaliphilic, extremophilic *Actinomycetes*, the *Thermomonospora* sp. synthesized gold nanoparticles. Manimaran and Kannabiran (2017) reported the production of gold nanoparticles by *Streptomyces glaucus* and silver nanoparticles by *Streptomyces* sp. ERI-3. According to a previous study, enzymes present in the cell walls and plasma membrane of the *Rhodococcus* sp. reduced gold ions to produce monodisperse gold nanoparticles in a size range of 5–15 nm.

Nanotechnology and sustainable agriculture

Nanotechnology plays a vital role in controlling nutrients for increased crop productivity and also have a significant role in monitoring pesticides and water quality for sustainable agriculture. For sustainable agriculture, nanotechnology is becoming a key factor. In the agricultural sector, applications of nanotubes, nanocapsules, nanofertilizers, nanopesticides, biosensors, controlled delivery systems, nanofiltration, etc. are prevalent. Nanotechnology has proved to be efficient in nutrient and water management of agricultural fields, drug delivery mechanisms, in plants, and also in maintaining soil health and fertility. Nowadays, nanosensors are extensively being used in agricultural areas for monitoring soil and water contamination due to their strength and fastness. Nanomaterials play a key role in degradation of wastes and toxic materials and also aid in improving a microbe's efficiency in waste and toxic material degradation. Nowadays, nano–nano interactions are considered exciting phenomena for removal of wastes and toxic materials from agricultural soils, thus ensuring sustainable development.

Nanofertilizers

Fertilizers play an essential role in plant growth and development through nutrition. Around 30%–40% of applied fertilizers are utilized for crop plants, and the remaining fertilizers are lost by evaporation, degradation, or leaching. According to previous reports, out of the total applied fertilizers, around 30%–40% of nitrogen, 80%–90% of phosphorus, and 50%–90% of potassium are lost in the environment. This leads to the reduction of fertilizer efficiency and increases the cost of cultivation. Excess use of fertilizers leads to environmental pollution including soil, water, and air pollution. To enhance the efficiency of fertilizers and to overcome the problem of environmental pollution, nanofertilizers might be a key alternative. Nanofertilizers are synthesized on the basis of their controlled release of nutrients, depending on the requirement of crops. They are believed to be more efficient than conventional fertilizers (Liu, 2006). Use of nanofertilizers leads to slow release of nutrients during crop growth, and, hence, plants are able to take up most of the nutrients without any loss. Encapsulation of fertilizers within nanoparticles produces nanofertilizers. Coating of fertilizers within nanoparticle membranes is responsible for the slow release of nutrients. According to the study by Chinnamuthu and Boopathi (2009), cereals when applied with nanocomposites containing macronutrients (N, P, K) and micronutrients showed enhanced uptake of nutrients. Leggo (2000) reported that zeolites have been used in agriculture for slow release of nitrogen. Fertilizers that are nanocoated play a key role in the reduction of soil, water, and air pollution. Nanocomposites with a layer of zinc aluminum lead to controlled release of plant growth regulators (DeRosa et al., 2010). The effects of fertilizers coated with nanoparticles have been studied in wheat, and the results showed that nanocomposites have significant effects on wheat. The results also showed that 99% of wheat seeds germinated with higher emergence and enhanced seedlings (Zhang et al., 2006). According to the study by Fernandez and Eichert (2009), carbon nanotubes penetrate tomato seeds and zinc oxide nanoparticles enter the root tissues of rye grass. Eichert et al. (2008) represented the novel drug delivery system with nanoscale porous material having wide applicability for plant surfaces.

Nanobiosensors

Nanosensors are used to monitor the proficient use of water and nutrients during farming. In the detection of pests, stresses, and contamination, nanosensors play a key role. Nanosensors have ability to allow farmers utilize fertilizers and pesticides more efficiently through indication of the water and nutrient status of crops on a temporal or spatial scale. This property of nanosensors helps farmers apply nutrients, pesticides, and fungicides only where necessary.

Crop protectors

Pesticide and herbicides having less water solubility, therefore needs a better delivery system for field application. Recently, encapsulation and the method of controlled release have revolutionized the practice of pesticides and herbicides. Pesticides and herbicides with nanoparticles within the size range of 100–250 nm dissolve in water more efficiently, which increases their activity.

Nanoherbicides

Nanoherbicides (due to their small size) are able to blend with soil and eradicate weeds without leaving any residues in an eco-friendly manner. Weeds, resistant to conventional herbicides, can be eradicated using nanoherbicides. Weeds spread and survive with the help of deep roots and tubers. Plowing weeds from infected areas by hand leads to the weeds spreading to uninfected areas. A combination of nano-sized active ingredients with a smart delivery system is highly effective because it is only applied to the required agricultural field. Crop productivity negatively affects or decreases due to weed growth. The efficacy of herbicides can be increased with the application of nanotechnology for higher crop production without any toxic effects on the field.

Nanopesticides

Nanopesticides are small particles exhibiting pesticidal activities (Bergeson, 2010). Nanopesticides have the ability to enhance the dispersibility of agricultural formulations, thus reducing the movement of harmful pesticides (Bergeson, 2010). Nanoparticles have various useful features such as solubility, permeability, rigidity, crystallinity, biodegradability, and thermal stability (Bordes et al., 2009); these properties of nanoparticles are essential for formulating nanopesticides. In the agriculture sector, subcategories of nanopesticides such as nanocapsules, nanoemulsions, nanocontainers, and nanocages have also been recently used (Bouwmeester et al., 2009). Nanomaterials can degrade faster in soils. Silver has a wide application in both its compound and metal forms due to its antimicrobial activity against pathogens or its nontoxicity toward humans (Elchiguerra et al., 2005). Nowadays, nanotechnology can be used to increase the efficiency of Ag NPs. Silver nanoparticles (Ag NPs) increase their microbial contact due to their large surface-to-volume ratio and also have the ability to permeate (Elchiguerra et al., 2005). Woo et al. (2009) evaluated the application of AgNPs in vitro aganist *Raffaelea* sp. a causal organism of oak wilt and found that AgNPs inhibited the growth of fungal hyphae and restricted the conidia germination. Kim et al. (2008) reported that Ag NPs inhibited *Rhizoctonia solani*, *Scleractinia*, and *S. minor* growth in vitro.

Applications of nanomaterials in disease management

Plant pathologists are working on solutions to protect crops from various diseases caused by bacteria, fungi, viruses, and nematodes. Nanoparticles have been prepared and can be used as pesticides, fungicides, herbicides and fertilizers more efficiently (Rai and Ingle, 2012). According to several studies, metallic nanoparticles are effective against various plant pathogens, pests, and insects. Eco-friendly nanofungicides act only inside the targeted pathogen by releasing their pathogen-killing properties (Liu, 2006). Nanomaterials exhibit antimicrobial activity, and silver nanomaterials have both antifungal and antibacterial properties. Kim et al. (2008) reported that nano silver showed fungicidal activities against various fungi including *Alternaria solani*, *A. alternata*, *A. brassicicola*, *Botrytis cinerea*, *Corynespora cassiicola*, *Cylindrocarpon destructans*, *Cladosporium cucumerinum*, *Didymella bryoniae*, *F. oxysporum*, *F. oxysporum* f. sp. *lycopersici*, *Fusarium oxysporum* f. sp. *cucumerinum*, *F. solani*, *Fusarium* sp., *Glomerella cingulata*, and a few others. Silver nanoparticles showed antimicrobial activity against *Bacillus megaterium*, *Erwinia* sp., *Fusarium graminearum*, *F. avenaceum*, *F. culmorum*, and *Pseudomonas syringae* (Gautam et al., 2020). According to the study by Mohanta et al. (2018), silver nanoparticles exhibited antibacterial activity against *P. aeruginosa*, *B. subtilis*, and *E. coli*. Ag NPs exerted antagonistic effects against *P. syringae* pv. *syringae* (Shahryari et al., 2020). According to the study by Hussein et al. (2019), silver nanoparticles exhibited antibacterial activity against *Staphylococcus aureus* and *Klebsiella pneumonia*.

Silver nanoparticles also showed antagonistic effects against various fungi including *Macrophomina phaseolina*, *Botrytis cinerea*, *Alternaria alternata*, *Rhizoctonia solani*, *Curvularia lunata*, and *Sclerotinia sclerotiorum* (Krishnaraj et al., 2012). According to Xing et al. (2016), nanoparticles of chitosan exerted antagonistic effects against *Aspergillus niger* and *Fusarium solani*. Divya et al. (2017) also reported the antagonistic activity of chitosan nanoparticles against *A. alternata* and *Macrophomina phaseolina*. According to Attar and Yapaoz (2018), ZnO and Au NPs exhibited antimicrobial activity against *E. coli*. Gold nanoparticles are reported to exert inhibitory effects against *Salmonella typhimurium*. Copper composites exert antibacterial activity against *Xanthomonas euvesicatoria*. Copper nanoparticles (CuO NPs) exhibited antifungal activity against *Penicillium digitatum*, *Magnaporthe oryzae*, *Botrytis cinerea*, *Rhizoctonia solani*, *Sclerotium rolfsii*, *Colletotrichum musae*, and *Colletotrichum graminicola* (Huang et al., 2015). According to the study by Sharma et al. (2015), MgO NPs exerted toxic effects against *Ralstonia solanacearum* and *Phomopsis vexans*. Derbalah et al. (2018) studied the antagonistic effects of silica nanoparticles on *Alternaria solani*. Akpinar et al. (2017) also found that SiO_2 nanoparticles possess antifungal properties against *F. oxysporum* f. sp. *radicis-lycopersici* and *Fusarium oxysporum* f. sp. *lycopersici*.

Si–Ag nanoparticles also exhibit antifungal activity against *P. syringae*, *Botrytis cinerea*, *Pythium ultimum*, *Colletotrichum gloeosporioides*, *Magnaporthe grisea*, *Rhizoctonia solani*, and *Xanthomonas campestris* pv. *vesicatoria*. ZnO NPs exhibit antifungal activity against *Aspergillus niger*, *A. alternata*, *Botrytis cinerea*, *Fusarium oxysporum*, and *Penicillium expansum* (Jamdagni et al., 2016). Navale et al. (2015) also found that ZnO NPs exhibited antimicrobial activity against *Aspergillus flavus* and *Aspergillus fumigates*. Rajiv et al. (2018) also studied the antifungal activity of ZnO NPs against *Aspergillus flavus*, *A. fumigates*, *A. niger*, *F. oxysporium*, and *Fusarium culmorum*. According to the study by Dimkpa et al. (2013), ZnO nanoparticles showed inhibitory effects against *Fusarium graminearum*.

Hamza et al. (2016) studied the antagonistic activity of TiO_2 nanoparticles against *Cercospora beticola*. TiO_2 exhibited inhibitory effects against *Fusarium oxysporum* f. sp. *Lycopersici* and *Fusarium oxysporum* f. sp. *radicis-lycopersici* (Sar et al., 2017). TiO_2 NPs showed nematicidal activity against *Meloidogyne incognita* (Ardakani, 2013). Cui et al. (2009) also studied the antibacterial activity of TiO_2 nanoparticles against *P. syringae* pv. *lachrymans* and *P. cubensis*.

Effects of nanoparticles on seed germination and plant growth

Nanomaterial application in seed germination and plant growth has been widely studied. According to the study by Zheng et al. (2005), utilization of TiO_2 nanoparticles on spinach increases its dry weight by 75%, photosynthetic rate by threefold, and chlorophyll content by 45% as compared to control plants over 30 days; this is due to enhanced nutrient absorption after application of TiO_2. Penetration of seeds by nanoparticles is the key mechanism for enhanced germination. Shah and Belozerova. (2009) reported that nanoparticles having a low concentration of palladium and gold and a higher concentration of silica, copper, and a combination gold and copper had significant effects on seed germination, plant growth, and increased the shoot–root ratio.

Nanotechnology in plant resistance

Plants are susceptible to various diseases caused by bacteria, fungi, viruses, and nematodes and various abiotic stresses such as drought, salinity, and heat stress. Nanotechnology plays a key role in the improvement of plant resistance using nanoparticles, nanocapsules, and nanofibers for gene multiplication, thus enhancing the resistance of plants (McKnight et al., 2003). A successful integration of plasmid DNA has been established through gene expression (Filipenko et al., 2007). Using nanotechnology, resistance genes can be introduced into a plant cell and thus form resistant varieties, which minimize the expense of agrochemicals used for disease control (Bouwmeester et al., 2009).

Conclusions and future perspectives

Nanotechnology has emerged as one of the promising technologies of the 21st century. It is now gradually being incorporated into the crop industry sector. Microbial nanotechnology possesses great potential to enhance agricultural yield by permitting the healthier management and conservation of inputs to the agricultural field. Nanoparticles can be synthesized by chemical, physical, and biological methods. Among these methods, the biological methods of nanoparticle synthesis are preferred due to their eco-friendly, nontoxic nature and also because these methods are cheaper as compared to other methods of NP synthesis. Recently, there has been more focus on the use of microbes such as fungi, bacteria, algae, and *Actinomycetes* for nanoparticle synthesis, which represents the green approach for nanoparticle biosynthesis. Fungi and bacteria represent the rapid and environmentally clean nanobiofactories for metallic nanoparticle synthesis. Nanomaterials have been shown to enhance seed germination, plant growth, and thus crop yield. They are also used for the detection and treatment of various plant diseases. Nanomaterials are also used for plant protection by acting as nanopesticides and nanoherbicides. Nanopesticides have a large surface area, which is responsible for increasing the affinity to a target. Various kinds of nanopesticides including nanoencapsulates, nanoemulsions, nanocontainers, and nanocages are available. Nanomaterials, when applied to agricultural fields, enhance the absorption of nutrients by plants, having great potential in the delivery of active ingredients to the target sites. Microbial nanotechnology is a fascinating and booming field for breakthrough nanoparticle synthesis. It exhibits great potential in agriculture, and this exciting area requires more analysis to understand the working mechanisms of nanoparticles to reap all the benefits of nanoparticles.

References

Ahmad, A., Senapati, S., Khan I., M., Kumar, S., Sastry, M., 2003a. Extracellular biosynthesis of monodisperse gold nanoparticle by a novel extremophilic Actinomycete, *Thermomonospora* sp. Langmuir 19 (8), 3550–3553.

Ahmad, A., Mukherjee, P., Senapati, S., Mandal, D., Khan, M.I., Kumar, R., Sastry, M., 2003b. Extracellular biosynthesis of silver nanoparticles using the fungus *fusarium oxysporum*. Colloids Surf. B: Biointerfaces 28 (4), 313–318.

Ahmad, A., Senapati, S., Khan I., M., Kumar, R., Sastry, M., et al., 2005. Extra-/ Intracellular biosyntheis of gold nanoparticles by an alkotolerant fungus *Trichothecium* sp. J. Biomed. Nanotechnol. 1 (1), 47–53.

Akpinar, I., Sar, T., Unal, M., 2017. Antifungal effects of silicon dioxide nanoparticles (SiO_2 NPs) against various plant pathogenic fungi. In: Tar, M., Bastas, K.K. (Eds.), International Workshop: Plant Health: Challenges and Solutions. Frontiers Media SA, Antalya, Turkey.

Ardakani, A.S., 2013. Toxicity of silver, titanium and silicon nanoparticles on the root-knot nematode, *Meloidogyne incognita*, and growth parameters of tomato. Nematology 15 (6), 671–677.

Attar, A., Yapaoz, M.A., 2018. Biomimetic synthesis, characterization and antibacterial efficacy of ZnO and au nanoparticles using echinacea flower extract precursor. Mater. Res. Exp. 5 (5), 055403.

Balaji, D.S., Basavaraja, S., Deshpande, R., Mahesh, D.B., Prabhakar, B.K., Venkataraman, A., 2009. Extracellular biosynthesis of functionalized silver nanoparticles by strains of *Cladosporium cladosporioides* fungus. Colloids Surf. B: Biointerfaces 68 (1), 88–92.

Bansal, V., Rautaray, D., Ahmad, A., Sastry, M., 2004. Biosynthesis of zirconia nanoparticles using the fungus fusarium oxysporum. J. Mater. Chem. 14 (22), 3303–3305.

Banu, A., Rathod, V., Ranganath, E., 2011. Silver nanoparticle production by *Rhizopus stolonifer* and its antibacterial activity against extended spectrum β-lactamase producing (ESBL) strains of enterobacteriaceae. Mater. Res. Bull. 46 (9), 1417–1423.

Bao, H., Hao, N., Yang, Y., Zhao, D., 2010. Biosynthesis of biocompatible cadmium telluride quantum dots using yeast cells. Nano Res. 3 (7), 481–489.

Basavaraja, S., Balaji, S., Lagashetty, A., Rajasab, A., Venkataraman, A., 2007. Extracellular biosynthesis of silver nanoparticles using the fungus *fusarium semitectum*. Mater. Res. Bull. 43 (5), 1164–1170.

Bergeson, L.L., 2010. Nanosilver pesticide products: what does the future hold? Environ. Qual. Manag. 19 (4), 73–82.

Beveridge, T.J., Murray, R.G., 1980. Sites of metal deposition in the cell wall of Bacillus subtilis. J. Bacteriol. 141 (2), 876–887.

Bhainsa, K.C., D'souza, S.F., 2006. Extracellular biosynthesis of silver nanoparticles using the fungus aspergillus fumigatus. Colloids Surf. B: Biointerfaces 47 (2), 160–164.

Bharde, A., Rautaray, D., Bansal, V., Ahmad, A., Sarkar, I., Yusuf, S.M., Sastry, M., 2006. Extracellular biosynthesis of magnetite using fungi. Small 2 (1), 135–141.

Birla, S.S., Tiwari, V.V., Gade, A.K., Ingle, A.P., Yadav, A.P., Rai, M.K., 2009. Fabrication of silver nanoparticles by *Phoma glomerata* and its combined effect against *Escherichia coli, Pseudomonas aeruginosa* and *Staphylococcus aureus.* Lett. Appl. Microbiol. 48, 173–179.

Bordes, P., Pollet, E., Averous, L., 2009. Nano-biocomposites: biodegradable polyester/nanoclay systems. Prog. Polym. Sci. 34 (2), 125–155.

Bouwmeester, H., Dekkers, S., Noordam, M.Y., Hagens, W.I., Bulder, A.S., De Heer, C., Sips, A.J., 2009. Review of health safety aspects of nanotechnologies in food production. Regul. Toxicol. Pharmacol. 53 (1), 52–62.

Brock, D.A., Douglsa, T.E., Queller, D.C., Strassmann, J.E., 2011. Primitive agriculture in social amoeba. Nature 469, 393–396.

Castro-Longoria, E., Vilchis-Nestor, A.R., Avalos-Borja, M., 2011. Biosynthesis of silver, gold and bimetallic nanoparticles using the filamentous fungus Neurospora crassa. Colloids Surf. B: Biointerfaces 83 (1), 42–48.

Chen, J.C., Lin, Z.H., Ma, X.X., 2003. Evidence of the production of silver nanoparticles via pretreatment of Phoma sp. 3.2883 with silver nitrate. Lett. Appl. Microbiol. 37 (2), 105–108.

Chinnamuthu, C.R., Boopathi, P.M., 2009. Nanotechnology and agroecosystem. Madras Agric. J. 96, 17–31.

Cui, H., Zhang, P., Gu, W., Jiang, J., 2009. Application of anatasa TiO_2 sol derived from peroxotitannic acid in crop diseases control and growth regulation. NSTI Nanotechnol. 2, 286–289.

Deepa, K., Panda, T., 2014. Synthesis of gold nanoparticles from different cellular fractions of *fusarium oxysporum*. J. Nanosci. Nanotechnol. 14, 3455–3463.

Derbalah, A., Shenashen, M., Hamza, A., Mohamed, A., El Safty, S., 2018. Antifungal activity of fabricated mesoporous silica nanoparticles against early blight of tomato. Egypt. J. Basic Appl. Sci. 5, 145–150.

DeRosa, M.R., Monreal, C., Schnitzer, M., Walsh, R., Sultan, Y., 2010. Nanotechnology in fertilizers. Nat. Nanotechnol. 5 (2), 91.

Devi, T.P., Kulanthaivel, S., Kamil, D., Borah, J.L., Prabhakaran, N., Srinivasa, N., 2013. Biosynthesis of silver nanoparticles from *Trichoderma* species. Indian J. Exp. Biol. 51, 543–547.

Dhillon, G.S., Brar, S.K., Kaur, S., Verma, M., 2012. Green approach for nanoparticle biosynthesis by fungi: current trends and applications. Crit. Rev. Biotechnol. 32 (1), 49–73.

Dimkpa, C.O., McLean, J.E., Britt, D.W., Anderson, A.J., 2013. Antifungal activity of ZnO nanoparticles and their interactive effect with a biocontrol bacterium on growth antagonism of the plant pathogen *fusarium graminearum.* Biometals 26, 913–924.

Divya, K., Vijayan, S., George, T.K., Jisha, M., 2017. Antimicrobial properties of chitosan nanoparticles: mode of action and factors affecting activity. Fibers Polym. 18, 221–230.

Duran, N., Marcato, P.D., Alves, O.L., DeSouza, G., Esposito, E., 2005. Mechanistic aspects of biosynthesis of silver nanoparticles by several *fusarium oxysporum* strains. J. Nanobiotechnol. 3, 8.

Eichert, T., Kurtz, A., Steiner, U., Goldbach, H.E., 2008. Size exclusion limits and lateral heterogeneity of the stomatal foliar uptake pathway for aqueous solutes and water-suspended nanoparticles. Plant Physiol. 134, 151–160.

El-Rafie, H., El-Rafie, M., Zahran, M., 2013. Green synthesis of silver nanoparticles using polysaccharides extracted from marine macro algae. Carbohydr. Polym. 96 (2), 403–410.

El-Sayed, A., Kamel, M., 2020. Advances in nanomedical applications: diagnostic, therapeutic, immunization, and vaccine production. Environ. Sci. Pollut. Res. 27, 19200–19213.

Elchiguerra, J.L., Burt, J.L., Morones, J.R., Camacho-Bragado, A., Gao, X., Lara, H.H., Yacaman, M.J., 2005. Interaction of silver nanoparticles with HIV-1. J. Nanobiotechnol. 3 (1), 1–10.

Fernandez, V., Eichert, T., 2009. Uptake of hydrophilic solutes through plant leaves: current state of knowledge and perspectives of foliar fertilization. Crit. Rev. Plant Sci. 28, 36–68.

Filipenko, E.A., Filipenko, M.L., Deineko, E.V., Shumnyi, V.K., 2007. Analysis of integration sites of T-DNA insertions in transgenic tobacco plants. Cytol. Genet. 41, 199–203.

Gaikwad, S., Birla, S.S., Ingle, A.P., Gade, A.K., Marcato, P.D., Rai, M.K., Duran, D., 2013. Screening of different *fusarium* species to select potential species for the synthesis of silver nanoparticles. J. Braz. Chem. Soc. 24, 1974–1982.

Gajbhiye, M., Kesharwani, J., Ingle, A., Gade, A., Rai, M., 2009. Fungus mediated synthesis of silver nanoparticles and its activity against pathogenic fungi in combination of fluconazole. Nanomedicine 5 (4), 282–286.

Gautam, N., Salaria, N., Thakur, K., Kukreja, S., Yadav, N., Yadav, R., Goutam, U., 2020. Green silver nanoparticles for phytopathogen control. Proc. Natl. Acad. Sci., India, Sect. B Biol. Sci. 90, 439–446.

Gericke, M., Pinches, A., 2006. Biological synthesis of metal nanoparticles. Hydrometallurgy 83, 132–140.

Govindaraju, K., Kiruthiga, V., Kumar, V.G., Singaravelu, G., 2009. Extracellular synthesis of silver nanoparticles by a marine alga, *Sargassum wightii* Grevilli and their antibacterial effects. J. Nanosci. Nanotechnol. 9 (9), 5497–5501.

Hamza, A., El-Mogazy, S., Derbalah, A., 2016. Fenton reagent and titanium dioxide nanoparticles as antifungal agents to control leaf spot of sugar beet under field conditions. J. Plant Protect. Res. 56, 270–278.

Huang, S., Wang, L., Liu, L., Hou, Y., Li, L., 2015. Nanotechnology in agriculture, livestock, and aquaculture in China. A review. Agron. Sustain. Dev. 35, 369–400.

Hussein, E.A.M., Al-Hajry, A.M., Harraz, F.A., Ahsan, M.F., 2019. Biologically synthesized silver nanoparticles for enhancing tetracycline activity against *Staphylococcus aureus* and *Klebsiella pneumonia*. Braz. Arch. Biol. Technol. 62, 25.

Ingle, A., Gade, A., Pierrat, S., Sonnichsen, C., Rai, M., 2008. Mycosynthesis of silver nanoparticles using the fungus *fusarium acuminatum* and its activity against some human pathogenic bacteria. Curr. Nanosci. 4, 141–144.

Ingle, A., Gade, A., Bawaskar, M., Rai, M., 2009. *Fusarium solani*, a novel biological agent for the extracellular synthesis of silver nanoparticles. J. Nanopart. Res. 11, 2079–2085.

Jamdagni, P., Khatri, P., Rana, J.S., 2016. Green synthesis of zinc oxide nanoparticles using flower extract of *Nyctanthes arbor-tristis* and their antifungal activity. J. King Saud Univ. Sci. 30 (2), 168–175.

Joshi, P., Bonde, S., Gaikwad, S., Gade, A., Abd-Elsalam, K.A., Rai, M., 2013. Comparative studies on synthesis of silver nanoparticles by *fusarium oxysporum* and *Macrophomina phaseolina* and its efficacy against bacteria and *Malassezia furfur*. J. Bionanosci. 7, 15.

Kannan, R.R.R., Arumugam, R., Ramya, D., Manivannan, K., Anantharaman, P., 2013. Green synthesis of silver nanoparticles using marine macroalga *Chaetomorpha linum*. Appl. Nanosci. 3 (3), 229–233.

Kathiraven, T., Sundaramanickam, A., Shanmugam, N., Balasubramanian, T., 2015. Green synthesis of silver nanoparticles using marine algae *Caulerpa racemosa* and their antibacterial activity against some human pathogens. Appl. Nanosci. 5 (4), 499–504.

Kathiresan, K., Manivannan, S., Nabeel, M.A., Dhivya, B., 2009. Studies on silver nanoparticles synthesized by a marine fungus, Penicillium fellutanum isolated from coastal mangrove sediment. Colloids Surf. B: Biointerfaces 71 (1), 133–137.

Khan, A.U., Malik, N., Khan, M., Cho, M.H., Khan, M.M., 2018. Fungi assisted silver nanoparticle synthesis and their applications. Bioprocess Biosyst. Eng. 41 (1), 1–20.

Kim, H., Kang, H., Chu, G., Byun, H., 2008. Antifungal effectiveness of nanosilver colloid against rose powdery mildew in greenhouses. Solid State Phenom. 135, 15–18.

Klaus-Joerger, T., Joerger, R., Olsson, E., Granqvist, C.G., 2001. Bacteria as workers in the living factory: metal-accumulating bacteria and their potential for materials science. Trends Biotechnol. 19 (1), 15–20.

Kour, D., Rana, K.L., Yadav, A.N., Yadav, N., Kumar, M., Kumar, V., Vyas, P., Dhaliwal, H.S., Saxena, A.K., 2020. Microbial biofertilizers: bioresources and eco-friendly technologies for agricultural and environmental sustainability. Biocatal. Agric. Biotechnol. 23, 101487.

Krishnaraj, C., Ramachandran, R., Mohan, K., Kalaichelvan, P.T., 2012. Optimization for rapid synthesis of silver nanoparticles and its effect on phytopathogenic fungi. Spectrochim. Acta A Mol. Biomol. Spectrosc. 93, 95–99.

Kumar, S.A., Abyaneh, M.K., Gosavi, S.W., Kulkarni, S.K., Pasricha, R., Ahmad, A., Khan, M.I., 2007. Nitrate reductase-mediated synthesis of silver nanoparticles from $AgNO_3$. Biotechnol. Lett. 29 (3), 439–445.

Kumar, R.R., Priyadharsani, P.K., Thamaraiselvi, K., 2012. Mycogenic synthesis of silver nanoparticles by the Japanese environmental isolate *aspergillus tamari*. J. Nanopart. Res. 14, 860–868.

Kuppusamy, P., Yusoff, M.M., Maniam, G.P., Govindan, N., 2016. Biosynthesis of metallic nanoparticles using plant derivatives and their avenues in pharmacological applications – an updated report. Saudi Pharm. J. 24 (4), 473–484.

Leggo, P.J., 2000. An investigation of plant growth in an organo-zeolitic substrate and its ecological significance. Plant Soil 219 (1), 135–146.

Li, G., He, D., Qian, Y., Guan, B., Gao, S., Cui, Y., Yokoyama, K., Wang, L., 2012. Fungus-mediated green synthesis of silver nanoparticles using *aspergillus terreus*. Int. J. Mol. Sci. 13, 466–476.

Liu, W.T., 2006. Nanoparticles and their biological and environmental applications. J. Biosci. Bioeng. 102, 107.

Manimaran, M., Kannabiran, K., 2017. Actinomycetes- mediated biogenic synthesis of metal and metal oxides nanoparticles nanoparticles: progress and challenges. Lett. Appl. Microbiol. 64 (6), 401–408.

McKnight, T.E., Melechko, A.V., Griffin, G.D., Guillorn, M.A., Merkulov, V.I., Serna, F., Hensley, D.K., Doktycz, M.J., Lowndes, D.H., Simpson, M.L., 2003. Intracellular integration of synthetic nanostructures with viable cells for controlled biochemical manipulation. Nanotechnology 14, 551–556.

Meyer, V., 2008. Genetic engineering of filamentous fungi progress, obstacles and future trends. Biotechnol. Adv. 26, 177–185.

Mishra, A.N., Bhadauria, S., Gaur, M.S., Pasricha, R., 2010. Extracellular microbial synthesis of gold nanoparticles using fungus *Hormoconis resinae*. J. Metals 62, 45–48.

Mohammad Fayaz, A., Balaji, K., Kalaichelvan, P.T., Venkateson, R., 2009. Fungal based synthesis of silver nanoparticles – an effect of temperature on the size of particles. Colloids Surf. B Biointerfaces 74 (1), 123–126.

Mohanpuria, P., Rana, N.K., Yadav, S.K., 2007. Biosynthesis of nanoparticles, technological concepts and future applications. J. Nanopart. Res. 7, 9275–9280.

Mohanta, Y.K., Nayak, D., Biswas, K.S., Abd-Allah, E.F., Hashem, A., Alqarawi, A.A., Yadav, D., Mohanta, T.P., 2018. Silver nanoparticles synthesized using wild mushroom show potential antimicrobial activities against food borne pathogens. Molecules 23 (3), 655.

Mukherjee, P., Roy, M., Mandal, B.P., Dey, G.K., Mukherjee, P.K., Ghatak, J., Tyagi, A.K., Kale, S.P., 2008. Green synthesis of highly stabilized nanocrystalline silver particles by a non- pathogenic and agriculturally important fungus *Trichoderma asperellum*. Nano 19, 075103.

Narayanan, K.B., 2010. Biological synthesis of metal nanoparticle by microbes. Adv. Colloid Interf. Sci. 156 (1), 1–13.

Navale, G.R., Thripuranthaka, M., Late, D.J., Shinde, S.S., 2015. Antimicrobial activity of ZnO nanoparticles against pathogenic bacteria and fungi. JSM Nanotechnol. Nanomed. 3, 10–33.

Nayak, R.R., Pradhan, N., Behera, D., Pradhan, K.M., Mishra, S., Sukla, L.B., Mishra, B.K., 2010. Green synthesis of silver nanoparticle by *Penicillium purpurogenum* NPMF, the process and optimization. J. Nanopart. Res. 13, 3129–3137.

Nithya, R., Ragunathan, R., 2009. Synthesis of silver nanoparticle using *Pleurotus sajor caju* and its antimicrobial study. Dig. J. Nanomater. Biostruct. 4, 623–629.

Noshad, A., Iqbal, M., Hetherington, C., Wahab, H., 2019. Biogenic AgNPs—a nano weapon against bacterial canker of tomato (BCT). Hindawi Adv. Agric. 2020.

Philip, D., 2009. Biosynthesis of au, ag and AuAg nanoparticles using edible mushroom extract. Spectrochim. Acta A 73, 374–381.

Prasad, T.N., Kambala, V.S.R., Naidu, R., 2013. Phyconanotechnology: synthesis of silver nanoparticles using brown marine algae *Cystophora moniliformis* and their characterisation. J. Appl. Phycol. 25 (1), 177–182.

Rai, M., Ingle, A., 2012. Role of nanotechnology in agriculture with special reference to management of insect pests. Appl. Microbiol. Biotechnol. 94, 287–293.

Rai, M., Yadav, A., Gade, A., 2009. Silver nanoparticles as a new generation of antimicrobials. Biotechnol. Adv. 27, 76–83.

Rai, M., Bonde, S., Golinska, P., Trzcinska-Wencel, J., Gade, A., Abd-Elsalam, K., A., Shende, S., Gaikwad, S., & Ingle, A. P., 2021. *Fusarium* as a novel fungus for the synthesis of nanoparticles: mechanism and applications. J. Fungi (Basel, Switz.) 7 (2), 139.

Rajiv, P., Vanathi, P., Thangmani, A., 2018. An investigation of using *Eicchornia* mediated zinc oxide nanoparticles on *Helianthus annus*. Biocatal. Agric. Biotechnol. 16, 419–424.

Raliya, R., Tarafdar, J.C., 2014. Biosynthesis and characterization of zinc, magnesium and titanium nanoparticles: an eco-friendly approach. Int. Nano Lett. 93, 310.

Rane, A.N., Baikar, V.V., Ravi kumar, V., & Deopurkar, R. L., 2017. Agro-industrial wastes for production of biosurfactant by *Bacillus subtilis* ANR 88 and its application in synthesis of silver and gold nanoparticles. Front. Microbiol. 8, 492.

Rashmi, K., Krishnaveni, T., Ramanamurthy, S., Mohan, P.M., 2004. Characterization of cobalt nanoparticle from a cobalt resistant strain of *Neurospora crassa*. In: International Symposium of Research Students on Materials Science and Engineering; December 2022. Chennai.

Riddin, T., Gericke, M., Whiteley, C., 2006. Analysis of inter and extracellular formation of platinum nanoparticles by *fusarium oxysporum* f. sp. *lycopersci* using response surface methodology. Nanotechnology 17, 3482–3489.

Salaam, A., Adebayo-Tayo, B., Ajibade, A., 2018. Phycosynthesis of silver nanoparticles using *Chlorella vulgaris* metabolites: its antibacterial, anti-biofilm and in-vitro cytotoxicity potential and effect of optimized conditions on biosynthesis. Afr. J. Biomed. Res. 23, 17–23.

Salem, S., Moustafa, M.G., Fouda, A., Mohamed, A., Ebtesam, M.A.-O., Allam, A.A., Shaheen, T.I., 2020. Antibacterial, cytotoxicity and larvicidal activity of green synthesized selenium nanoparticles using *Penicillium corylophilum*. J. Clust. Sci. 32 (2), 351–361.

Sanghi, R., Verma, P.A., 2009. A facile green extracellular biosynthesis of CdS nanoparticles by immobilized fungus. Chem. Eng. J. 155, 886–891.

Sar, T., Akpinar, I., Unal, M., 2017. Inhibitory effect of antifungal activity of titanium dioxide (TiO2) nanoparticles on some pathogenic *fusarium* isolates. In: Proceedings of the International Workshop Plant Health: Challenges and Solutions, Antalya, Turkey, 23–28 April 2017, p. 78.

Saravanan, M., Arokiyaraj, S., Lakshmi, T., Pugazhendhi, A., 2018. Synthesis of silver nanoparticles from *Phenerochaete chrysosporium* (MTCC-787) and their antibacterial activity against human pathogenic bacteria. Microb. Pathog. 117, 68–72.

Sastry, M., Ahmad, A., Khan, M.I., Kumar, R., 2003. Biosynthesis of metal nanoparticles using fungi and actinomycete. Curr. Sci. 85, 162–170.

Shah, V., Belozerova., 2009. Influence of metal nanoparticles on the soil microbial community and germination of lettuce seeds. Water Air Soil Pollut. 197 (1), 143–148.

Shahi, S.K., Patra, M., 2003. Biotechnological aspect for the synthesis of bioactive nanoparticle and their formulation active against human pathogenic fungi. Rev. Adv. Mater. Sci. 5, 501–509.

Shahryari, F., Rabiei, Z., Sadighian, S., 2020. Antibacterial activity of synthesized silver nanoparticles by sumac aqueous extract and silver-chitosan nanocomposite against *Pseudomonas syringae* pv. *Syringae*. J. Plant Pathol. 102, 469–475.

Shakiba, S., Astete, C.E., Paudel, S., Sabliov, C.M., Rodrigues, D.F., Louie, S.M., 2020. Emerging investigator series: polymeric nanocarriers for agricultural applications: synthesis, characterization, and environmental and biological interactions. Environ. Sci. Nano 7, 37–67.

Shaligram, N.S., Bule, M., Bhambure, R., Singhal, R.S., Singh, S.K., Szakacs, G., Pandey, A., 2009. Biosynthesis of silver nanoparticles using aqueous extract from the compactin producing fungal strain. Process Biochem. 44, 939–943.

Sharma, N., Jandaik, S., Kumar, S., Chitkara, M., Sandhu, I.S., 2015. Synthesis, characterisation and antimicrobial activity of manganese and iron-doped zinc oxide nanoparticles. J. Exp. Nanosci. 11, 54–71.

Shende, S., Rajput, V., Gade, A., Minikana, T., Rai, M., Singh, S.P., 2020. Impact of nanomaterials on human health through/via food chain. In: Intellectual Property Issues in Nanotechnology. CRC Press.

UNFAO, 2018. RUAF and Laurier Centre for Sustainable Food Systems. https://ruaf.org/partner/lcsfs-canada/.

United Nations, 2014. Questions about nanotechnology. Aug. 21.2012. Available from https://www.epa.gov./chemical-research-/research-nanomaterials.

Verma, V.C., Kharwar, R.N., Gange, A.C., 2010. Biosynthesis of antimicrobial silver nanoparticles by the endophytic fungus *aspergillus clavatus*. Nanomedicine 5, 3340.

Vigneshwaran, N., Kathe, A.A., Varadarajan, P.V., Nachane, R.P., Balasubramanya, R.H., 2006. Biomimetics of silver nanoparticles by white rot fungus, *Phaenerochaete chrysosporium*. Colloids Surf., B 53, 55–59.

Wang, J., Chen, C., 2009. Biosorbents for heavy metals removal and their future. Biotechnol. Adv. 7, 195–226.

Woo, K.S., Kim, K.S., Lamsal, K., Kim, Y.J., Kim, S.B., Jung, M., Sim, S.J., Kim, H.S., Chang, S.J., Kim, J.K., Lee, Y.S., 2009. An in vitro study of the antifungal effect of silver nanoparticles on oak wilt pathogen *Raffaelea* sp. J. Microbiol. Biotechnol. 19, 760–764.

Xing, K., Shen, X., Zhu, X., Ju, X., Miao, X., Tian, J., Feng, Z., Peng, X., Jiang, J., Qin, S., 2016. Synthesis and in vitro antifungal efficacy of oleoyl-chitosan nanoparticles against plant pathogenic fungi. Int. J. Biol. Macromol. 82, 830–836.

Zhang, F., Wang, R., Xiao, Q., Wang, Y., Zhang, J., 2006. Effects of slow/controlled-release fertilizer cemented and coated by nano-materials on biology. II effects of slow/controlled-release fertilizer cemented and coated by nano-materials on plants. Nano 11, 18–26.

Zheng, L., Hong, F., Lu, S., Liu, C., 2005. Effect of nano-TiO_2 on strength of naturally aged seeds and growth of spinach. Biol. Trace Elem. Res. 104, 83–91.

CHAPTER 11

Microbial cross talk: Below and above ground

Sandhya Bind[a], Sudha Bind[a], Anand Kumar[a,b], and Dinesh Chandra[a,c]

[a]Department of Biological Sciences, College of Basic Sciences and Humanities, Govind Ballabh Pant University of Agriculture and Technology Pantnagar, US Nagar, Uttarakhand, India, [b]Department of Botany, D.S.B. Campus, Kumaun University, Nainital, Uttarakhand, India, cGIC Chamtola, Almora, Uttarakhand, India

Chapter outline

Introduction

Plant growth and development depends on accurate cell division, cell expansion, and cell differentiation. Soil is a region below the earth's surface composed of both living and nonliving factors. Biotic factors such as plant roots and microbial communities enhance soil fertility by increasing organic matter and by regulating the pH and other properties of soil. A soil microbiome is considered to be the microbial seed bank of a rhizospheric microbiome (Lennon and Jones, 2011). The physicochemical properties of soils such as soil moisture, pH, nutrient status, and temperature shape microbial communities and also determine the assembly of rhizospheric microbial communities. Anthropogenic activities like agricultural

Unravelling Plant-Microbe Synergy. https://doi.org/10.1016/B978-0-323-99896-3.00006-0

management practices intensively affect the physicochemical properties of soils, which have a direct impact on rhizospheric microbial communities (Lekberg et al., 2021). Research studies based on amplicon sequencing of rhizospheric microbial communities have revealed that soil type is considered the primary driving force for influencing the structure of bacterial communities in the rhizosphere, though for fungal communities, climate plays a key role (Lundberg et al., 2012; Thiergart et al., 2020).

Plants provide space and nutrients for the growth and proliferation of various microbial communities such as bacteria, fungi, nematodes, viruses, algae, and protists (Lundberg et al., 2012; Coleman-Derr et al., 2016). These microbes play a vital role in crop health and productivity by establishing a complex interaction with the host plant (Chandra et al., 2020; Chandra and Sharma, 2020; Fitzpatrick et al., 2018). Specifically, plants establish a 'holobiont', which is an association of microbes with plants and evolutionary selection between microbes and plants contributes to stability of ecosystem (Hamonts et al., 2018; Xu et al., 2018). The shoot-to-root signal communication encounters different abiotic and biotic stress factors. Plant roots produce root exudates, which are composed of organic acids, vitamins, sugars, and phenolic compounds used by microbial populations as nutrients and for signaling. In order to do this, microbes produce different kinds of volatile compounds and phytohormones that activate the immune system of the plant or help in controlling the morphogenesis and plant growth (Ortiz-Castro et al., 2009). Plants compete with other plants and microbes for space, nutrients, and water (Baetz and Martinoia, 2013). Interactions of microbes with plants might be advantageous, deleterious, or neutral. One gram of rhizospheric soil is reported to contain 10^{11} microbial cells. Microbial populations affect plant interactions with the soil environment (Mendes et al., 2013). Plant roots having a high population of microbial communities are generally referred to as the root microbiome (Hacquard et al., 2015). Plants have the potential to shape the rhizospheric microbiome structure by attracting beneficial microbes such as fungi and bacteria through secretion of root exudates (Berendsen et al., 2012). The host genotype is a key factor influencing the shape, composition, and pattern of a microbiome (Bulgarelli et al., 2015). Interactions between plant roots and rhizospheric microbes enhance plant growth and development (Panke-Buisse et al., 2015). Plant roots directly influence the microbial communities in close proximity to them through secretion of root exudates, which promotes beneficial microbes or might restrict other competitive species (Nardi et al., 2000). Establishment of a microbial community and structure in rhizospheric soil depends on parameters such as soil type and soil pH, which brings about changes in the composition of the microbial community or may cause alteration in their functions, more precisely in the rhizosphere (Cristina et al., 2017). Rhizobacteria in the soil may contribute to plant growth and development by providing growth-promoting substances, thus helping make the soil structure firm, releasing nutrients from organic compounds, and also forming a symbiotic association with plant roots. Bacteria and AMF play a vital role in the maintenance and improvement of soil structure, soil health, and soil fertility (Babalola and Glick, 2012).

The rhizosphere is defined as a narrow region of soil, directly influenced by the plant roots and activities of microorganisms (Hiltner, 1904). This area is a highly populated and competitive zone, where plant roots compete with the inhabitant root systems of adjacent plants for mineral nutrients, water, and space as well as with other soil-borne organisms, including bacteria, fungi, and insects that feed on the ample source of organic substances (Bais et al., 2004). About 5%–21% of total photosynthetically fixed carbon is secreted by plants as root exudates (Marschner, 1995). The organic compounds present in root exudates are influenced by periodic changes in the environment in a natural ecosystem (Whipps and Lynch, 1986). Microbial community identification is relevant to the effects of climate change on agricultural productivity (Alexandratos and Bruinsma, 2012).

Microorganisms also produce different types of chemical signals like phytohormones, antibiotics, volatile organic compounds (VOCs), quorum-sensing molecules, e.g., *N*-acyl homoserine lactones (AHLs), and diffusible signal factors (DSFs). These molecules function as interkingdom signals, and this leads to their interactions with plant hormonal signaling via salicylic acid, jasmonic acid, and ethylene. These signals are responsible for the symbiotic relationship between plants and beneficial microbes. A symbiotic relationship between microbes and plants is initiated by the coordination of specific plant signals such as flavonoids and strigolactones and microbial signals such as Nod factors and Myc. For appropriate functioning, health, stability, and ecosystem function, interspecies or interkingdom cross talk in plant–plant or plant–microbe interaction is vital (Pellegrino and Bedini 2014).

Plant roots are exceedingly essential for productivity and mineral nutrient uptake, and, therefore, there is need to continue to conduct more research on the interactions of plant roots with soil microbes. Similarly, more focus must be given to the application of these microorganisms as biopesticides or biofertilizers in field conditions to reduce the adverse impacts of conventional chemical fertilizers and pesticides, resulting in increased crop productivity to ensure global food security. With the evolution of omics technologies, it is now practical to successfully observe modifications in the microbial communities of the rhizosphere while simultaneously assessing their effects on plant growth and development. This may guide scientists in the discovery of various microbial species, which can effectively interact with plants (Igiehon and Babalola, 2018).

Beneficial functions of rhizospheric microbiomes

Beneficial microbes reside in the rhizosphere of plants and are crucial for growth, development, and plant health. Nowadays, various studies describe the taxonomic composition of rhizospheric microbiomes. The structure of a microbiome inside the rhizosphere is pivotal to the improvement of plant fitness. Dissection of plant–microbiome interactions is highly complex because of the vast diversity of rhizospheric microbiomes. It is essential to understand the mechanism that drives the assembly of communities and interactions among different members of the gnotobiotic system (Bengtsson-Palme, 2020). In spite of conferring health benefits on plants directly through nutrient acquisition, rhizospheric microbiomes also confer plant health in an indirect manner, which includes suppression of various plant pathogens and increasing stress tolerance. Various members belonging to rhizospheric microbiomes have been shown to increase the growth and development of plants (Backer et al., 2018).

Nutrient acquisition

Interactions between plants and microbes activate the important functions of a rhizospheric microbiome, which improves plant nutrition. Plant growth-promoting bacteria (PGPR) can improve the plant nutrient status by increasing the availability of soil nutrients or modulating the root system architecture of host plants. Subsequently, plant rhizospheric microbiomes can also enhance iron (Fe) and phosphorous (P) nutrition under iron and phosphate deprivation conditions, respectively. Under phosphate-limiting conditions, the master transcriptional regulator PHR1-mediated PSR (phosphate starvation response) in *A. thaliana* is devoted to the assembly of the rhizospheric microbiome, and it was also reported that the rhizospheric microbiome can increase PSR gene transcription in plants, thus enhancing inorganic phosphate uptake (Castrillo et al., 2017). Plant-adaptive and plant-assertive strategies might enclose PSR-induced rhizospheric microbiome assembly, in which assertive colonizers build to intensify phosphate starvation of the host plant (Finkel et al., 2019). Likewise, root-specific MYB72-regulated biosynthesis and an iron chelator BGLU42-dependent secretion of the coumarin scopoletin changes the rhizosphere microbiome composition by its different antimicrobial activity against different bacterial taxa (Harbort et al., 2020) and has essential roles in both the regulation of immunity under iron-limiting conditions and rhizobacteria-mediated iron acquisition (Stringlis et al., 2018).

Stress tolerance

Host plants are capable of gathering a beneficial rhizospheric microbial community under biotic or abiotic stress conditions, which, in turn, alter the plants' responses to environmental stresses in which drought stress is the most common (Chandra et al., 2019a,b). An integrated study of genome-resolved metagenomics and comparative genomics showed that in the sorghum rhizosphere, bacterial metabolism functionality and iron transport are highly correlated with their enhancement under drought stress (Xu et al., 2021). Particularly, drought stress resulted in the loss of an iron-phytosiderophore transporter and led to the proliferation of *Actinobacteria*; this enhancement can be disturbed by exogenous application of iron.

The drought-induced enhancement of endospheric *Actinobacteria*, particularly *Streptomyces*, may have a stable and beneficial influence on rice and wheat (Santos-Medellín et al., 2021). According to Lian et al. (2020), application of some microbial species such as *Rhizobium*, *Dyella*, and *Thiomonas* can control the salt tolerance gene in rice under salt stress (Lian et al., 2020). In particular, Li et al. (2021) proposed that the application of a rhizospheric bacterial consortium provides eternal resistance against salt stress than do individual members. Moreover, the flowering time of plants has been demonstrated to be modified by the rhizospheric microbiome (Lu et al., 2018). Formation of root diffusion barriers such as suberin deposition on the endodermis is crucial for maintaining the nutrient balance of plants and seems to be mediated by root microbial repression of abscisic acid responses in roots (Salas-Gonzalez et al., 2021). Regulation of flowering time and development of the root diffusion barriers have revealed manifold opportunities to develop crops that are resistant to various adverse environmental stresses.

Pathogen suppression

Plant pathogens secrete effector molecules during colonization of their host for disease development (Rovenich et al., 2014). For example, *Verticillium dahliae*, a fungal pathogen, can abolish antagonistic bacteria by secreting the virulence effector VdAve1 and manipulate the rhizospheric microbiome, thus facilitating colonization of cotton and tomato (Snelders et al., 2020). Pathogen invasion and disease progression occur because of a reduction in the diversity of the protective Gram-positive bacterial community (Lee et al., 2021). Rhizospheric microbes can alter plant responses to pathogen invasions through multiple pathways (Vannier et al., 2019). Primarily, rhizospheric microbes provide direct protection against various soil-borne pathogens through secretion of antimicrobial compounds that inhibit the growth of pathogens (Chen et al., 2018). Rhizospheric microbes have the ability to evoke or lessen MTI responses to alter the susceptibility of the host to pathogens (Ma et al., 2021). Interestingly, soil management and organic crops activate the enrichment of micrococcales in the rhizosphere, which produces a higher concentration of salicylic acid, thus improving the plants' resistance to herbivorous insects (Blundell et al., 2020). Eventually, rhizospheric microbiomes have the ability to control the growth defense trade-off of the host plant (Heil and Baldwin, 2002). According to Hu et al. (2018), cereal roots produce secondary metabolites, benzoxazinoids, which have a defensive property, through enhancement of jasmonate signaling for plant defense, though plant growth can be constrained and composition of rhizospheric microbes can also be altered for improvement of resistance against herbivorous insects, even in future plant generations (Hu et al., 2018). The soil-borne legacy theory proposes a suppressive memory for future plant generations growing in the same soil (Hannula et al., 2021). This suppressive memory to improve plant health is believed to be helpful in attracting microbes for modification of plant root exudates (Yuan et al., 2018).

Signal molecule-mediated communication between microorganisms and plants

Roots are the key sites for plant and microbial interactions and its surroundings (Bais et al., 2006). Different types of organic compounds such as amino acids, organic acids, sugars, polysaccharides, proteins, fatty acids, alkaloids, phenolics, flavonoids, and glucosinolates that act as signals or carry out other activities are released by the roots in the rhizospheric soil (Zhang et al., 2014). Root exudates also have the ability to control biotic and abiotic stresses. Modulation of the microbial composition by the phytochemicals secreted by *Arabidopsis thaliana* was reported by Badri et al. (2013). To attract *Rhizobia* that fixes nitrogen, legumes release flavonoids. They also secrete nodulation factors that are important for a symbiotic relationship, and, in return, the *Rhizobia* receive nutrition and shelter (Zhang and Cheng, 2006). Symbiotic associations exceed by up to $50\text{–}70 \times 10^6$ tons of fixed nitrogen per annum in soils, which reduces the requirement of N fertilizers (Herridge et al., 2008). Arbuscular mycorrhizal fungi (AMF) form symbiotic relationships with plants; they form a network of hyphae, which increases the root surface area for better absorption of water and nutrients (especially phosphorus) from the soil (Jeffries et al., 2003). A better grasp of the essential mechanisms along with ecological balances might generate a healthier utilization and management of these signals for protection and crop production (Rasmann and Turlings, 2016).

Cross talk between plants and microbes

Signal molecules produced by the microbes, intra domain signal-mediated bacterial communications may also act on plants. This effect was reported in some bacterial and fungal strains from bean and maize rhizospheres (Prikyrl et al., 1985). This signal-mediated communication also leads to the release of carbohydrates from plant cell walls. Bacterial signals, such as formation and release of indole compounds, are responsible for cooperative activities like formation of biofilms, production of virulence factors, and, through interference of auxin signaling, plant root development (Bailly et al., 2014). Research studies on below- and above-ground microbial communities have discovered innumerable interactions inside the rhizosphere, which promote secretion of root exudates (Philippot et al., 2013). In the rhizosphere, root exudates play a crucial role in determining biotic and abiotic functions. The quality and quantity of root exudates depend on the age of the individual, species of the plant, and external abiotic and biotic factors (Jones et al., 2004). Root exudates are divided into two classes of compounds: (1) low-molecular-weight compounds such as organic acids, sugars, amino acids, and other secondary metabolites, which make up much of the root exudate diversity and (2) high-molecular-weight compounds like proteins and mucilages (polysaccharides), which in terms of mass make up a larger section of root exudates (Badri and Vivanco, 2009), as shown in Fig. 11.1.

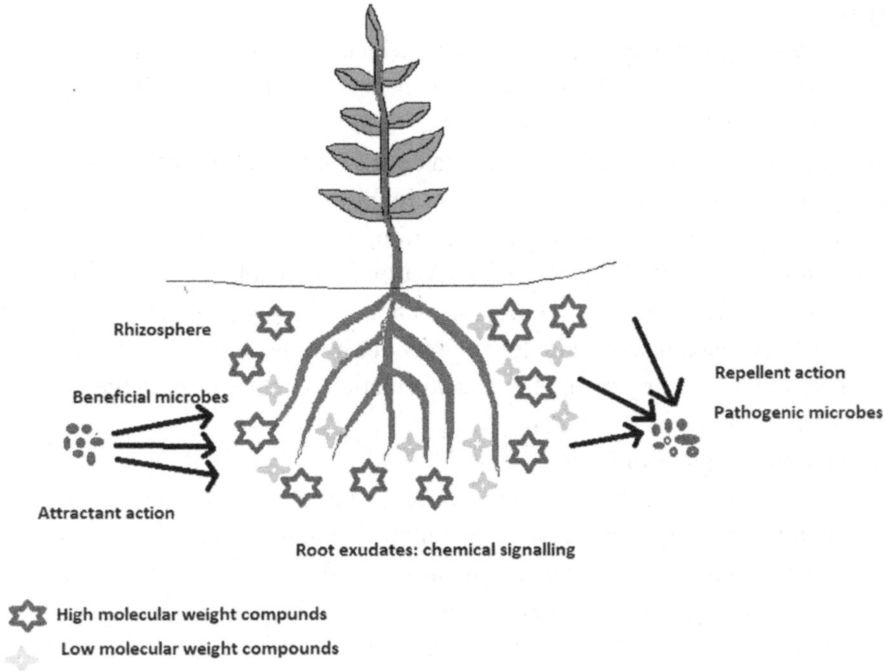

Fig. 11.1
Chemical signaling involved in a plant-microbe interaction.

(1) **Low-molecular-weight exudates**

The functions of root exudates include changes in the physicochemical properties of soil, determining the structure of the microbial community, and restricting the propagation of competing plants. Symbiotic interactions within soils are mediated by root exudate compounds. These include monosaccharides like glucose, disaccharides such as sucrose, and polysaccharides, various types of amino acids, e.g., arginine, organic acids, and benzoic acids.

(i) **Organic acids**

Different types of low-molecular-weight organic acids like citrate and malate are produced by the roots, which attract various kinds of microorganisms. These organic acids are beneficial for chelating cations. Acids such as citrate and malate produced by the roots are beneficial for reducing metal toxicity, especially phosphorus and aluminum ions (Al^{3+}), respectively. Lack of nutrients also provokes organic acid production by the roots. Roots of the *Poaceae* plant species secrete mugineic (Mas) acid under iron-deficient conditions that are normal in alkaline soils (Ma, 2005; Takagi et al., 1984). The effects of organic acids on P dissolution in soils have been reported.

(ii) **Sugars**

Sugars are extremely important molecules, which are involved in morphological and physiological cross talk. According to the study by Currier (1980), with one trefoil strain, maltose, mannose, and ribose were found to be strong attractants, fructose, galactose, and sucrose were weak attractants, and five other sugars including glucose was found to be ineffective. According to Koo et al. (2005), root exudates, by increasing the bond strength between soil particles, enhance the stabilization of the soil structure in the rhizosphere. In addition to their role in plant nutrition, root exudates may affect vital soil processes, such as soil organic matter transformation, cycling of nutrients in soils, and development of a stable soil structure. Root exudates play a crucial role in organic matter transformation, cycling of nutrients, and formation of a stable soil structure.

(2) **High-molecular-weight exudates**

Secretion of high-molecular-weight compounds such as alkaloids, tannins, fatty acids, nucleotides, and vitamins plays an essential role in the interaction of AMF and rhizobacteria with plant roots in soils (Rasmann and Turlings, 2016). The chief roles of these interactions are growth stimulation, nutrient acquisition, phytopathogen inhibition, and maintenance of soil health and fertility for plant growth and development.

Impact of positive and negative interactions on plants and microbial diversity

A relationship between rhizospheric microorganisms and plants is vital and complex (Philippot et al., 2013; Turner et al., 2013). Substances produced in the above-ground biomass exert significant positive or negative effects on the below-ground microbial diversity and vice versa. However, below-ground interactions enhance above-ground productivity. Plant immunity plays a role in shaping microbial structures. The absence of acquired immunity in the mutant strain of *A. thaliana* showed a microbial community structure different from that of the wild-type; however, activation of acquired immunity through chemicals did not bring about any distinguishable changes in the rhizospheric bacterial community. The defense mechanism activated by salicylic acid in the phyllosphere of *A. thaliana* led to the depletion of the abundance and richness of endophytes; meanwhile, plants showed significant diversity of epiphytes without the jasmonate-activated defense mechanism. These results indicate that a definite plant immune mechanism differentially influences the structure of the soil microbial community, particularly certain groups of bacteria (Philippot et al., 2013). Rhizobacteria produce various plant growth-promoting hormones such as indole-3-acetic acid (IAA), gibberellins, and cytokinins. Pathogens gain entry into plants through their stomatal opening by the compounds released by *Pseudomonas syringae*, which are similar to the hormones that block ethylene and jasmonate signaling (Philippot et al., 2013). A growth-specific interaction in certain microbial groups occurs through various chemical compounds produced by plants.

For example, flavonoids produced by plants cause various responses in rhizobacteria, AMF, and root pathogens, and, at the same time, plant roots that produce strigolactones at low concentrations stimulate AMF hyphal branching and seed germination of parasitic plants, e.g., *Orobanche* spp. (Turner et al., 2013; Philippot et al., 2013).

Microbial communities in the roots and rhizosphere are modified by glucosinolates generated by the transgenic *Arabidopsis*. Phenolics, alkaloids, and terpenoids are generally produced by plants. *Arabidopsis* and the *Brassicales* order produce glucosinolates; however, *Avena strigosa* produces avenacins that exhibit antifungal activity against a wide range of fungi. The mutant *A. strigosa*, low in avenacins content exhibited highrisk of disease from pathogenic fungi than the wild-type (Turner et al., 2013). An analytical study showed that, surprisingly, there exists no significant variation in the fungal diversity in the rhizosphere of wild-type and mutant *A. strigosa*, although the mutant genotype with insufficient avenacins critically affects the communities of *Amoeba* as well as *Alveolata* but has no effect on other bacterial communities (Turner et al., 2013). This study supported the view that specific changes in plant metabolism exert significant influence on the microorganism community associated with plants.

Understanding the below- and above-ground microbial interactions via omics studies

Plants establish interactions with a number of different microorganisms, and a large percentage of these tremendously complex microbial groups have not yet been characterized. According to a recent report, only 7% of all terrestrial fungi have been characterized (Mora et al., 2011), and a noteworthy fraction of unknown fungi interact with plants in several ways. Exploring this huge and unknown microbial diversity is a big task. For that reason, studying rhizospheric microbes using omics technology helps in discovering novel plant–microbe interactions and it may be convenient to reduce drought stress in plants. Real-time application of metatranscriptomics, transcriptomics, and metagenomics to rhizospheric microorganisms provides more insights into the intricate microbial interactions (Kakirde et al., 2010).

Culture-based transcriptomic analysis combined with PhyloChip metagenomics of rhizospheric microbiomes of plants cultivated in disease-suppressive soil reveals the presence of 33,000 bacterial and archaeal species (Mendes et al., 2011). In addition, to make the projection via statisticalmodeling and combining the data from structural genes, functional genes, protein analysis and metabolomic investigations help to detect the various crucial biological functions. This approach was used for understanding the interactions involving ectomycorrhizal *Populus tremuloides* and *Laccaria bicolor* roots. With the help of next-generation sequencing techniques, transcriptomic profiling of matured ectomycorrhizal fungi resulted in the identification of transcripts and was used to create a fungal metabolome model. This model was also used for the estimation of metabolites such as glycine and glutamate produced by *L. bicolor* and may be used by *P. tremuloides* that, in turn, provides carbon sources, e.g., glucose and fructose, to AMF. Omics technology is highly effective in determining the role of different genes, RNA, proteins, and metabolites engaged in plant–microbe interactions and would eventually detect key processes that build up these interactions. For accomplishing such an approach, one simply needs to establish 'genome-scale models' that are suitable for analysis of metabolic flux. In this model, microbial groups present themselves as a single superorganism during interactions with plants' 'genomic-scale models' (de Oliveira Dal'Molin et al., 2010). This kind of a model has been prepared for *Arabidopsis* and some bacterial species.

Influence of below-ground microbiota on above-ground interactions

Various reports have shown how rhizospheric interactions affect above-ground interactions with decomposers, pathogens, mutualists, herbivores, and carnivores (Van der Heijden et al., 1998). Modifications of the water and nutrient uptake or the plant defense mechanism can trigger above- and below-ground interactions in plants (Staley et al., 2007). Rhizospheric AMF and rhizobacterial species such as *Bacillus*, *Rhizobium*, and *Pseudomonas* can enhance the systemic immune responses of the host to various pathogens and pests from the above-ground plant biomass (Zamioudis and Pieterse, 2012). Responses of plants to above-ground herbivores is particularly dependent on AMF and plant interactions at a below-ground level. Induced defense responses may spread from the phyllosphere to the roots and influence rhizospheric microorganisms (De Roman et al., 2011). Activation of defense responses in plants by rhizobacteria leads to enhancement of systematic resistance upon a pathogen attack, secondary metabolite production, oxidative burst due to swelling of cell walls, and immune responses stabilization by plant-derived hormones such as ethylene, jasmonic acid, and salicylic acid. For example, the concentration and type of secondary metabolite deleterious to herbivores are found to be regulated in plants by activities of rhizospheric microbes (Joosten et al., 2009). Rhizobacterial interactions with plant help in the production of defense metabolites such as glucosinolates; in addition, they also amplify secondary metabolite production in plants having unknown structures and functions (van de Mortel et al., 2012).

Conclusions and future perspectives

The below- and above-ground cross talk is vital for any plant. This signaling promotes plant growth and development by accomplishing availability of various types of phytohormones and nutrients such as nitrogen, phosphorus, potassium, iron, etc. It also provides resistance to different abiotic stresses such as droughts, heat, salt stress, and oxidative stress, water logging conditions, and biotic stresses like pathogen infestation. The interactions between plants and microbial species are not completely understood even after a large amount of research in this area. To gain more useful insights that can be used for sustainable agricultural development, more such studies on plant–microbe interactions are required. This is also important for developing microbial inoculation with various plant growth-promoting properties through various direct and indirect mechanisms. A number of factors are involved in achieving plant microbial interactions, which include the nature of the host plant, microorganism properties, soil matrix, and other environmental constituents. Application of microbial inoculation as biofertilizers or biopesticides reduces the deleterious effects of chemical fertilizers, pesticides, and insecticides on the environment. Advanced studies on plant–microbe interactions might also have a significant role in developing sustainable agriculture by controlling the spread of a wide variety of pathogens. Beneficial plant–microbe interactions play a crucial role in crop health, productivity, and yield for achieving global food security. Nevertheless, more such studies on physiology are needed to understand the useful interactions of microorganisms, microbial inoculant introduction, and microbial–cultivar acclimatization relationships. It also brings to attention knowledge about beneficial microbial populations, activities, and how they are maintained in the rhizospheric soil. So, this ultimately leads to the development of novel, more useful, and favorable compounds from microbial interactions that could be compatible with the agricultural field. Similarly, more research is required to understand the whole mechanisms involved in the interactions between host plants and microorganisms by establishing in vivo and in vitro techniques or models of great importance.

References

Alexandratos, N., Bruinsma, J., 2012. World Agriculture towards 2030/2050: The 2012 Revision. ESA Working paper No. 12-03, FAO, Rome.

Babalola, O.O., Glick, B.R., 2012. Indigenous African agriculture and plant associated microbes: current practice and future transgenic prospects. Sci. Res. Essays 7 (28), 2431–2439.

Backer, R., Rokem, J.S., Ilangumaran, G., Lamont, J., Praslickova, D., Ricci, E., 2018. Plant growth-promoting rhizobacteria: context, mechanisms of action, and roadmap to commercialization of biostimulants for sustainable agriculture. Front. Plant Sci. 9, 1473.

Badri, D.V., Vivanco, J.M., 2009. Regulation and function of root exudates. Plant Cell. Environ. 32, 666–681.

Badri, D.V., Chaparro, J.M., Zhang, R., Shen, Q., Vivanco, J.M., 2013. Application of natural blends of phytochemicals derived from the root exudates of *Arabidopsis* to the soil reveal that phenolic-related compounds predominantly modulate the soil microbiome. J. Biol. Chem. 288 (7), 4502–4512.

Baetz, U., Martinoia, E., 2013. Root exudates: the hidden part of plant defense. Trends Plant Sci. 19 (2), 90–98.

Bailly, A., Groenhagen, U., Schulz, S., Geisler, M., Eberl, L., Weisskopf, L., 2014. The interkingdom volatile signal indole promotes root development by interfering with auxin signalling. Plant J. 80 (5), 758–771.

Bais, H.P., Park, S.W., Weir, T.L., Callaway, R.M., Vivanco, J.M., 2004. How plants communicate using the underground information superhighway. Trends Plant Sci. 9, 26–32.

Bais, H.P., Weir, T.L., Perry, L.G., Gilroy, S., Vivanco, J.M., 2006. The role of root exudates in rhizosphere interactions with plants and other organisms. Annu. Rev. Plant Biol. 57, 233–266.

Bengtsson-Palme, J., 2020. Microbial model communities: to understand complexity, harness the power of simplicity. Comput. Struct. Biotechnol. J. 18, 3987–4001.

Berendsen, R.L., Pieterse, C.M., Bakker, P.A., 2012. The rhizosphere microbiome and plant health. Trends Plant Sci. 17, 478–486.

Blundell, R., Schmidt, J.E., Igwe, A., Cheung, A.L., Vannette, R.L., Gaudin, A.C.M., 2020. Organic management promotes natural pest control through altered plant resistance to insects. Nat Plants 6 (5), 483–491.

Bulgarelli, D., Garrido-Oter, R., Munch, P.C., Weiman, A., Droge, J., Pan, Y., McHardy, A.C., Schulze Lefert, P., 2015. Structure and function of the bacterial root microbiota in wild and domesticated barley. Cell Host Microbe 17 (3), 392–403.

Castrillo, G., Teixeira, P.J.P.L., Paredes, S.H., Law, T.F., de Lorenzo, L., Feltcher, M.E., 2017. Root microbiota drive direct integration of phosphate stress and immunity. Nature 543 (7646), 513–518.

Chandra, D., Sharma, A.K., 2020. Field evaluation of consortium of bacterial inoculants producing ACC deaminase on growth, nutrients and yield components of rice and wheat. J. Crop. Sci. Biotechnol. 24 (3), 293–305.

Chandra, D., Srivastava, R., Gupta, V.V., Franco, C.M., Sharma, A.K., 2019a. Evaluation of ACC-deaminase-producing rhizobacteria to alleviate water-stress impacts in wheat (*Triticum aestivum* L.) plants. Can. J. Microbiol. 65 (5), 387–403.

Chandra, D., Srivastava, R., Gupta, V.V., Franco, C.M., Paasricha, N., Saifi, S.K., Tuteja, N., Sharma, A.K., 2019b. Field performance of bacterial inoculants to alleviate water stress effects in wheat (*Triticum aestivum* L.). Plant Soil 441 (1), 261–281.

Chandra, D., Srivastava, R., Glick, B.R., Sharma, A.K., 2020. Rhizobacteria producing ACC deaminase mitigate water-stress response in finger millet (*Eleusine coracana* (L.) Gaertn.). 3 Biotech 10, 65.

Chen, Y., Wang, J., Yang, N., Wen, Z., Sun, X., Chai, Y., 2018. Wheat microbiome bacteria can reduce virulence of a plant pathogenic fungus by altering histone acetylation. Nat. Commun. 9, 3429.

Coleman-Derr, D., Desgarennes, D., Fonseca-Garcia, C., Gross, S., Clingenpeel, S., Woyke, T., North, G., Visel, A., Partida-Martinez, L.P., Tringe, S.G., 2016. Plant compartment and biogeography affect microbiome composition in cultivated and native *Agave* species. New Phytol. 209, 798–811.

Cristina, C., Gouveia, C., Dias, T., Varma, A., Babalola, O.O., 2017. How to disentangle changes in microbial function from changes in microbial community. In: Modern Tools and Techniques to Understand Microbes. Springer, pp. 149–158.

Currier, W., 1980. Chemotaxis of a birds foot trefoil strain of *Rhizobium* to simple sugars. FEMS Microbiol. Lett. 8, 43–46.

de Oliveira Dal'Molin, C.G., Quek, L.E., Palfreyman, R.W., Brumbley, S.M., Nielsen, L.K., 2010. AraGEM, a genome-scale reconstruction of the primary metabolic network in *Arabidopsis*. Plant Physiol. 152 (2), 579–589.

De Roman, M., Fernandez, I., Wyatt, T., Sahrawy, M., Heil, M., Pozo, M.J., 2011. Elicitation of foliar resistance mechanisms transiently impairs root association with arbuscular mycorrhizal fungi. J. Ecol. 99 (1), 36–45.

Finkel, O.M., Salas-Gonzalez, I., Castrillo, G., Spaepen, S., Law, T.F., Teixeira, P.J.P.L., 2019. The effects of soil phosphorus content on plant microbiota are driven by the plant phosphate starvation response. PLoS Biol 17 (11). e, 3000534.

Fitzpatrick, C.R., Copeland, J., Wang, P.W., Guttman, D.S., Kotanen, P.M., Johnson, M.T., 2018. Assembly and ecological function of the root microbiome across angiosperm plant species. Proc. Natl Acad. Sci. USA 115, 1157–1165.

Hacquard, S., Garrido-Oter, R., Gonzalez, A., Spaepen, S., Ackermann, G., Lebeis, S., McHardy, A.C., Dangl, J.L., Knight, R., Ley, R., Schulze-Lefert, P., 2015. Microbiota and host nutrition across plant and animal kingdoms. Cell Host Microbe 17 (5), 603–616.

Hamonts, K., Trivedi, P., Garg, A., Janitz, C., Grinyer, J., Holford, P., Botha, F.C., Anderson, I.C., Singh, B.K., 2018. Field study reveals core plant microbiota and relative importance of their drivers. Environ. Microbiol. 20, 124–140.

Hannula, S.E., Heinen, R., Huberty, M., Steinauer, K., De Long, J.R., Jongen, R., 2021. Persistence of plant-mediated microbial soil legacy effects in soil and inside, roots. Nat. Commun. 12, 5686.

Harbort, C.J., Hashimoto, M., Inoue, H., Niu, Y., Guan, R., Rombola, A.D., 2020. Root secreted coumarins and the microbiota interact to improve iron nutrition in *Arabidopsis*. Cell Host Microbe 28 (6), 825–837.

Heil, M., Baldwin, I.T., 2002. Fitness costs of induced resistance: emerging experimental support for a slippery concept. Trends Plant Sci. 7, 61–67.

Herridge, D.F., Peoples, M.B., Boddey, R.M., 2008. Global inputs of biological nitrogen fixation in agricultural systems. Plant Soil 311, 1–18.

Hiltner, L., 1904. Uber neuere Erfahrungen und Problem auf dem gebiet der Bod enbakteriologie und unter besonderer Berücksichtigung der Gründüngung und Brachte. Arb. Dt. Landw. 98, 59–78.

Hu, L., Robert, C.A.M., Cadot, S., Zhang, X., Ye, M., Li, B., 2018. Root exudate metabolites drive plant-soil feedbacks on growth and defense by shaping the rhizosphere microbiota. Nat. Commun. 9, 2739.

Igiehon, N.O., Babalola, O.O., 2018. Rhizosphere microbiome modulators: contributions of nitrogen fixing bacteria towards sustainable agriculture. Int. J. Environ. Res. Public Health 15 (4), 574–598.

Jeffries, P., Gianinazzi, S., Perotto, S., Turnau, K., Barea, J.M., 2003. The contribution of arbuscular mycorrhizal fungi in sustainable maintenance of plant health and soil fertility. Biol. Fertil. Soils 37, 1–16.

Jones, D.L., Angela, Hodge, A., Kuzyakov, Y., 2004. Plant and mycorrhizal regulation of rhizodeposition. New Phytol. 163, 459–480.

Joosten, L., Mulder, P.P., Klinkhamer, P.G., Van Veen, J.A., 2009. Soil-borne microorganisms and soil-type affect pyrrolizidine alkaloids in *Jacobaea vulgaris*. Plant Soil 325 (1–2), 133.

Kakirde, K.S., Parsley, L.C., Liles, M.R., 2010. Size does matter: application-driven approaches for soil metagenomics. Soil Biol. Biochem. 42 (11), 1911–1923.

Koo, B.J., Adriano, D.C., Bolan, N.S., Barton, C.D., 2005. Root exudates and microorganisms. In: Encylopedia of Soils in the Environment, pp. 421–428.

Lee, S.M., Kong, H.G., Song, G.C., Ryu, C.M., 2021. Disruption of Firmicutes and Actinobacteria abundance in tomato rhizosphere causes the incidence of bacterial wilt disease. Int. Soc. Microb. Ecol. J. 15 (1), 330–347.

Lekberg, Y., Arnillas, C.A., Borer, E.T., Bullington, L.S., Fierer, N., Kennedy, P.G., 2021. Nitrogen and phosphorus fertilization consistently favor pathogenic over mutualistic fungi in grassland soils. Nat. Commun. 12, 3484.

Lennon, J., Jones, S., 2011. Microbial seed banks: the ecological and evolutionary implications of dormancy. Nat. Rev. Microbiol. 9 (2), 119–130.

Li, H., La, S., Zhang, X., Gao, L., Tian, Y., 2021. Salt-induced recruitment of specific root associated bacterial consortium capable of enhancing plant adaptability to salt stress. Int. Soc. Microb. Ecol. J., 1–18.

Lian, T., Huang, Y., Xie, X., Huo, X., Shahid, M.Q., Tian, L., 2020. Rice SST variation shapes the rhizosphere bacterial community, conferring tolerance to salt stress through regulating soil metabolites. MSystems 5 (6). e00721-20.

Lu, T., Ke, M., Lavoie, M., Jin, Y., Fan, X., Zhang, Z., 2018. Rhizosphere microorganisms can influence the timing of plant flowering. Microbiome 6, 231.

Lundberg, D.S., Lebeis, S.L., Paredes, S.H., Yourstone, S., Gehring, J., Malfatti, S., Tremblay, J., Engelbrektson, A., Kunin, V., del Rio, T.G., 2012. Defining the core *Arabidopsis thaliana* root microbiome. Nature 488 (7409), 86–90.

Ma, J.F., 2005. Plant roots responses to three abundant soil minerals: silicon, aluminium and iron. Crit. Rev. Plant Sci. 24, 267–281.

Ma, K.W., Niu, Y., Jia, Y., Ordon, J., Copeland, C., Emonet, A., 2021. Coordination of microbe–host homeostasis by crosstalk with plant innate immunity. Nat. Plants 7 (6), 814–825.

Marschner, H., 1995. Mineral Nutrition of Higher Plants, second ed. Academic, London.

Mendes, R., Kruijt, M., de Bruijn, I., 2011. Deciphering the rhizosphere microbiome for disease-suppressive bacteria. Science 332 (6033), 1097–1100.

Mendes, R., Garbeva, P., Raaijmakers, J.M., 2013. The rhizosphere microbiome: significance of plant beneficial, plant pathogenic and human pathogenic microorganisms. FEMS Microbiol. Rev. 37, 634–663.

Mora, C., Tittensor, D.P., Adl, S., Simpson, A.G., Worm, B., 2011. How many species are there on Earth and in the ocean? PLoS Biol. 9 (8). e,1001127.

Nardi, S., Concheri, G., Pizzeghello, D., Sturaro, A., Rella, R., Parvoli, G., 2000. Soil organic matter mobilization by root exudates. Chemosphere 41 (5), 653–658.

Ortiz-Castro, R., Contreras-Cornejo, H.A., Macias-Rodriguez, L., Lopez-Bucio, J., 2009. The role of microbial signals in plant growth and development. Plant Signal. Behav. 4 (8), 701–712.

Panke-Buisse, K., Poole, A., Goodrich, J., Ley, R., Kao-Kniffin, J., 2015. Selection on soil microbiomes reveals reproducible impacts on plant function. Int. Soc. Microb. Ecol. J. 9, 980–989.

Pellegrino, E., Bedini, S., 2014. Enhancing ecosystem services in sustainable agriculture: biofertilization and biofortification of chickpea (*Cicer arietinum* L.) by arbuscular mycorrhizal fungi. Soil Biol. Biochem. 68, 429–439.

Philippot, L., Raaijmakers, J.M., Lemanceau, P., van der Putten, W.H., 2013. Going back to the roots: the microbial ecology of the rhizosphere. Nat. Rev. Microbiol. 11 (11), 789–799.

Prikryl, Z., Vancura, V., Wurst, M., 1985. Auxin formation by rhizosphere bacteria as a factor of root growth. Plant Biol. 27, 159–163.

Rasmann, S., Turlings, T.C.J., 2016. Root signals that mediate mutualistic interactions in the rhizosphere. Curr. Opin. Plant Biol. 32, 62–68.

Rovenich, H., Boshoven, J.C., Thomma, B.P., 2014. Filamentous pathogen effector functions: of pathogens, hosts and microbiomes. Curr. Opin. Plant Biol. 20, 96–103.

Salas-Gonzalez, I., Reyt, G., Flis, P., Custodio, V., Gopaulchan, D., Bakhoum, N., 2021. Coordination between microbiota and root endodermis supports plant mineral nutrient homeostasis. Science 371 (6525), 0695.

Santos-Medellin, C., Liechty, Z., Edwards, J., Nguyen, B., Huang, B., Weimer, B.C., 2021. Prolonged drought imparts lasting compositional changes to the rice root microbiome. Nat. Plants 7 (8), 1065–1077.

Snelders, N.C., Rovenich, H., Petti, G.C., Rocafort, M., van den Berg, G.C.M., Vorholt, J.A., 2020. Microbiome manipulation by a soil-borne fungal plant pathogen using effector proteins. Nat. Plants 6 (11), 1365–1374.

Staley, J.T., Mortimer, S.R., Morecroft, M.D., Brown, V.K., Masters, G.J., 2007. Summer drought alters plant-mediated competition between foliar-and root feeding insects. Glob. Chang. Biol. 13 (4), 866–877.

Stringlis, I.A., Yu, K., Feussner, K., de Jonge, R., Van Bentum, S., Van Verk, M.C., 2018. MYB72-dependent coumarin exudation shapes root microbiome assembly to promote plant health. Proc. Natl Acad. Sci. USA 115 (22), 5213–5222.

Takagi, S., Nomoto, K., Takemoto, T., 1984. Physiological aspects of muigneic acid, a possible phytosiderophore of graminaceous plant. J. Plant Nutr. 7, 469–477.

Thiergart, T., Duran, P., Ellis, T., Vannier, N., Garrido-Oter, R., Kemen, E., 2020. Root microbiota assembly and adaptive differentiation among European *Arabidopsis* populations. Nat. Ecol. Evol. 4 (1), 122–131.

Turner, T.R., James, E.K., Poole, P.S., 2013. The plant microbiome. Genome Biol. 14 (6), 209.

van de Mortel, J.E., de Vos, R.C., Dekkers, E., 2012. Metabolic and transcriptomic changes induced in *Arabidopsis* by the rhizobacterium *Pseudomonas fluorescens* SS101. Plant Physiol. 160 (4), 2173–2188.

Van der Heijden, M.G., Klironomos, J.N., Ursic, M., 1998. Mycorrhizal fungal diversity determines plant biodiversity, ecosystem variability and productivity. Nature 396 (6706), 69–72.

Vannier, N., Agler, M., Hacquard, S., Zipfel, C., 2019. Microbiota-mediated disease resistance in plants. PLoS Pathog. 15 (6), e1007740.

Whipps, J.M., Lynch, J.M., 1986. The influence of the rhizosphere on crop productivity. In: Advances in Microbial Ecology. Springer, New York, pp. 187–244.

Xu, J., Zhang, Y., Zhang, P., Trivedi, P., Riera, N., Wang, Y., Liu, X., Fan, G., Tang, J., Coletta-Filho, H.D., 2018. The structure and function of the global citrus rhizosphere microbiome. Nat. Commun. 9, 4894.

Xu, L., Dong, Z., Chiniquy, D., Pierroz, G., Deng, S., Gao, C., 2021. Genome-resolved metagenomics reveals role of iron metabolism in drought-induced rhizosphere microbiome dynamics. Nat. Commun. 12, 3209.

Yuan, J., Zhao, J., Wen, T., Zhao, M., Li, R., Goossens, P., 2018. Root exudates drive the soil-borne legacy of aboveground pathogen infection. Microbiome 6, 156.

Zamioudis, C., Pieterse, C.M., 2012. Modulation of host immunity by beneficial microbes. Mol. Plant-Microbe Interact. 25 (2), 139–150.

Zhang, X.S., Cheng, H.P., 2006. Identification of *Sinorhizobium meliloti* early symbiotic genes by use of a positive functional screen. Appl. Environ. Microbiol. 72 (4), 2738–2748. https://doi.org/10.1128/AEM.72.4.2738-2748.2006.

Zhang, N., Wang, D., Liu, Y., Li, S., Shen, Q., Zhang, R., 2014. Effects of different plant root exudates and their organic acid components on chemotaxis, biofilm formation and colonization by beneficial rhizosphere-associated bacterial strains. Plant Soil 374, 689–700.

CHAPTER 12

Arbuscular mycorrhizal fungi symbiosis and food security

Fokom Raymond[a,b,d], Eke Pierre[c,d], Adamou Souleymanou[b,e], Ngo Oum Therese[b], Fekam Boyom Fabrice[d], and Nwaga Dieudonne[b]

[a]Department of Food Processing and Quality Control, Institute of Fisheries and Aquatic Sciences, University of Douala, Douala, Cameroon, [b]Soil Microbiology Laboratory, The Biotechnology Centre, Department of Microbiology, Faculty of Sciences, University of Yaoundé I, Yaoundé, Cameroon, [c]Department of Crop Production Technology, College of Technology, University of Bamenda, North-West Region, Cameroon, [d]Antimicrobial & Biocontrol Agents Unit (AmBcAU), Laboratory for Phyto Biochemistry and Medicinal Plants Studies, Faculty of Science, University of Yaoundé I, Yaoundé, Cameroon, [e]Faculty of Agronomy and Agricultural Sciences, Department of Agriculture, University of Dschang, Dschang, Cameroon

Chapter outline

Unravelling Plant-Microbe Synergy. https://doi.org/10.1016/B978-0-323-99896-3.00012-6

Introduction

Food security is a current worldwide concern that aims to provide everyone with consistent access to a sufficient quantity of nutritious and inexpensive food, subject to both qualitative and quantitative demands (Capone et al., 2014). In contrast, food insecurity is a state in which there is a scarcity of economic, nutritionally adequate, and secure food or the ability to access sufficient food in a socially acceptable manner is constrained (FAO, 2021; Samim et al., 2021). Food insecurity is therefore defined as insufficient food intake, insufficient food access, and susceptibility to subsistence mechanisms that provide sufficient food daily but are not resilient under shock (Samim et al., 2021). Despite considerable progress in eradicating global malnutrition and hunger, food security continues to be the primary concern of people and families worldwide (Gebre and Rahut, 2021). The overall global population afflicted by hunger kept increasing in 2020, despite the COVID-19 pandemic. According to the Food and Agricultural Organization's most current estimates from 2018, 26.4% of the global population is food-insecure. Regional disparities persist and are worrying, according to statistics. In 2020, around 21% (one in five) of the people in Africa faced hunger, more than twice the number in any other continent. This indicates a three-point improvement in 1 year. More than half (418 million) of the world's 768 million malnourished people live in Asia, 8% in Latin America and the Caribbean (60 million), and nearly a third (282 million) in Africa. In all, 46 million more people in Africa, more than 57 million more in Asia, and almost 14 million more in Latin America and the Caribbean were affected by hunger in 2020 as compared to that in 2019. According to the Food Insecurity Experience Scale (FIES), the proportion of individuals worldwide who face moderate or severe food insecurity has gradually increased from 22.6% in 2014 to 26.6% in 2019. Then, in 2020, the year of the COVID-19 outbreak, it jumped to 30.4%, almost as much as it had in the previous 5 years put together. Thus, more than a third of the world's population was food-insecure in 2020, an increase of 320 million individuals from 2.05 billion to 2.37 billion in only 1 year. Roughly 40% of those people (11.9% of the world's population or nearly 928 million individuals) faced severe food insecurity (FAO, 2021).

Challenges to agricultural development: The driving force behind food security

Agricultural output has more than tripled between the years 1960 and 2015, due in part to increased productivity, enhanced green revolution technology, and a considerable increase in natural resource, water, and agricultural land usage (Food and Agriculture Organization of the United Nations, 2017). Nonetheless, widespread hunger and malnutrition continue to be a significant burden in many regions of the world. At the present pace of advancement, hunger will not be eradicated by 2030 or even by 2050 as forecasted earlier in the millennia. Conversely, the FAO forecasts that global food production would need to grow by 70% by 2050 to feed the 9 billion people who will occupy the planet by then; 90% of this increase will need to occur in developing nations, with an overall 80% resulting from increased agricultural yields. Meanwhile, food systems have evolved in response to and as a result of shifting dietary choices and patterns of overconsumption, as seen by the startling rise in the incidence of overweight and obesity worldwide (Food and Agriculture Organization of the United Nations, 2017). Increasing food production and economic development have oftentimes come at a high environmental cost. Agriculture in the 21st century faces an array of problems, including soil erosion, acidification, water shortages, insect pests, climate change, phytopathogens, pesticide-resistant weeds, etc., all of which pose significant threats to the current agricultural paradigm. As a result, approximately half of the trees that covered the earth in the past have been lost. Groundwater resources are dwindling rapidly. Biodiversity has been degraded systematically. Each year, fossil fuel burning emits billions of tons of greenhouse gases into the atmosphere, contributing to global warming and climate change. These developments jeopardize the world's ability to satisfy its food demands substantially. Even minor changes in the climate, such as tiny variations in yearly rainfall or seasonal precipitation patterns, are believed to have a significant impact on crop productivity (Food and Agriculture Organization of the United Nations, 2017).

Agricultural sustainability as a viable option

Any food or fiber production system that systematically pursues objectives including the comprehensive integration of natural processes such as nutrient cycle, nitrogen fixation, and pest and predator interactions is referred to as sustainable agriculture. The former minimizes the use of external and nonrenewable inputs that harm the environment or harm the health of farmers and consumers. As a result of the significant, complicated, and contentious issues raised by the widespread use of xenobiotic chemicals in intensive agriculture, the so-called 'sustainable goals' have emerged. No doubt, using agrochemicals in agricultural systems poses a significant threat to food security, human health, and environmental sustainability, which has been hotly debated by the entire global scientific community (Shorette, 2012). Contamination concerns (e.g., soil acidification, soil infertility, hormone and antibiotic residues, wildlife poisoning, and pollution of groundwater, surface water, and the atmosphere) have resulted in ongoing environmental degradation, hurting human and animal health (Guo et al., 2010; Tian and Niu, 2015).

Currently, sustainability in agricultural models promotes the concepts of persistence (the ability of systems to continuously turn over) and resilience (the capacity of systems to buffer effects on the economy, society, and the environment) (Pretty, 2008). Pesticides are being replaced by natural enemies, and plowing is being replaced by zero tillage in many traditional agricultural sustainability models. This means that these large system models require drastic reductions in inputs (e.g., moving away from nitrogen-fixing legumes to zero tillage) (Pretty and Bharucha, 2014). Amidst all of this, the recent years have seen significant progress in reducing the negative environmental impacts of conventional agriculture while still meeting the current demand without endangering the ability of future generations to meet their own needs (Pretty and Bharucha, 2014). Eco-environmentally sustainable methods that use fewer agrochemicals and do not exhaust natural resources are therefore being promoted as a means of reducing environmental impact, safeguarding and improving soil health, and ensuring food security (Azcón-Aguilar and Barea, 2015). It is necessary to promote plant and soil health properties resulting from interactions between physicochemical and biological factors to achieve these goals (Altieri, 2004; Meena et al., 2016; Yadav and Sidhu, 2016; Saha et al., 2016) and to allow farmers to maintain healthy, productive soils for crops without degrading the agroecosystem; the diverse genetic soil macro-, meso-, and microbiota that make up organic residue decomposition and the biogeochemical cycling of plant nutrients are just two of the many functions that these soil biological components perform in agroecosystems (Azcón-Aguilar and Barea, 2015). Sustainable, low-input cropping systems rely more on the soil microbiota than does conventional agriculture to maintain soil health and control phytopathogens biologically (Johansson et al., 2004). Mycorrhizal fungi have long upheld their promise of improving crop health, nutritional status, and thus production in a wide range of habitats. AMF are a critical component of the soil microbiota (Eke et al., 2016). To ensure that our planet's food supply will continue to meet the needs of future generations, use and holistic application of AMF in the sustainable intensification of agriculture are essential (Fig. 12.1).

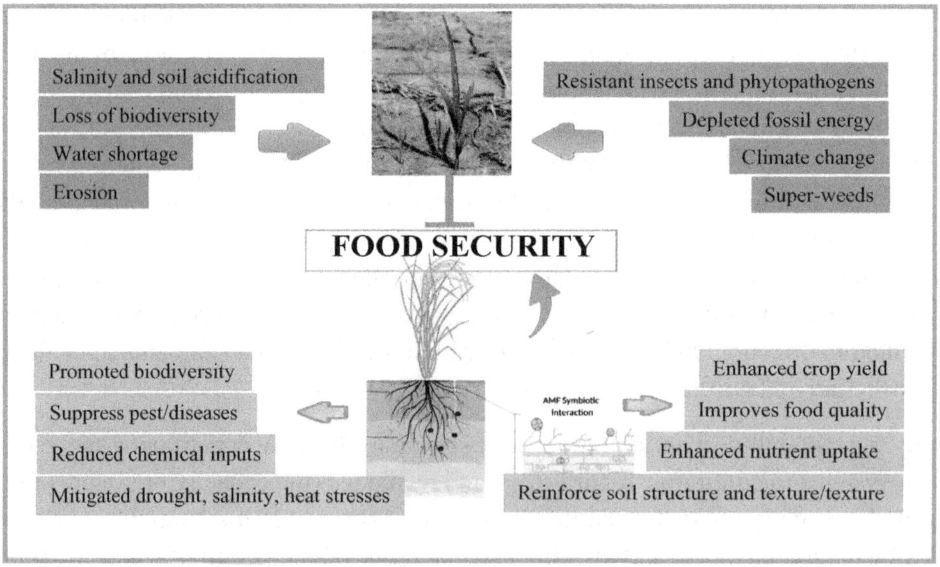

Fig. 12.1
Major factors hampering the current food security and mitigating measures mediated by arbuscular mycorrhizal fungi.

The general concept of arbuscular mycorrhizal symbiosis

Microorganisms are our planet's most ancient living organisms, dating back to around 3.5 billion years (Burchell et al., 2010). They have evolved features and adaptability to cohabit with other living creatures throughout the evolutionary process, developing various interactions in terms of both function and form with their hosts (Taylor et al., 2012). Symbioses between plants and heterotrophic microbes are the most well-known of the diverse biological interactions that exist between living things on Earth. Symbioses between soil fungi and plant roots are the most well-known mutualist associations between plant roots and soil fungi (Zuccaro et al., 2014). As it turns out, AMF are a form of soil-dwelling fungi that symbiotically interact with more than 90% of the vascular plant species on the planet. AMF are soil-borne fungi that are required for plant nutrient uptake and for providing plant resistance to a range of abiotic and biotic challenges. AMF are beneficial to plants in several ways (Sun et al., 2018).

The majority of the mycorrhizal species belongs to the subphylum *Glomeromycotina*, which is also a part of the phylum *Mucoromycota*. This subphylum contains four orders of AMF, namely, *Archaeosporales*, *Paraglomerales*, *Diversisporales*, and *Glomerales* (Redecker et al., 2013; Spatafora et al., 2016). The latter are biotrophs that benefit from plant photosynthates and other macromolecules to bring their life cycle to an end (Jiang et al., 2017). As compared to root hairs, mycorrhizal fungi have a greater absorptive surface area, which facilitates the absorption of immobilized ions (such as copper, zinc, and phosphate) in the soil outside of the depleted zones (Pichardo et al., 2012). Apart from promoting plant growth by boosting water and mineral nutrient absorption from the nearby soil, AMF also safeguard plants against fungal, bacterial, and viral pathogens (Smith and Read, 2008; Jung et al., 2012). Furthermore, it has been established that mycorrhizal fungi may minimize the impacts of abiotic variables such as salt, drought, and heavy metal toxicity, hence supporting plant growth. Subsequently, AMF are essential endosymbionts that significantly contribute to both plant production and ecological function. They are fundamental for long-term agricultural improvement under the changing climate and depleted soil nutrients (Gianinazzi et al., 2010). Spore germination and degree of advancement of the resulting fungal hyphae during root colonization differentiate AMF symbiosis into several stages (Gutjahr and Paszkowski, 2013). The 'presymbiotic' stage stands for mutual recognition of the plant roots and AMF spores.

It has been well demonstrated that in the vicinity of each other, a chemical dialogue between AMF quiescent spores and root cells initiates the presymbiotic phase, characterized by the release of root exudates composed of strigolactones and flavonoids. The latter triggers the germination and early branching of AMF spores in the rhizosphere (Parniske, 2008). The 'symbiotic' phase or formation of the hyphopodium refers to the formation of the appressorium and large, intracellular, tree-like structures called arbuscules. Hence, appressoria are swollen structures formed at the tips of the hyphae, which exert physical

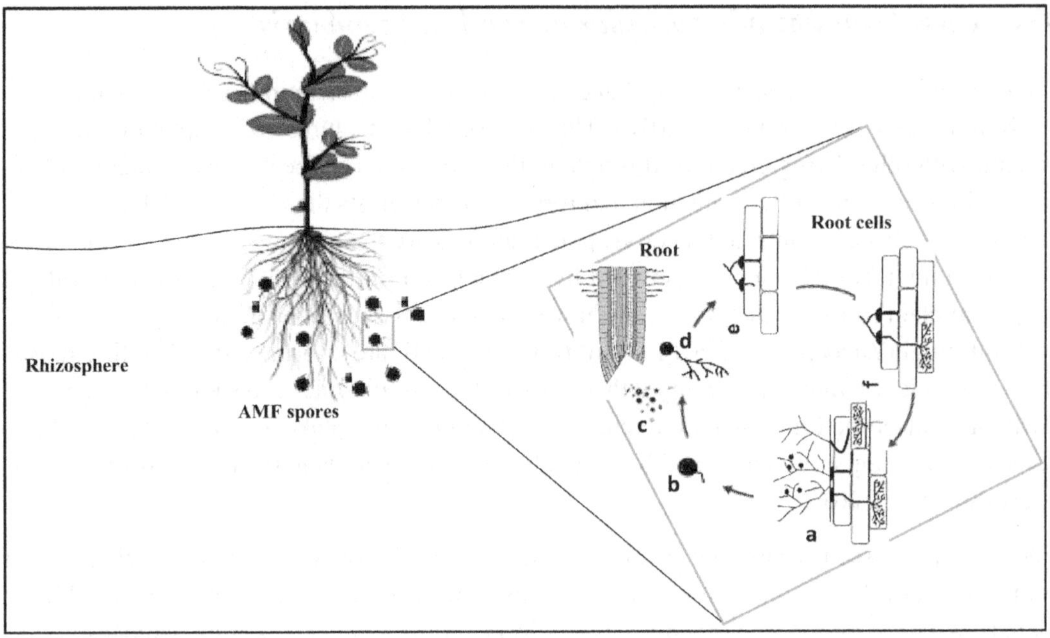

Fig. 12.2
Development of arbuscular mycorrhiza and soil recolonization. Mycorrhizal developmental phases: (A), arbuscule establishment, external hyphae growth, and spore formation at hyphal tips; (B) AMF spore; (C) root exudates (flavonoids and strigolactones); (D) germination of AMF spores and initiation of the hyphal network formation; (E) establishment of physical pressure on the root tissues through appressoria formation; and (F) tissue (inter- and intracellular) colonization and arbuscule formation.*Adapted from Marsh and Schultze (2001).*

pressure on the root epidermis to allow the entry of the AMF into the root cortex and adjacent tissues. Then, the AMF hyphal network migrates intercellularly within the fine root tissues (the cortex and endodermis) and colonizes the whole root system. Likewise, the exchanges of photosynthates (from the host) with a plethora of soil nutrients by the hyperbranched intracellular hyphal network termed mycorrhiza have been reported (Akiyama et al., 2005; Harrison, 2005, 2012; Genre et al., 2008; Parniske, 2008; Hocher et al., 2011; Maillet et al., 2011) (Fig. 12.2). The following stage of the AM association's growth is known as 'root colonization,' which is characterized by the expansion of the fungal hyphae within nearly all the root systems of the host and undergoes profound alterations in plant biochemistry and physiology (Hause et al., 2007).

Numerous alterations are associated with defense systems, which are likely responsible for the plant's ability to retain control over its symbiotic partner (Ludwig-Muller and Guther, 2007; Liu et al., 2007; Hause et al., 2007; Schliemann et al., 2008; Fiorilli et al., 2009;

Hause and Schaarschmidt, 2009; Bari and Jones, 2009; García-Garrido et al., 2010). Finally, the formation of a mycorrhizal network is realized (Bever et al., 2010; Gutjahr and Paszkowski, 2013). Following the establishment of all of these stages, the plant must regulate the degree of fungal invasion within its roots to avoid excessive colonization and carbon drainage, hence preserving mutualistic interactions in time (Jung et al., 2012).

Direct benefits of arbuscular mycorrhizal symbiosis

AMF symbiosis improves nutritional status and crop growth

The overwhelming pressure placed on agricultural lands by the green revolution frequently has significant negative or converse repercussions on biodiversity, which may thus damage ecosystem functions (Wagg et al., 2015). AMF inoculation has an impact on the metabolism of the host plant, with repercussions on the concentrations of different macro- and micronutrients, resulting in an increase in photosynthate production and, as a result, enhanced plant yields, qualitatively and quantitatively (Mitra et al., 2019). Plant nutrient absorption, particularly phosphorus uptake, is suggested to be stimulated upon AMF colonization as it can increase inorganic nutrient intake in nearly all hosts (Nell et al., 2010). Mycorrhizal fungi generate arbuscules, which contributes to the interchange of carbon, minerals, and phosphorus molecules, giving host plants a boost in vigor (Prasad et al., 2017).

Moreover, mycorrhizae formation is particularly useful for supporting plants in the absorption of essential nutrients in all soil types including nutrient-poor soils (Kayama and Yamanaka, 2014). AMF have also been shown to promote the availability of copper and zinc in addition to macronutrients (Smith and Read, 1997). The ability of the host roots to absorb surface water is also improved in mycorrhizal plants (Bielcski, 1973). The Pi absorption rate, for example, was significantly enhanced in mycorrhizal maize seedlings in a study conducted by Garcés-Ruiz et al. (2017). Higher photosynthetic activity and other leaf functions are closely related to greater AMF inoculation and plant growth frequency, which is correlated with the intake of P, N, and carbon, which moves to roots and promotes tuber development in root crops. Under diverse irrigation regimes, AMF have been proven to maintain N and P absorption, thereby promoting plant development under high and low P levels and stressful circumstances (Liu et al., 2014, 2018). Under drought stress, for example, mycorrhizal symbiosis improved Fe, P, and N concentrations in *Pelargonium graveolens* L. (Amiri et al., 2017). Instant fungal attachment enhanced Ca, K, and P levels in *Euonymus japonica* under salinity stress, according to Gomez-Bellot et al. (2015). Another study indicated that under drought stress, AMF-inoculated pistachio trees exhibited high levels of P, Mn, Zn, and K (Bagheri et al., 2012). Furthermore, AMF inoculation increased seedling weight by increasing water content and intercellular CO_2, N, and P contents in *Chrysanthemum morifolium* plant tissues (Wang et al., 2018).

AMF symbiosis improves crop yield

Beneficial rhizospheric microorganisms not only boost crop nutritional status but may also improve crop quality. As multipurpose beneficial organisms, AMF have been used in the field production of yam (Lu et al., 2015), potato (Hijri, 2016), and maize (Sabia et al., 2015), demonstrating their high potential in crop yield enhancement. Furthermore, AMF-colonized strawberries had greater proportions of synthesized metabolites, leading to enhanced antioxidant properties (Castellanos-Morales et al., 2010). By doing so, AMF may improve crop nutritional quality by influencing and producing carotenoids and other health beneficial volatile chemicals. Indeed, Bona et al. (2017) discovered that AMF had a positive influence on tomato quality in terms of fruit quantitative yield and carotenoid content. In another research conducted by Zeng et al. (2014), *Glomus versiforme* was shown to boost the levels of organic acids, sugars, vitamin C, minerals, and flavonoids, resulting in better citrus fruit quality. AMF may also boost the synthesis of beneficial biochemicals in edible plants (Rouphael et al., 2015). Extensive research has also shown the ability of AMF to affect metabolic levels and boost metabolite production in medicinal plants (Sbrana et al., 2014). By doing so, mycorrhization alters plant secondary metabolism, thus sometimes resulting in favorable changes in their therapeutic potentials (Chaudhary et al., 2008). Several attempts have been made to understand the effects of mycorrhizal inoculation on the secondary metabolism in several host medicinal plants, such as *Artemisia annua* L. (Domokos et al., 2018), *Anadenanthera colubrina* (Vell.) Brenan (Kapoor et al., 2007), *Cynara cardunculus* L. var. sco (Smith et al., 2011), and *Valeriana officinalis* L. (Nell et al., 2010). Similarly, Fokom et al. (2019) demonstrated that in field conditions, lemongrass plants inoculated with a Cameroonian AMF inoculum displayed higher biomass production and triggered essential oil output, indicating a change in both primary and secondary metabolic pathways (Fig. 12.3).

Moreover, the analysis of cassava tubers and leaves further highlighted the role of mycorrhizal symbiosis in decreasing their concentration in the antinutrients saponin, phytate, oxalate, and cyanide (Table 12.1).

Indirect benefits of arbuscular mycorrhizal symbiosis

AMF affect plant defense and disease resistance

It is not well understood how plants tolerate AMF colonization without activating a defense mechanism, given that AMF contain a variety of molecular signals on their membranes that are recognized by plants and have been shown to initiate innate immunity in a variety of plant species (Boller and Felix, 2009). As a consequence, it is proposed that AMF repress the defense system of their host. Indeed, when infected with AM fungi, plant mutants, defective in genes important in symbiotic signaling and mycorrhizal establishment, usually display recognizable defense responses, demonstrating that these fungi contain powerful signaling

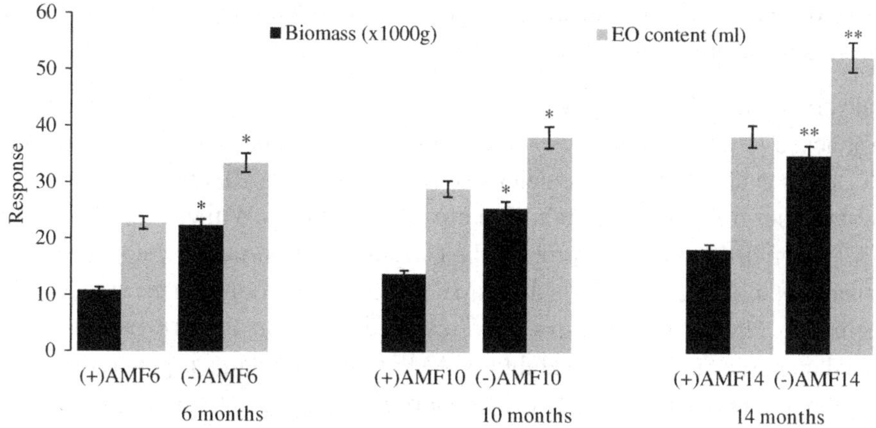

Fig. 12.3
Biological yield and essential oil production by the mycorrhizal *Cymbopogon citratus* (lemongrass) plant and its nonmycorrhizal counterpart under field conditions at 6, 10, and 14 months of age. Bar charts bearing different numbers of asterisks are significantly different at $P \leq 0.05$.

Table 12.1: Antinutritional content of cassava tubers and leaves under AM fungi inoculation in the field.

	Tuber		Leaf	
	AM fungi	NAM fungi	AM fungi	NAM fungi
Saponin (%)	16.16^{d}	18.5^{c}	20.83^{b}	24^{a}
Phytate (%)	0.40^{d}	0.53^{c}	1.10^{b}	1.70^{a}
Oxalate (%)	0.30^{d}	0.64^{c}	6.4^{b}	9.9^{a}
Cyanide (%)	0.0174^{c}	0.0308^{b}	0.0081^{d}	0.0510^{a}

AMF samples are obtained from plants inoculated with arbuscular mycorrhizal fungi; NAM: sample from nonarbuscular mycorrhizal fungi.
The mean values in each column, superscripted with the different letters (a,b,c and d) are significantly different at the statistical threshold of $P \leq 0.05$.

molecules that trigger a defense, which is muted during normal AM growth. Typically, infections produce defense inhibitors (referred to as effectors), and, recently, numerous effectors have been predicted to exist in AMF genomes (Sędzielewska Toro and Brachmann, 2016). Despite this, only a few have been functionally studied (Kloppholz et al., 2011). While host defenses must be reduced to allow AMF invasion and root colonization, the overall defense must remain active to counteract rhizospheric diseases. Indeed, mycorrhizal plants retain their overall disease resistance. On the other hand, mycorrhizal plants often exhibit increased resistance to disease (Jung et al., 2012; Cameron et al., 2013). Experiments with divided root systems show that this effect is often systemic, safeguarding the whole plant from infection. This may be done by either an overall improvement in plant health as a consequence of increased nutrients or through systemic induction of the defense responses, known as systemic acquired resistance (SAR).

Additionally, mycorrhizal plants may be prepared to react more rapidly and consistently to a pathogen attack, a process referred to as induced systemic resistance (ISR) or immune system priming (Conrath et al., 2006). These defensive capabilities of AMF are critical for developing long-term strategies for plant protection (Solaiman et al., 2014). While systemic priming occurs, AMF are mostly utilized to protect plants against soil-borne illnesses (Jung et al., 2012; Cameron et al., 2013). Additionally, AMF and other microorganisms associated with their mycelia may directly fight for space and resources with rhizospheric pathogens or release antimicrobial compounds. While the ability of AMF to protect plants is generally recognized, it is worth noting that in certain situations, mycorrhizal crops may not benefit from AMF and may even demonstrate decreased growth and fitness. Therefore, the above investigation, emphasizing the process of ISR induction or mycorrhiza-induced resistance (MIR) is extensively reported (Jacott et al., 2017). Within the current scenario, whereby pest and disease pressures pose a significant threat to food production systems, AMF could effectively serve to dwarf the suppressive effects of phytopathogens and major pests on crop productivity. It is widely alleged that roughly 35% of crop failure or yield shortage is ascribed to pest and pathogen attacks. Hence, a holistic adoption and implementation of the AMF technology in modern agriculture would not only repress chemical inputs significantly but also assure improved quantity and quality of food to meet the demands of the ever-increasing world population.

Quality of process products from AMF plants

AM fungi not only improve the physiological characteristics of plants but also impact their various life cycles. Three groups of hen feed with different food including (a) the standard ration, (b) ration containing cassava leaves from nonmycorrhizal plants, and (c) ration containing cassava leaves from mycorrhizal plants have different reactions. Chicken fed food fortified with cassava leaves (b and c) have remarkable liveliness, with wings that adhere better to their body compared to chicken fed standard rations. When the cassava leaves used are from mycorrhizal plants, the chicken show better characteristics as mentioned above. A histological section of their striated muscles shows finely sized twitching fibers in hens fed a ration containing cassava leaves from mycorrhizal plants compared to those of hens fed food containing cassava leaves from nonmycorrhizal plants and standard food (Fig. 12.4).

Similarly, three groups of fish (*Clarias gariepinus*) feed with different food rations including the standard ration (control), ration containing cassava leaves from nonmycorrhizal plants (nonmycorrhizal), and ration containing cassava leaves from mycorrhizal plants (mycorrhizal) showed differential reactions with regard to weight gain. The group of fish fed food fortified with cassava leaves from mycorrhizal plants gained more weight than did the control group and those receiving food enriched with cassava leaves from nonmycorrhizal plants (Fig. 12.5).

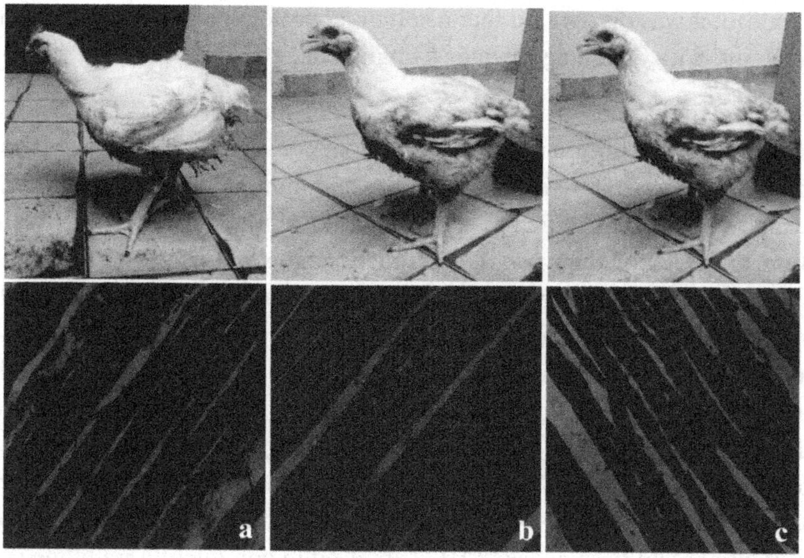

Fig. 12.4
Influence of food quality on the shape and structure of the striated muscles of a hen at 45 days of growth. (A) Feed with standard food. (B) Feed with food enriched with cassava leaves from nonmycorrhizal plants. (C) Feed with food enriched with leaves from mycorrhizal plants.

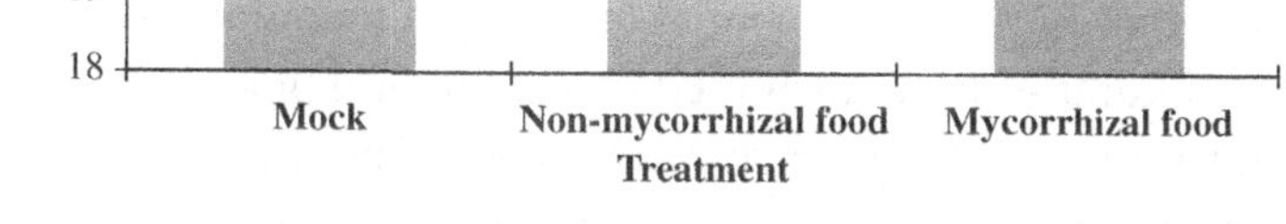

Fig. 12.5
Influence of food quality on the weight gain of fish (*Clarias gariepinus*) at 60 days of age. Control related to fish feed with a standard died. Nonmycorrhiza is related to fish feed with food enriched with cassava leaves from nonmycorrhizal plants. Mycorrhiza is related to fish feed with food enriched with cassava leaves from mycorrhizal plants.

AMF improve soil quality and reduce soil erosion

One of the most important functions performed by AMF in both natural and agricultural contexts is the adjustment of the soil structure to its advantage (Leifheit et al., 2014). Factually, the dense hyphal network of the highly branched AMF mycelia forms a three-dimensional matrix that interconnects soil particles while preventing them from being compacted. Indeed, it was discovered that a soil glycoprotein known as glomalin is critical for maintaining the integrity of soil aggregates (Singh et al., 2013).

It is known as glomalin because it is believed to be generated by AMF. Glomalin is neither a well-defined gene product nor a chemically homogenous molecular species; rather, it is a soil fraction characterized by its extractability and immunoreactivity (Rillig, 2004). Glomalin-related soil proteins (GRSPs) and glomalin itself have lately seen a resurgence in popular culture, even though their origin and function are still unclear. However, despite their short half-lives (estimated to range from several years to decades), they have a substantial influence on soil quality and a relatively stable source of carbon. Grasse-rich soil particles (GRSPs) are believed to account for a significant portion (2%–5%) of total organic soil carbon. Moreover, because they protect other forms of organic carbon from degradation due to enhanced soil particle aggregation, they may make a significant contribution to soil carbon sequestration (Rillig et al., 2001; Wilson et al., 2009). Wind and water erosions are reduced when the hyphal network of AMF and its favorable effects on plant growth and root system development (Gutjahr and Paszkowski, 2013) are used in conjunction with one another to protect soil from wind and water erosions. The cumulative impacts of AM fungi on soil properties result in greater water retention capacity, which is beneficial to plant growth in addition to improved nutrient availability. When it comes to plants growing on dry sandy soils in arid climes, the benefits of AMF are critical. These soils are often infertile and are highly prone to erosions by wind and rain. In such cases, mycorrhizal plantings may prove to be a cost-effective and long-term solution for erosion control and soil fertility improvement.

AMF, in addition to improving soil structure, also aid in the prevention of nutrient loss from soils (Cavagnaro et al., 2015). In addition to causing soil fertility loss and pollution of ground and surface water, nutrient leaching is a major public health concern (lakes, rivers). As a result of the efficiency with which roots and soil bacteria adsorb and retain nutrients in intact ecosystems, they have great potential for nutrient retention (including AMF). On the other hand, agricultural soils are disturbed by agricultural practices (especially plowing) and receive huge quantities of fertilizers, typically in the form of nitrogen, phosphorus, and potassium. Without an adequate nutrient retention system in place, these nutrients, notably the highly mobile nitrate, are susceptible to being washed away from the soil and into groundwater (Cameron et al., 2013). The favorable effects of AM fungi on nutrient leaching occur at several different levels in the soil. The following are some of the benefits of an improved soil structure: first, increased nutrient sequestration in mycorrhizal soil micro- and macroaggregates, second, AMF take up nutrients from the soil solution (Clark and Zeto, 2000; George, 2000), and third, mycorrhizal soils have a higher retention capacity for the soil solution (Querejeta, 2017), thereby increasing the availability of nutrients and water to the plant at the same time. A comprehensive description of the positive impact of AMF on tomato plants under drought stress has just been published in the journal *Plant and Soil* (Bitterlich et al., 2018). Reductions in leaching from mycorrhizal soils have been documented for phosphorus and nitrogen in particular, while this reduction might apply to other mineral nutrients as well. AMF, when considered as a whole, contribute to the integration of soil nutrient fluxes by generating closed nutrient cycles, which help sustain long-term soil fertility (Cavagnaro et al., 2015).

Single versus multiple species-based AMF inoculants for efficiency assurance

Due to a lack of understanding of the specificity of AMF to host plants at intraspecific and interspecific levels, as well as interactions among AMF species and between AMF and other microorganisms, the use of microorganisms in novel agriculture is still modest (Eke et al., 2016). AMF strains and their interactions with one host are the focus of the majority of research, which ignores the tremendous microbial variety within these functional groups that coexist in, on, and around plant roots in natural systems due to the limited number of studies available (Wehner et al., 2010). According to research, AMF species differ in both the expression of features (Pozo and Azcón-Aguilar, 2007) and their abilities to guard their hosts against infection ingresses (Sikes et al., 2009). As a result, assemblages of AMF originating from many species may have a stronger capacity to protect host plants against diseases than a single AMF species. This idea might contribute to the synergistic relationship between the rich variety of AMF in assemblages and the development of individual plants or the diversity and productivity of plant communities associated with these assemblages (Maherali and Klironomos, 2007).

There is evidence of varied responses of mycorrhizal preparations based on the specific fungi–plant relationships that have been observed. The symbiotic behavior of the *Scutellospora* and *Glomus* species differs due to the distinct strategies of nutrient acquisition from the rhizosphere used by the two species. On the other hand, it is hypothesized that combining various species of *Glomus* that have the same positive characteristics would have a detrimental influence on the other partners. The authors also speculated that the variety of AMF infected by the same host might result in higher nutrient intake, which they hypothesized was due to synergistic and/or additive interactions between the partners. These impacts might result in divergence of features such as the ability of fungal mycelia to develop and colonize the host root, their extension out of the root system to colonize the rhizosphere, and the ability of the fungi to take up nutrients from the soil, among others characteristics (Jansa et al., 2005). On the other hand, it is believed that the interaction between AMF inoculum and soil-associated saprophytic fungi would result in unpredictable effects. Such interactions may produce results that are paradoxical among species of the same genus and even between strains of the same species. In the end, because mycorrhizal symbiosis is known to alter soil microbial population composition (Linderman, 1988; Linderman and Paulitz, 1990), proper investigations or the establishment of an artificial consortia of compatible bioagents including AMF and saprophytic organisms or the interactions between the two are essential to predict plausible additive or synergistic effects when applied in a multifaceted environment such as the rhizosphere. Consequently, full knowledge of the interactions between AM fungi and beneficial saprophytic fungi is required for their efficient use in crop promotion while reducing the need for chemical inputs.

Conclusions

Food security is a huge concern to mankind nowadays. In contrast to the current scenario, the earlier boom in world population after the Industrial Revolution was successfully tackled with mechanization, using agrochemicals, modern breeding techniques, genetic engineering, improved planting materials, and enhanced processing methods. However, the drawbacks perpetrated by these so-called methods have imposed a shift in crop production strategies, advocating sustainable measures as they are known to be eco-friendly and assuring lasting climatic resilient measures to withstand modern-day human conflicts. In this regard, arbuscular mycorrhizal fungi have upheld their promise of sustainable crop growth and yield as well as enhanced plant resistance to a plethora of biotic and abiotic stresses. In the current scenario in which the ratio of the population suffering from food insecurity is increasing dramatically, coupled with the COVID-19 pandemic, AMF could constitute a prominent alternative or a key tool to achieving the food security prospects of agricultural practitioners. Nonetheless, complexes rather than individual AMF-based inoculum are likely to be more efficient as they are believed to reinforce each other in a synergistic or additive manner. Efforts should then be taken to develop artificial and compatible bioagent mixtures aimed at improving food security in the near future.

References

Akiyama, K., Matsuzaki, K.I., Hayashi, H., 2005. Plant sesquiterpenes induce hyphal branching in arbuscular mycorrhizal fungi. Nature 435 (7043), 824–827.

Altieri, M.A., 2004. Linking ecologists and traditional farmers in the search for sustainable agriculture. Front. Ecol. Environ. 2 (1), 35–42.

Amiri, R., Ali, N., Nematollah, E., Mohammad, R.S., 2017. Nutritional status, essential oil changes and water-use efficiency of rose geranium in response to arbuscular mycorrhizal fungi and water deficiency stress. Symbiosis 73, 15–25.

Azcón-Aguilar, C., Barea, J.M., 2015. Nutrient cycling in the mycorrhizosphere. J. Soil Sci. Plant Nutr. 25 (2), 372–396.

Bagheri, V., Shamshiri, M.H., Shirani, H., Roosta, H., 2012. Nutrient uptake and distribution in mycorrhizal pistachio seedlings under drought stress. J. Agric. Sci. Technol. 14, 1591–1604.

Bari, R., Jones, J.D.G., 2009. Role of plant hormones in plant defense responses. Plant Mol. Biol. 69 (4), 473–488.

Bever, J.D., Dickie, I.A., Facelli, E., Facelli, J.M., Klironomos, J., Moora, M., Rillig, M.C., Stock, W.D., Tibbett, M., Zobel, M., 2010. Rooting theories of plant community ecology in microbial interactions. Trends Ecol. Evol. 25, 468–478.

Bieleski, R.L., 1973. Phosphate pools, phosphate transport, and phosphate availability. Annu. Rev. Plant Physiol. 24 (1), 225–252.

Bitterlich, M., Franken, P., Graefe, J., 2018. Arbuscular Mycorrhiza improves substrate hydraulic conductivity in the plant available moisture range under root growth exclusion. Front. Plant Sci. 9, 301.

Boller, T., Felix, G.A., 2009. Renaissance of elicitors: perception of microbe-associated molecular patterns and danger signals by pattern-recognition receptors. Annu. Rev. Plant Biol. 60, 379–406.

Bona, E., Cantamessa, S., Massa, N., Manassero, P., Marsano, F., Copetta, A., Lingua, G., D'Agostino, G., Gamalero, E., Berta, G., 2017. Arbuscular mycorrhizal fungi and plant growth-promoting pseudomonads improve yield, quality and nutritional value of tomato: a field study. Mycorrhiza 27 (1), 1–11.

Burchell, M.J., Parnell, J., Bowden, S.A., Crawford, I.A., 2010. Hypervelocity impact experiments in the laboratory relating to lunar astrobiology. Earth Moon Planet. 107 (1), 55–64.

Cameron, D.D., Neal, A.L., Van Wees, S.C.M., Ton, J., 2013. Mycorrhiza-induced resistance: more than the sum of its parts? Trends Plant Sci. 18, 539–545.

Capone, R., Bilali, H.E., Debs, P., Cardone, G., Driouech, N., 2014. Food system sustainability and food security: connecting the dots. J. Food Sec. 2 (1), 13–22.

Castellanos-Morales, V., Villegas, J., Wendelin, S., Vierheiling, H., Eder, R., Cardenas-Navarro, R., 2010. Root colonization by the arbuscular mycorrhizal fungus *Glomus intraradices* alters the quality of strawberry fruit (*Fragaria ananassa* Duch.) at different nitrogen levels. J. Sci. Food Agric. 90, 1774–1782.

Cavagnaro, T.R., Bender, S.F., Asghari, H.R., van der Heijden, M.G.A., 2015. The role of arbuscular mycorrhizas in reducing soil nutrient loss. Trends Plant Sci. 20, 283–290.

Chaudhary, L.V., Kapoor, R., Bhatnagar, K.A., 2008. Effectiveness of two arbuscular mycorrhizal fungi on concentrations of essential oil and artemisinin in three accessions of *Artemisia annua* L. Appl. Soil Ecol. 40 (1), 174–181.

Clark, R.B., Zeto, S.K., 2000. Mineral acquisition by arbuscular mycorrhizal plants. J. Plant Nutr. 23, 867–902.

Conrath, U., Beckers, G.J., Flors, V., García-Agustín, P., Jakab, G., Mauch, F., Newman, M.A., Pieterse, C.M., Poinssot, B., Pozo, M.J., Pugin, A., 2006. Priming: getting ready for battle. Mol. Plant Microbe Interact. 19 (10), 1062–1071.

Domokos, E., Jakab-Farkas, L., Darkó, B., Bíró-Janka, B., Mara Gy, A.C., Balog, A., 2018. Increase in *Artemisia annua* plant biomass content and guaiacol peroxidase activity using the arbuscular mycorrhizal fungus *Rhizophagus irregularis*. Front. Plant Sci. 9, 478.

Eke, P., Chatue, G.C., Wakam, L.N., Kouipou, R.M.T., Fokou, P.V.T., Boyom, F.F., 2016. Mycorrhiza consortia suppress the *fusarium* root rot (*fusarium solani* f. sp. *Phaseoli*) in common bean (*Phaseolus vulgaris* L.). Biol. Control 103, 240–250.

FAO, IFAD, UNICEF, WFP, & WHO, 2021. The state of food security and nutrition in the world 2021. In: Transforming Food Systems for Food Security, Improved Nutrition and Affordable Healthy Diets for all. FAO, Rome.

Fiorilli, V., Catoni, M., Miozzi, L., Novero, M., Accotto, G.P., Lanfranco, L., 2009. Global and cell-type gene expression profiles in tomato plants colonized by an arbuscular mycorrhizal fungus. New Phytol. 184, 975–987.

Fokom, R., Adamou, S., Essono, D., Ngwasiri, D.P., Eke, P., Teugwa, M.C., Tchoumbougnang, F., Fekam, B.F., Amvam, Z.P.H., Nwaga, D., Sharma, A.K., 2019. Growth, essential oil content, chemical composition and antioxidant properties of lemongrass as affected by harvest period and arbuscular mycorrhizal fungi in field conditions. Ind. Crop Prod. 138, 111477.

Food and Agriculture Organization of the United Nations, 2017. The Future of Food and Agriculture—Trends and Challenges. FAO, Rome.

Garcés-Ruiz, M., Calonne-Salmon, M., Plouznikoff, K., Misson, C., Navarrete-Mier, M., Cranenbrouck, S., Declerck, S., 2017. Dynamics of short-term phosphorus uptake by intact mycorrhizal and nonmycorrhizal maize plants grown in a circulatory semi-hydroponic cultivation system. J. Front. Plant Sci. 8, 1–8.

García-Garrido, J.M., León-Morcillo, R.J., Martín-Rodríguez, J.A., Ocampo-Bote, J.A., 2010. Variations in the mycorrhization characteristics in roots of wild-type and ABA-deficient tomato are accompanied by specific transcriptomic alterations. Mol. Plant Microbe Interact. 23, 651–664.

Gebre, G.G., Rahut, D.B., 2021. Prevalence of household food insecurity in East Africa: linking food access with climate vulnerability. Clim. Risk Manag. 33, 100333.

Genre, A., Chabaud, M., Faccio, A., Barker, D.G., Bonfante, P., 2008. Prepenetration apparatus assembly precedes and predicts the colonization patterns of arbuscular mycorrhizal fungi within the root cortex of both Medicago truncatula and Daucus carota. Plant Cell 20 (5), 1407–1420.

George, E., 2000. Nutrient uptake—contributions of arbuscular mycorrhizal fungi to plant mineral nutrition. In: Kapulnik, Y., Douds, D.D. (Eds.), Arbuscular Mycorrhizas: Physiology and Function. Kluwer Academic Publishers, Dordrecht, pp. 307–343.

Gianinazzi, S., Golotte, A., Binet, M.N., Van Tuinen, D., Redecker, D., Wipf, D., 2010. Agroecology: the key role of arbuscular mycorrhizas in ecosystem services. Mycorrhiza 20, 519–530.

Gomez-Bellot, M.J., Ortuño, M.F., Nortes, P.A., Vicente-Sánchez, J., Bañón, S., Sánchez Blanco, M.J., 2015. Mycorrhizal euonymus plants and reclaimed water: biomass, water status and nutritional responses. Sci. Hortic. 186, 61–69.

Guo, M., Rupe, M.A., Danilevskaya, O.N., Yang, X., Hu, Z., 2010. Genome-wide mRNA profiling reveals heterochronic allelic variation and a new imprinted gene in hybrid maize endosperm. Plant J. Cell Mol. Biol. 36, 30–44.

Gutjahr, C., Paszkowski, U., 2013. Multiple control levels of root system remodelling in arbuscular mycorrhizal symbiosis. Front. Plant Sci. 4, 204.

Harrison, M., 2005. Signalling in the arbuscular mycorrhizal symbiosis. Annu. Rev. Microbiol. 59, 19–42.

Harrison, M.J., 2012. Cellular programs for arbuscular mycorrhizal symbiosis. Curr. Opin. Plant Biol. 15, 691–698.

Hause, B., Schaarschmidt, S., 2009. The role of jasmonates in mutualistic symbioses between plants and soil-borne microorganisms. Phytochemistry 70, 1589–1599.

Hause, B., Mrosk, C., Isayenkov, S., Strack, D., 2007. Jasmonates in arbuscular mycorrhizal interactions. Phytochemistry 68, 101–110.

Hijri, M., 2016. Analysis of a large dataset from field mycorrhizal inoculation trials on potatoes showed a highly significant increase in yield. Mycorrhiza 2, 209–214.

Hocher, V., Alloisio, N., Auguy, F., Fournier, P., Doumas, P., Pujic, P., Gherbi, H., Queiroux, C., Da Silva, C., Wincker, P., Normand, P., Bogusz, D., 2011. Transcriptomics of actinorhizal symbioses reveals homologs of the whole common symbiotic signaling cascade. Plant Physiol. 156 (2), 700–711.

Jacott, C.N., Murray, J.D., Ridout, C.J., 2017. Trade-offs in arbuscular mycorrhizal symbiosis: disease resistance, growth responses and perspectives for crop breeding. Agronomy 7, 75.

Jansa, J., Mozafar, A., Frossard, E., 2005. Phosphorus acquisition strategies within the arbuscular mycorrhizal fungal community of a single field site. Plant and Soil 276, 163–176.

Jiang, Y., Wang, W., Xie, Q., Liu, N., Liu, L., Wang, D., Zhang, X., Yang, C., Chen, X., Tang, D., Wang, E., 2017. Plants transfer lipids to sustain colonization by mutualistic mycorrhizal and parasitic fungi. Science 356 (6343), 1172–1175.

Johansson, J.F., Paul, L.R., Finlay, R.D., 2004. Microbial interactions in the mycorrhizosphere and their significance for sustainable agriculture. FEMS Microbiol. Ecol. 48, 1–13.

Jung, S.C., Martinez-Medina, A., Lopez-Raez, J.A., Pozo, M.J., 2012. Mycorrhiza-induced resistance and priming of plant defences. J. Chem. Ecol. 38, 651–664.

Kapoor, R., Chaudhary, V., Bhatnagar, A.K., 2007. Effects of arbuscular mycorrhiza and phosphorus application on artemisinin concentration in *Artemisia annua* L. Mycorrhiza 17, 581–587.

Kayama, M., Yamanaka, T., 2014. Growth characteristics of ectomycorrhizal seedlings of *Quercus glauca*, *Quercus salicina*, and *Castanopsis cuspidata* planted on acidic soil. Trees 28, 569–583.

Kloppholz, S., Kuhn, H., Requena, N., 2011. A secreted fungal effector of Glomus intraradices promotes symbiotic biotrophy. Curr. Biol. 21 (14), 1204–1209.

Leifheit, E.F., Veresoglou, S.D., Lehmann, A., Morris, E.K., Rillig, M.C., 2014. Multiple factors influence the role of arbuscular mycorrhizal fungi in soil aggregation-a meta-analysis. Plant and Soil 374, 523–537.

Linderman, R.G., 1988. Mycorrhizal interactions with the rhizosphere microflora–the mycorrhizosphere effect. Phytopathology 78, 366–371.

Linderman, R.G., Paulitz, T.C., 1990. Mycorrhizal-rhizobacterial interactions. In: Hornby (Ed.), Biological control of soil-borne plant pathogens. CAB International, Redwood Press Ltd., Melksham, Wiltshire, Great Britain, pp. 261–283.

Liu, J.Y., Maldonado-Mendoza, I., Lopez-Meyer, M., Cheung, F., Town, C.D., Harrison, M.J., 2007. Arbuscular mycorrhizal symbiosis is accompanied by local and systemic alterations in gene expression and an increase in disease resistance in the shoots. Plant J. 50, 529–544.

Liu, L.Z., Gong, Z.Q., Zhang, Y.L., Li, P.J., 2014. Growth, cadmium uptake and accumulation of maize *Zea mays* L. under the effects of arbuscular mycorrhizal fungi. Ecotoxicology 23, 1979–1986.

Liu, C., Ravnskov, S., Liu, F., Rubæk, G.H., Andersen, M.N., 2018. Arbuscular mycorrhizal fungi alleviate abiotic stresses in potato plants caused by low phosphorus and deficit irrigation/partial root-zone drying. J. Agric. Sci. 156, 46–58.

Lu, F., Lee, C., Wang, C., 2015. The influence of arbuscular mycorrhizal fungi inoculation on yam (*Dioscorea* spp.) tuber weights and secondary metabolite content. Peer J 3, 12–66.

Ludwig-Muller, J., Guther, M., 2007. Auxins as signals in arbuscular mycorrhiza formation. Plant Signal. Behav. 2, 194–196.

Maherali, M., Klironomos, J.N., 2007. Influence of phylogeny on fungal community assembly and ecosystem functioning. Science 316, 1746–1748.

Maillet, F., Poinsot, V., André, O., Puech-Pagès, V., Haouy, A., Gueunier, M., Cromer, L., Giraudet, D., Formey, D., Niebel, A., Martinez, E.A., Dénarié, J., 2011. Fungal lipochitooligosaccharide symbiotic signals in arbuscular mycorrhiza. Nature 469 (7328), 58–63.

Marsh, J., Schultze, M., 2001. Analysis of arbuscular mycorrhizas using symbiosis-defective plant mutants. New Phytol. 150, 525–532.

Meena, V.S., Maurya, B.R., Verma, J.P., Meena, R.S., 2016. Potassium Solubilizing Microorganisms for Sustainable Agriculture. Springer, New Delhi, pp. 1–20.

Mitra, D., Navendra, U., Panneerselvam, U., Ansuman, S., Ganeshamurthy, A.N., Divya, J., 2019. Role of mycorrhiza and its associated bacteria on plant growth promotion and nutrient management in sustainable agriculture. Int. J. Life Sci. Appl. Sci. 1, 1–10.

Nell, M., Wawrosch, C., Steinkellner, S., Vierheilig, H., Kopp, B., Lössl, A., 2010. Root colonization by symbiotic arbuscular mycorrhizal fungi increases sesquiterpene acid concentrations in *Valeriana officinalis* L. Planta Med. 76, 393–398.

Parniske, M., 2008. Arbuscular mycorrhiza: the mother of plant root endosymbiosis. Nat. Rev. Genet. 6, 763–775.

Pichardo, S.T., Su, Y., Han, F.X., 2012. The potential effects of arbuscular mycorrhizae (AM) on the uptake of heavy metals by plants from contaminated soils. J. Bioremed. Biodegr. 3 (10), 1–4.

Pozo, M.J., Azcón-Aguilar, C., 2007. Unravelling mycorrhiza-induced resistance. Curr. Opin. Plant Biol. 10, 393–398.

Prasad, R., Bhola, D., Akdi, K., Cruz, C., Sairam, K.V.S.S., Tuteja, N., Varma, A., 2017. Introduction to mycorrhiza: historical development. In: Mycorrhiza-Function, Diversity, State of the Art. Springer, Cham, pp. 1–7.

Pretty, J., 2008. Agricultural sustainability: concepts, principles and evidence. Philos. Trans. R. Soc. B 363, 447–466.

Pretty, J., Bharucha, Z.P., 2014. Sustainable intensification in agricultural systems. Ann. Bot. 114, 1571–1596.

Querejeta, J.I., 2017. Soil water retention and availability as influenced by mycorrhizal symbiosis: consequences for individual plants, communities, and ecosystems. In: Mycorrhizal Mediation of Soil. Elsevier, pp. 299–317.

Redecker, D., Schüssler, A., Stockinger, H., Stürmer, S.L., Morton, J.B., Walker, C., 2013. An evidence-based consensus for the classification of arbuscular mycorrhizal fungi (Glomeromycota). Mycorrhiza 23 (7), 515–531.

Rillig, M.C., 2004. Arbuscular mycorrhizae, glomalin and soil quality. Can. J. Soil Sci. 84, 355–363.

Rillig, M.C., Wright, S.F., Nichols, K.A., Schmidt, W.F., Torn, M.S., 2001. Large contribution of arbuscular mycorrhizal fungi to soil carbon pools in tropical forest soils. Plant and Soil 233, 167–177.

Rouphael, Y., Franken, P., Schneider, C., Schwarz, D., Giovannetti, M., Agnolucci, M., 2015. Arbuscular mycorrhizal fungi act as bio-stimulants in horticultural crops. Sci. Hortic. 196, 91–108.

Sabia, E., Claps, S., Morone, G., Bruno, A., Sepe, L., Aleandri, R., 2015. Field inoculation of arbuscular mycorrhiza on maize (*Zea mays* L.) under low inputs: a preliminary study on quantitative and qualitative aspects. Ital. J. Agron. 10, 30–33.

Saha, M., Maurya, B.R., Meena, V.S., Bahadur, I., Kumar, A., 2016. Identification and characterization of potassium solubilizing bacteria (KSB) from indo-Gangetic Plains of India. Biocatal. Agric. Biotechnol. 38 (8), 202–209.

Samim, S.A., Hu, Z., Stepien, S., Amini, S.Y., Rayee, R., Niu, K., Mgendi, G., 2021. Food insecurity and related factors among farming families in Takhar region, Afghanistan. Sustainability 13 (18), 10211.

Sbrana, C., Avio, L., Giovannetti, M., 2014. Beneficial mycorrhizal symbionts affecting the production of health-promoting phytochemicals. Electrophoresis 35 (11), 1535–1546.

Schliemann, W., Ammer, C., Strack, D., 2008. Metabolite profiling of mycorrhizal roots of *Medicago truncatula*. Phytochemistry 69 (1), 112–146.

Sędzielewska Toro, K., Brachmann, A., 2016. The effector candidate repertoire of the arbuscular mycorrhizal fungus *Rhizophagus clarus*. BMC Genomics 17 (101), 1–17.

Shorette, K., 2012. Outcomes of global environmentalism: Longitudinal and cross-national trends in chemical fertilizer and pesticide use. Soc. Forces 91 (1), 299–325.

Sikes, B.A., Cottenie, K., Klironomos, J.N., 2009. Plant and fungal identity determine pathogen protection of plant roots by arbuscular mycorrhizas. J. Ecol. 97 (6), 1274–1280.

Singh, B., Sharma, A., Joshi, N., Mittal, P., Singh, S., 2013. Combining ability for grain yield and its components in barley (*Hordeum vulgare*). Indian J. Agric. Sci. 83, 96–98.

Smith, S.E., Read, D.J., 1997. Mycorrhizal Symbiosis, second ed. Academic Press, London.

Smith, S.E., Read, D.J., 2008. Mycorrhizal Symbiosis, third ed. Academic Press, London.

Smith, S.E., Jakobsen, I., Grnlund, M., Smith, F.A., 2011. Roles of arbuscular mycorrhizas in plant phosphorus nutrition: interactions between pathways of phosphorus uptake in arbuscular mycorrhizal roots have important implications for understanding and manipulating plant phosphorus acquisition. Plant Physiol. 156 (3), 1050–1057.

Solaiman, Z.M., Abbott, L.K., Varma, A., 2014. Mycorrhizal fungi: use in sustainable agriculture and land restoration. In: Soil Biology. vol. 41. Springer-Verlag, p. 415.

Spatafora, J.W., Chang, Y., Benny, G.L., Lazarus, K., Smith, M.E., Berbee, M.L., Bonito, G., Corradi, N., Grigoriev, I., Gryganskyi, A., James, T.Y., Stajich, J.E., 2016. A phylum-level phylogenetic classification of zygomycete fungi based on genome-scale data. Mycologia 108 (5), 1028–1046.

Sun, Z., Song, J., Xin, X., Xie, X., Zhao, B., 2018. Arbuscular mycorrhizal fungal proteins 14-3-3- are involved in arbuscular formation and responses to abiotic stresses during AM symbiosis. Front. Microbiol. 5, 9–19.

Taylor, L.L., Banwart, S.A., Valdes, P.J., Leake, J.R., Beerling, D.J., 2012. Evaluating the effects of terrestrial ecosystems, climate and carbon dioxide on weathering over geological time: a global-scale process-based approach. Philos. Trans. R. Soc. B 367, 565–582.

Tian, D., Niu, S., 2015. A global analysis of soil acidification caused by nitrogen addition. Environ. Res. Lett. 10 (2), 024019.

Wagg, C., Barendregt, C., Jansa, J., Heijden, M.G.A., 2015. Complementarity in both plant and mycorrhizal fungal communities are not necessarily increased by diversity in the other. J. Ecol. 103, 1233–1244.

Wang, Y., Wang, M., Li, Y., Wu, A., Huang, J., 2018. Effects of arbuscular mycorrhizal fungi on growth and nitrogen uptake of *Chrysanthemum morifolium* under salt stress. PLoS One 13 (4), e0196408.

Wehner, J., Pedro, M.A., Powell, J.R., Mazukatow, J., Rillig, C.M., 2010. Plant pathogen protection by arbuscular mycorrhizas: a role for fungal diversity. Pedobiologia 53 (3), 197–201.

Wilson, M.R., Wood, G., Vine, S.J., 2009. Anxiety, attentional control, and performance impairment in penalty kicks. J. Sport Exerc. Psychol. 31 (6), 761–775.

Yadav, B.K., Sidhu, A.S., 2016. Dynamics of potassium and their bioavailability for plant nutrition. In: Potassium Solubilizing Microorganisms for Sustainable Agriculture. Springer, New Delhi, pp. 187–201.

Zeng, L., Li, J., Liu, J., Wang, M., 2014. Effects of arbuscular mycorrhizal (AM) fungi on citrus fruit quality under nature conditions. Southwest China J. Agric. Sci. 27 (5), 2101–2105.

Zuccaro, A., Lahrmann, U., Langen, G., 2014. Broad compatibility in fungal root symbioses. Curr. Opin. Plant Biol. 20, 135–145.

CHAPTER 13

Microbe-mediated abiotic stress management for sustainable agriculture

Satish Chandra Pandey[a], Veni Pande[a], Diksha Sati[a], Amir Khan[b], Ajay Veer Singh[b], Arjita Punetha[c], Yogita Martoliya[d], and Mukesh Samant[a]

[a]*Cell & Molecular Biology Laboratory, Department of Zoology, Soban Singh Jeena University, Almora, Uttarakhand, India,* [b]*Department of Microbiology, College of Basic Sciences and Humanities, Govind Ballabh Pant University of Agriculture and Technology, Pantnagar, Uttarakhand, India,* [c]*CSIR-Central Institute of Medicinal & Aromatic Plants (CIMAP), Research Centre, Pantnagar, Uttarakhand, India,* [d]*School of Life Sciences, Jawaharlal Nehru University, New Delhi, India*

Chapter outline

Unravelling Plant-Microbe Synergy. https://doi.org/10.1016/B978-0-323-99896-3.00015-1

Introduction

Plants that are anchored to the soil are exposed to a variety of abiotic challenges during their lifetime, and they must evolve systems to cope with these pressures in order to live. High and low temperatures, droughts, floods, salt, cold, pH shifts, intense light, UV, and heavy metals are all particularly damaging to agriculture. Such unfavorable environmental circumstances have a severe effect on crop productivity, which has the capability to become a serious food security issue, especially in tropical areas. Abiotic stress management in plants should be prioritized, taking into consideration that the methodology used should be both environmentally friendly and cost-effective. Agriculture is facing significant problems. To this end, substantial research is being conducted globally to create techniques to deal with abiotic pressures, such as the introduction of heat- and drought-tolerant varieties, crop calendar shifts, and resource management practices, among others (Venkateswarlu and Shanker, 2009). While most of these technologies may not reach farmers due to their high cost, there is another technique that holds high promise for helping plants endure abiotic stresses while also being extremely environmentally safe and cost-effective. This technique entails combining the multifaceted features of various microorganisms that have a history of promoting plant growth, nutrient management, and disease control. Some reports on the use of such microorganisms for inducing tolerance to abiotic conditions have surfaced in the last two decades. The term 'induced systemic tolerance' (IST) was coined by Yang et al. (2009) to describe changes in plants caused by plant growth-promoting rhizobacteria (PGPR), which result in increased abiotic stress tolerance. Plants have a close symbiotic relationship with a group of fungi called arbuscular mycorrhizal fungi (AMF), which provide plants with enhanced nutritional absorption as well as increased tolerance to abiotic and biotic stresses, while the fungi gain a protected ecological niche and plant photosynthates (Ruíz-Lozano, 2003). This plant–AMF interaction can be effectively used to construct abiotic stress control methods. In this chapter, we cover research findings in the field of abiotic stress relief by microorganisms as well as the processes behind the reported alleviation.

Abiotic stresses and their impact on plant growth and development

Unfavorable conditions are commonly imposed on plants by living or naturally occurring nonliving entities. Temperatures, heavy metals, droughts, salt, and UV radiation are examples of nonliving factors that have significant negative consequences on plant survival, biomass production, and yields and are known as abiotic stresses (as well as a division of plant species in diverse habitats) (Boyer, 1982; Khan et al., 2020). In excellent environmental conditions, plants develop at an exponential rate, but when plants experience any abiotic stress, many severe consequences are exerted, such as ion and nutrient imbalance and cell organelle disruption. In addition, plants respond to mild stress, trying to adapt to or sustain themselves under those conditions, which results in the commencement of a variety of transcription factors such as activation and protein formation that aid in combating the mild stress effects. However, under high stress, most crops cannot survive and their survival, yield, and biomass production are reduced by ~70%, posing a threat to the world's food supply (Khalid et al., 2019). In a large number of woody plants, abiotic stress suppresses leaf extension, internode lengthening, and leaf initiation, resulting in premature leaf senescence (Vinocur and Altman, 2005).

Temperature

Plant physiology is modulated by temperature in different ways, including heating, chilling, and freezing. Long-lasting coverage of plants by extremely high temperatures can cause significant harm and even death. Plant growth stages and cellular processes define the vulnerability that directs the nature of the injury. Extreme heat does more than just harm a plant's cells; it also disrupts a slew of intricate processes and structures, ultimately leading to its demise. At high temperatures, various cellular changes occur, including alteration of lipoic acids and reactive oxygen species (ROS) production, which disrupt the metabolism of cellular membranes. Proteins are most effective at certain temperatures, but when that temperature is exceeded, they become inactive, changing the action potential of enzymes and boosting reactive oxygen species (ROS) production (Kumar, 2020). Menopausal dysfunction, pollen tube growth, pollen germination, and megagametophyte surrender are a few outcomes of high-temperature stress (Song et al., 2015). Rice spikes/florets are also reduced through heat stress. A modest reduction in carotenoids, including β-carotene, has been reported after high-temperature treatments (Khalid et al., 2019). When a plant is exposed to active oxygen species (AOS), it suffers from diminished photosynthesis, increased oxidative stress, and altered assimilate transport, all of which are fatal to the plant (Khalid et al., 2019). During plant responses, ROS can be combated using nonenzymatic and enzymatic antioxidants (such as proline). However, antioxidants have little influence on AOS activity, which is frequent in environments where temperatures are particularly high (Hasanuzzaman et al., 2020). In response to abiotic stress incidence, different types of mechanisms and physiological changes are adopted by plants, for example, when plants are exposed to extreme temperatures, they create specific proteins to help them deal with the stress. These proteins are called heat-shock proteins (HSPs) (Boston et al., 1996). Mitigating heat stress via diverse pathways involves HSPs as the only component of heat tolerance.

Low temperatures also have a negative impact on plant development and physiology. The integrity of the cellular membranes is a prime target for cold stress. A decrease in temperature alters membrane shape, protoplasmic streaming, and electrolyte leakage/plasmolysis. When plants like maize, cotton, soybean, rice, mango, and tomato are planted in tropical or subtropical areas, they are more liable to freezing damage. The introduction of plants into a low-temperature environment changes the state of cell membrane lipids from liquid to solid, primarily owing to the amplitude of the unsaturated fatty acids that they contain (Waqas et al., 2019). In response to cooling or frost stress, various phenomena in plant cell membranes occur, which change the chemical and physical states that increase membrane permeability and pH imbalances that reduce the amount of ATP (Levitt, 1980). Frosty temperatures also have an impact on mineral uptake, photosynthetic respiration, and metabolism. A shift in the transcription of genes associated with the creation of many low-molecular-weight proteins is linked to plants' responses to low temperatures. During the progression known as 'cold acclimation,' plants engage a variety of physiological and biochemical systems to deal with freezing. Physical, biological, and genetic factors all play a role in cold acclimation. Freezing temperatures cause changes in plants. Nevertheless, freezing tolerance processes are not linked to a single gene that is able to deal with the cold. Low temperatures can cause freezing injury (damage to the cell membrane). Thus, plants create several types of osmoprotectants and osmolytes like proline, lipids, sugars, and glycine betaine to prevent low temperature impact on plant development and physiology (Waqas et al., 2019).

Salinity

Salinity stress occurs when the salt content rises in the soil milieu owing to the evaporation/transpiration rates becoming higher than the precipitation rates. Agricultural productivity and crop yields are being restricted worldwide due to salt stress, which is a serious, severe issue in semiarid and arid countries. Salinity decreases the crop output and productivity of about 20% of arable lands and 50% of irrigated lands (Flowers, 2004). The principal center of fundamental research is currently on the way to improve plant development and yield in salt-affected soils. Because salt stress is also lethal to plant life as early phases of plant development are more exposed to salinity compared to later stages, lower humidity negatively impacts plant growth and production. Seed germination is hampered by various factors like drought sensitivity and salt stress. Salt-stressed seeds take a longer time to germinate, and germination takes place at lower rates and speeds, which reduces plant growth, incorrect development, and lower yields in the end. It is possible that soil salinity shapes seed germination, either by increasing the osmotic potential outside the seeds and so preventing water intake or by toxically affecting developing seeds with Na^+ and Cl^- ions (Khajeh-Hosseini et al., 2003). Salt reduces shoot development by stopping leaf creation and internode growth and hastening leaf abscission (Singh et al., 2021). Decreased growth may be connected to chloride accumulation, which may cause leaf shedding in all plants, including gymnosperms and angiosperms (Tanveer and Yousaf, 2020). Cell development, membrane function, and cytosolic metabolism are all harmed by salt stress. To combat the upshot of salinity, plants respond in diverse ways. Plant salinity adjustment involves phenological and physiological reflexes that are obligatory in plant health management in a salty environment but could reduce production (Tanveer and Yousaf, 2020). Various tomato species, mostly wild species and primitive cultivars, hold genes responsible for salt tolerance (Cuartero et al., 1992).

Drought stress

Water is critical to plants since it serves multiple purposes. According to Kramer and Boyer (1995), herbaceous plants contain 90% water of their fresh weight. Water helps in the respiration process by keeping plant cells turgor-free. Water is used as a solvent because it has a number of biophysical properties (such as high-temperature vaporization and higher surface tension). It acts as an electron donor during photosynthesis and hence is crucial in metabolic processes. Water limiting conditions are a challenge to plant expansion and production, declining plant health, productivity, and microbial diversity (Waqas et al., 2019; Khan and Singh, 2021). Roughly 35% of the earth's crust surface is either completely dry or semidry. The only way to get water is to wait for the rain. Rainfall that is sufficient but unevenly disseminated across all areas also causes water shortages that have a negative impact on crop production. The amount of moisture conserved in soils has an impact on seedling germination and growth. Lack of water during grain loading reduces the amylose levels of wheat starch, which has been explained by the restriction of granule-bound starch synthase (GBSS) activity (an enzyme involved in amylose synthesis). Soil moisture can limit an embryo's access to oxygen, causing seed germination to occur faster or later (Ali and Elozeiri, 2017). Droughts are responsible for about half of all global yield losses. Different processes such as root establishment and branching or enlargement of mature roots are the ways by which soil moisture reduces root nutrient uptake and plant development in diverse ecosystems. Drought stress impacts plants through numerous means including hampering cell division, reducing turgor pressure, causing hormonal imbalance, increasing Na^+ uptake, and accumulating ABA and ethylene. It also disturbs the membrane integrity, reduces immune response and carbohydrate translocation, and lowers internal CO_2 concentration and chlorophyll and carotenoid content, which ultimately affect the germination, seed weight, growth, and yield (Cramer et al., 2011).

Flooding occurs when there is an excessive quantity of water in a particular area. Waterlogging and submergence are the two forms of excessive water conditions or floods. They both interfere with a plant's capacity to get enough oxygen, resulting in stumpy oxygen content or hypoxia. Generally, land floods are caused by overirrigation, poor drainage, and heavy rainfall. Waterlogging is a major issue nowadays; it not only occurs in locations with heavy rainfall but also in those places where high irrigation is practiced. Pitiable drainage and water leakage via water channels cause flooding in 0.7 million acres and waterlogging in 60,000 acres of land every year (Khalid et al., 2019). In some soils, sodium can also produce waterlogging due to sodicity, and this can lead to infiltration. An augmentation in the number of Na^+ ions in the soil inhibits soil pores, preventing free movement of air and water, resulting in waterlogging (Martínez-Alvarez et al., 2018). A wide range of impacts on plants occur when exposed to flood stress: decreased photosynthesis, decrement in stomatal conductance, and reduced chlorophyll and Rubisco (or RuBisCO). Plants respond differently, based on their developmental stage, to waterlogging circumstances. Waterlogging circumstances significantly lessen a plant's yield and productivity. Dormant plants have a minor impact that is only noticeable for a short period. Under waterlogging stress, plant characteristics such as seed germination resistance, reduced vegetative and reproductive growth, and altered structures all occur more quickly (Cramer et al., 2011).

Heavy metal stress

Metals with a high atomic number, weight, and density are called heavy metals. To grow and develop, plants use most of the nutrients from the soil. There are many essential metals like Zn Cu, Se, Co, Ni, and Fe that are needed by plants to perform at their best. Such heavy metals can be harmful in high concentrations. Excessive accumulation in soil makes them hazardous. The discharge of industrial wastes, burning of solid/liquid fuels, disposal of household wastes, and other activities in developing countries pollute the surroundings with toxic heavy metals. One of the foremost causes of crop yield declines is heavy metal pollution of soil. The presence of certain nonessential constituents such as cadmium, silicon (Si), and lead (Pb) in soil reduces crop plant development and yield. Heavy metals greatly impact seed germination, dry weight, root and shoot length, and total soluble protein levels. Other impacts comprise membrane damage, oxidative damage, sugar and protein metabolic changes, and nutrient loss, all of which contribute to the toxicity of seeds and the consequent loss of yield (Chibuike and Obiora, 2014).

Heavy metal pollution in soil increases the risk to animals and people's health because it accumulates in plant parts, which, upon consumption, is transported to animals and humans. Nickel (Ni) inhibits enzyme activity, e.g., amylase, protease, and ribonuclease, which slows seed germination and plant growth in numerous crops. The impacts of Ni stress include reduced yield, disruption of pigments, and the accretion of sodium, potassium, and calcium ions. Numerous proteins require copper, including photosynthetic pigments such as plastocyanin, and the cytochrome oxidase enzyme in the electron transport chain (Asati et al., 2016). Overconcentration of copper in soils has a cytotoxic effect, which increases plant stress, slowing down plant development, causing chlorosis on the leaves, and, finally, damaging the plants (Asati et al., 2016). Plants exposed to excessive Cu experience oxidative stress and produce reactive oxygen species (ROS) (Waqas et al., 2019). Oxidative stress disrupts metabolic pathways and damages macromolecules. Plant cells are extremely poisonous when they encounter high concentrations of Hg^{2+}. A toxic quantity of Hg^{2+} in plants can cause visual damage and physiological problems (Azevedo and Rodriguez, 2012). Water channel proteins binding to Hg^{2+} can result in the closure of the plant stomata and physical restriction of water flow. A high magnitude of Hg^{2+} interferes with mitochondrial action and causes oxidative stress. This causes harm to biological membrane lipids and results in altered cellular metabolism in plants (Waqas et al., 2019). Seed shape and physiology have been reported to be adversely affected by exposure to lead (Pb). Pb exposure hampers root extension, seed germination and development, transpiration, plant growth, chlorophyll production, and protein content, which alter chloroplast functioning, obstruct the electron transportation chain, inhibit Calvin cycle enzyme, and reduce uptake of vital elements like Mg and Fe (Yadav, 2010).

UV radiation

About 8–9% of solar radiation is UV light, which belongs to the nonionizing area of the electromagnetic spectrum. The wavelengths of ultraviolet light are generally split into three categories: UV-A (320–400 nm), UV-B (280–320 nm), and UV-C (200–280 nm). UV-C is exceptionally injurious to organisms, and UV-B has gained interest because although it represents only about 1.5% of the total spectrum, it can cause damaging effects on plants, which are lesser than those of UV-C. However, UV-A occupies around 6.3% of the whole incoming solar light and is characterized as the less dangerous fraction of UV radiation. Plants must capture light for the synthesis of food via photosynthesis, and thus they are constantly exposed to ultraviolet radiation coming from the sun. It is reported that contact with UV light, particularly in its UV-B band (280–320 nm), regulates plant development. However, UV-B radiations are substantial stressors for plants, with potentially harmful effects. Tissue necrosis and pathogen defense pathway activation, changes in photosynthesis and transpiration, morphology and development, and plant reproductive processes have been reported in many systems as plant responses during UV-B exposure (Hollósy, 2002). UV radiation can cause mutations in DNA during replication since DNA is the foremost recognized target. UV radiation can cause breakage of DNA strands, DNA and protein cross-linking, and base-pair deletions or insertions (Rastogi et al., 2010). Tyrosine could be photooxidized to 4,4-dihidroxyphenylalanine and to dityrosine upon UV-B exposure (Creed, 1984); all of these cause mutations in the plants and result in abnormal growth. Cysteine's disulfide bonding also splits into sulfhydryl group reactive forms upon UV exposure, which has significant implications in protein folding and functions (Creed, 1984). Ultraviolet radiation not only modifies or degrades amino acid residues but also renders proteins and enzymes inactive. A study found that exposing *Petunia hybrida* protoplasts to UV light condensed the organization of cortical microtubules (Staxén et al., 1993). Tubulin contains a high concentration of aromatic amino acids, making it a highly vulnerable target for UV radiation (Zaremba et al., 1984). Ultraviolet exposure slows cell cycle progression, particularly in the S, G1, and G2 phases. UV absorption by lipids is the reason for modified lipids. Unsaturated fatty acids are found in phospholipids and glycolipids that are crucial components of plant membranes. When exposed to UV-B irradiation in the presence of oxygen, these unsaturated fatty acids are destroyed. The large amplitude of UV-B radiation can harm both chlorophylls and carotenoids. The plants' epidermis serves as an important barrier against the potentially lethal effects of ultraviolet light. In response to UV exposure, the accretion of flavonoids and other phenolics in epidermal layers is enhanced and can be used in UV-absorbing sunscreens. Moreover, another inherent enzyme, i.e., photolyase, splits cyclobutane pyrimidine dimers directly through photoreactivation repair and protects plants from mutations (Siipola et al., 2015).

Alleviation of abiotic stress in plants by microorganisms

Agriculture plays an extremely important role in society as it fulfills our basic need for food among others. In India, about 58% of the population is directly dependent on agriculture for its livelihood. So, the growth and development of major crops is a concern for everyone. Increases in population and industrialization are the main reasons for depletion of natural resources and also for environmental stresses (Wan et al., 2012). These environmental stresses are termed 'abiotic stress' in plants. The major abiotic stresses are temperatures, floods, salinity, heavy metals, pH, etc., which severely affect plant growth and productivity. Climate change is one of the problems that hamper crop productivity worldwide. Some microorganisms that are associated with plants have an important role in fighting against these abiotic stresses. Microbes are those that are present in close proximity to plants and help them in many different ways (McNear, 2013). Some are important for biocontrol and some for the accomplishment of nutrient availability. Microorganisms can undergo a quick metabolic change and help plant growth in adverse conditions. They produce and secrete a variety of biomolecules that help in plant growth and productivity. Microbes produce the ACC deaminase enzyme, secondary metabolites, siderophores, EPSs, microbial volatile compounds, etc., which help plants withstand stressful circumstances (Inbaraj, 2021).

Production of plant hormones

Many microorganisms produce phytohormones, which are used by plants to grow their roots and increase the root surface area so that they can get enough nutrition for their growth and development. Kang and So (2016) reported an increase in the content of GA and ABA and a reduction in the content of jasmonate and salicylate in pepper plants inoculated with gibberellin-producing PGPR. This resulted in an increase in plant growth under low-temperature stress conditions. Some auxin-producing microbes are also able to produce IAA and GA3 that induce plant growth and help in the development of the plant root system (Atzorn et al., 1988;

Bottini et al., 2004). Production of these phytohormones by a variety of bacteria helps plants in salinity and drought conditions (Arkhipova et al., 2007; Chang et al., 2014).

Production of ACC deaminase

A class of microbes, i.e., PGPR, produce an enzyme called 1-aminocyclopropane-1-carboxylate (ACC) deaminase to lower the production of ethylene in plants by cleaving plant-produced ACC (Saleem et al., 2007). Ethylene is a phytohormone synthesized in higher plants in various abiotic and biotic stresses. A decreased level of ethylene allows plants to be more resistant to a wide variety of environmental stresses. Rhizobial strains that produce the ACC deaminase enzyme are more efficient in producing nitrogen-fixing nodules than others (Ma et al., 2003a, b). Production of the ACC deaminase protects plants from various pathogens and also helps them in fighting against many abiotic stresses, i.e., floods, droughts, salinity, etc.

Production of exopolysaccharides (EPSs)

Microbial exopolysaccharides are a form of carbohydrate polymers that have multiple roles in the rhizosphere. EPSs have a unique water-holding capacity that is extremely helpful in water-deficit conditions for plants in drought and salinity conditions. This property of EPSs ensures proper moisture supply in the rhizosphere. EPSs also have many reactive sites by which they are to trap micronutrients and then release them slowly into the surroundings (Hepper, 1975; Selvakumar et al., 2012; Alami et al., 2000). EPSs have an affinity toward NA^+ ions, and, so, they help plants survive under salinity stress. Many studies have reported the production of EPSs in drought and salinity conditions, which helps plants tolerate stressful conditions (Sen and Chandrasekhar, 2014). *Pseudomonas fluorescence* has been reported to have the highest ability to produce ACC deaminase and EPSs.

Production of microbial volatile organic compounds (MVOCs)

Many rhizospheric bacteria and fungi produce volatile organic compounds, which help plants in the development of the root system and growth. These MVOCs are complex mixtures with low molecular weight and are vital infochemicals that mediate essential communication systems in the entire plant kingdom (Dweck et al., 2015; Bitla et al., 2017). Some of these compounds trigger induced systemic resistance (ISR) in plants, which helps them tolerate drought and salinity conditions (Cho et al., 2008).

Nutrient cycle management

As we all know, plants need several macro- and micronutrients for their growth and development. In harsh environmental conditions, plants do not get enough of these nutrients, which leads to less productivity and even death. Here, also, rhizospheric microbes play an extremely important role as they can provide the much-needed nutrients to plants by various methods. pH change is an abiotic stress factor, which leads to lower bioavailability of many nutrients. Alkaline soil converts iron from Fe^{2+} to the Fe^{3+} form; this form is not available to plants for use, and, hence, in that condition, microbes produce siderophores, which provide iron in its Fe^{2+} form and help the plants take up enough iron for growth and development (Hao et al., 2012; Supanekar et al., 2013). Other nutrients are N_2 and P that are also not available to plants in environmental stresses. Some rhizobacteria, i.e., *Rhizobium* sp., are well-known for their ability to fix nitrogen for plants in a symbiotic system. Phosphorus is also solubilized by microorganisms so that it is available to plants. The enzyme phytase has also been shown to have a significant role in phosphorus mobilization by microorganisms (Kohler et al., 2007). Abiotic stress management by microbes is shown in Table 13.1.

Rhizosphere management to improve soil and plant productivity

Rhizosphere management is the effective management of microbes, soils, and plants with the goal of enhancing nutrient utilization capacity, soil quality, and crop yield (Ryan et al., 2009). Our overdependence on the green revolution has deteriorated soil quality and crop yield, and its insufficiency to further enhance the production of food requires some modern-day agricultural, soil, and biological management practices (Bender et al., 2016). These can be grouped under various management practices and utilized at multiple levels to enhance their benefits. These management approaches work by altering certain biological populations and increasing the overall richness of the rhizospheric soils (Dessaux et al., 2016).

Crop management

Crop management encompasses both single plant-level and large heterogeneous plant community-level management, with both being adapted and enhanced for optimal utilization of soil and plant resources. The single plant-level root system could be modified for superior rhizospheric activity. By boosting soil organic matter content, plants can greatly increase microbial activity in the soil (Lange et al., 2015; Sati et al., 2020). Plants influence soil biology and hence play an essential role in rhizosphere design. According to Bender et al. (2016), soil health can be improved by developing effective, versatile, and integrated strategies like crop rotations and crop species selection. It can boost crop growth in a specific soil ecosystem by strengthening soil biology (Deguchi et al., 2007). The co-utilization of assorted selective crops and their varieties provides the benefit of manipulating the soil biota, their characteristics, and services (Vandermeer et al., 1998). Improvement of soil quality also hastens the activities of soil microorganisms in a range of cropping systems. By employing crops, we can alter soil biological communities both chronologically (e.g., crop rotation and

Table 13.1: Abiotic stress management by microbes.

Abiotic stress	Major effect on plants	Helpful microbes	Mechanisms	References
Drought	Poor root growth, decreased water and nutrient availability, poor stem elongation, reduction of leaf expansion	*Achromobacter xylosoxidans*, *Bacillus pumilus*, *Klebsiella variicola*, *Pseudomonas fluorescens*, *Raoultella planticola*	Delay of maturity, modulation of the rate of transpiration, production of EPSs and VOCs, induction of HSPs, increase of photosynthesis	Khan et al. (2019), Gou et al. (2015)
Salinity	Ion toxicity, root growth damage, reduced rate of seed germination, decreased vegetative growth	*Bacillus licheniformis*, *Aneurinibacillus aneurinilyticus*, *Paenibacillus* sp., *Rhizoglomus intraradices*, *Funneliformis mosseae*, *Rhizoglomus fasciculatum*, and *Gigaspora* sp.	Production of EPSs and ACC deaminase, N fixation, P solubilization, and siderophore formation, increased accumulation of proline, K+, carbohydrates, etc.	Parihar et al. (2020), Lim et al. (2011), Gupta and Pandey (2019)
Chilling	Affects nitrogen fixation and the process of nodulation	*Rhizophagus irregularis*, *Pseudomonas vancouverensis*, *Pseudomonas frederiksbergensis*, *Funneliformis mosseae*	Increased antioxidants, reduction in cellular lipid peroxidation, production of ABA and GA	Subramanian et al. (2015), Pasbani et al. (2020)
Heavy metals (Pb, Cd, Cr, Zn, AS, Cu)	Inhibition of cell division, chlorosis	*Pseudomonas aeruginosa*, *Burkholderia gladioli*, *B. siamensis*, *B pumilus*, *B. firmus*	Phosphate solubilization, acidification, siderophore production, ACC deaminase and IAA production, metal-chelating molecules	Yang et al. (2015), Abdallah et al. (2015), Bilal et al. (2020)
Flooding	Stomata closure, oxygen deprivation	*Bacillus* sp., *Agrobacterium rhizogenes*, *Pseudomonas putida*, *Fusarium solani*	Antioxidant defense system, ROS regulation, compatible solute accumulation, EPS production, biofilms	Farwell et al. (2007), Kavroulakis et al. (2018)

cover cropping) and geographically (e.g., mixed cropping and intercropping), resulting in a favorable impact on the agricultural environment (Li et al., 2007). They could further assist in cropping system management and reduce the requirement of exogenous components (Bender et al., 2016). Furthermore, crop system management has the ability to accelerate the biotic potential, health, and crop performance of the soil.

Soil management

Soil management is carried out with the purpose of improving tillage, reducing weeds, preparing the seedbed, adjusting the biotic potential of soil, and, consequently, declining the economic assets. Improper tilth can interfere with soil biology, although good tillage practices stimulate decomposition and mineralization of inorganic and organic matter, thereby augmenting nutrient usage efficiency (Zimmerman et al., 2011). Advanced soil management strategies, e.g., zero tillage, minimum tillage, and strip tillage, with the inclusion of cover crops and manures can enhance the biotic potential of the soil (Pittelkow et al., 2014). Employing soil management strategies promotes soil biota proliferation, minimizes the chances of infection by soil-borne insects, and checks weed population (Mader et al., 2002). Soil additives have been proven to improve soil quality and crop yield in the last 50 years (Ryan et al., 2009). Biochar can significantly speed up nitrogen fixation and vesicular-arbuscular mycorrhizae (VAM) activity (Guerena et al., 2015). By neutralizing the acidic pH of soils, increasing water retention capacity, and accelerating the breakdown of soil organic matter, biochar effectively boosts the bioactivity of soil (Zimmerman et al., 2011). According to Dessaux et al. (2016), carbon-rich materials, e.g., calcium silicate, coal fly ash, organic residues, and organic compost, can increase soil bioactivity, carbon mineralization and availability, and overall soil health. In conclusion, the application of all the abovementioned soil amendments has aided in increasing soil quality and plant viability.

Microbiological management

Soil biological populations such as PGPR and VAM aid in optimizing the functioning of plants by hydrolyzing and releasing various organic and inorganic nutrients and providing them to plant roots (Ceballos et al., 2013). Crop output is enhanced by inoculating the soil and seedlings with a specific biotic population. Application of the *Rhizobia* spp. in legumes led to a reduction in plant exogenous nitrogen requirement owing to nitrogen fixation (Vargas et al., 2000). These perks, however, are largely sensitive to soil type, plant species, and ecological conditions. In sustainable agricultural practices, biocontrol agents (*Trichoderma*, *Pseudomonas*, and *Bacillus*) that establish plant systemic resistance to bacterial pathogen infestations have also been used to manage crop pests (Pieterse et al., 2014). According to Ryan et al. (2009), the soil microbiota contribute to the synthesis of stress hormones, enzymes, and other antibiotics that assist plants in enduring multiple stressors. Hence, amendments in soil microbial communities can improve plant productivity while also ensuring environmental protection.

Rhizospheric biota management through a holobiont approach

Plants and the soil biota work together in shaping the rhizosphere, according to research (Bulgarelli et al., 2012). Hence, it often becomes crucial in research to breed the plant community, which can assist in improving the rhizospheric diversity with the desired crop plant performance. Integrating plants with the desired rhizospheric communities by various breeding programs might be adequate to address agricultural sustainability requirements (Chaparro et al., 2012; Sati et al., 2021). Root secretions and carbon distribution within the rhizosphere, for example, provide energy to symbionts present in the plant roots (Walder and Van Der Heijden, 2015). Crops that exude a higher level of carbon enhance the rhizospheric communities more efficiently, along with their functioning. However, a particular plant microbiome can be used to change the plant phenotype, limit infections (Mendes et al., 2011), regulate plant flowering period, etc. (Panke-Buisse et al., 2015), for instance, a *Bacillus* spp. can be genetically manipulated with the nitrogen fixation mechanism in order to synthesize higher amounts of plant hormones (Arkhipova et al., 2005; Kim and Timmusk, 2013). A mixed consortium of enhanced nitrogen fixers, such as *Bacillus* spp., *Pseudomonas*, *Rhizobium*, and *Bradyrhizobium*, could generate complex and dynamic native rhizospheric biofunctions (Ahkami et al., 2017). The integration of plants with rhizospheric microbial communities can be a necessary stage in rhizospheric management for improved utilization efficiency of plants. In the continually changing climatic conditions, the research concentrated on the development and selection of genotypes with multiple desirable traits such as root developmental plasticity, water use efficiency (WUE), and root nutrient uptake that could boost agricultural output (Ahkami et al., 2017). Transgenic approaches can also be used to strengthen plants and the rhizospheric microbial communities, which can facilitate the growth of stress-tolerant varieties or microbial stains that aid in crop improvement (Bender et al., 2016). All of these factors must be considered in designing future plant microbial approaches with the effort to expand the rhizospheric responsiveness to plant communities.

Conclusions and future perspectives

It is evident that a variety of environmental factors, including excessive climate variability, water shortages, urbanization, overpopulation, salinization, global warming, and so on, have put a huge strain on plant survival and productivity in general and on crop plants in particular. Therefore, as abiotic stress levels increase, proper management approaches will be required to provide enough crop output to feed millions of hungry mouths. Application of versatile microorganisms is a tempting option among the possibly useful management methods based on cost-effective, low-cost technologies for improving the growth and production of crops as well as for the mitigation of abiotic/biotic stresses. Numerous studies have demonstrated the effectiveness of specific bacteria and fungi not only in the protection of crops but also in growth enhancement. Their application mechanism has been operated in various agricultural systems, both in soils and plants. It is envisaged that, in the near future, such microorganisms will be commonly utilized in the field for sustainable farming and that these low-input methods will be offered to farmers. In the field of PGPR biofertilizer technology, great improvements have been made globally, since they exhibit efficient plant growth promotion and have the ability to lessen various environmental pressures, enhance soil fertility and nutritional quality of food, and increase agricultural production. The application of PGPR as stress relievers, biofertilizers, biocontrol agents, and biofortifiers provides a cost-effective substitute to chemical fertilizers for global crop development.

References

Abdallah, E.F., Abeer, H., Alqarawi, A.A., Alwathnani, H.A., 2015. Alleviation of adverse impact of cadmium stress in sunflower (*Helianthus annuus* L.) by arbuscular mycorrhizal fungi. Pak. J. Bot. 47, 785–795.

Ahkami, A.H., White III, R.A., Handakumburaa, P.P., Jansson, C., 2017. Rhizosphere engineering: enhancing sustainable plant ecosystem productivity. Rhizosphere 3 (2), 233–243.

Alami, Y., Achouak, W., Marol, C., Heulin, T., 2000. Rhizosphere soil aggregation and plant growth promotion of sunflowers by an exopolysaccharide-producing *Rhizobium* sp. strain isolated from sunflower roots. Appl. Environ. Microbiol. 66 (8), 3393–3398.

Ali, A.S., Elozeiri, A.A., 2017. Metabolic processes during seed germination. Adv. Seed Biol., 141–166.

Arkhipova, T.N., Veselov, S.U., Melentiev, A.I., Martynenko, E.V., Kudoyarova, G.R., 2005. Ability of bacterium *Bacillus subtilis* to produce *cytokinins* and to influence the growth and endogenous hormone content of lettuce plants. Plant and Soil 272, 201–209.

Arkhipova, T.N., Prinsen, E., Veselov, S.U., Martinenko, E.V., Melentiev, A.I., Kudoyarova, G.R., 2007. Cytokinin producing bacteria enhance plant growth in drying soil. Plant and Soil 292 (1), 305–315.

Asati, A., Pichhode, M., Nikhil, K., 2016. Effect of heavy metals on plants: an overview. Int. J. Appl. Innov. Eng. Manag 5 (3), 56–66.

Atzorn, R., Crozier, A., Wheeler, C.T., Sandberg, G., 1988. Production of gibberellins and indole-3-acetic acid by *Rhizobium phaseoli* in relation to nodulation of *Phaseolus vulgaris* roots. Planta 175 (4), 532–538.

Azevedo, R., Rodriguez, E., 2012. Phytotoxicity of mercury in plants: a review. J. Bot. 2012.

Bender, S.F., Wagg, C., van der Heijden, M.G., 2016. An underground revolution: biodiversity and soil ecological engineering for agricultural sustainability. Trends Ecol. Evol. 31 (6), 440–452.

Bilal, S., Shahzad, R., Imran, M., Jan, R., Kim, K.M., Lee, I.J., 2020. Synergistic association of endophytic fungi enhances Glycine max L. resilience to combined abiotic stresses: Heavy metals, high temperature and drought stress. Ind. Crop Prod. 143, 111931.

Bitla, U.M., Sorty, A.M., Meena, K.K., Singh, N.P., 2017. Rhizosphere signaling cascades: fundamentals and determinants. In: Singh, D.P., Singh, H.B., Prabha, R. (Eds.), Plant-microbe interactions in agro-ecological perspectives. Springer, Singapore, pp. 211–226.

Boston, R.S., Viitanen, P.V., Vierling, E., 1996. Molecular chaperones and protein folding in plants. Plant Mol. Biol. 32 (1–2), 191–222.

Bottini, R., Cassán, F., Piccoli, P., 2004. Gibberellin production by bacteria and its involvement in plant growth promotion and yield increase. Appl. Microbiol. Biotechnol. 65 (5), 497–503.

Boyer, J.S., 1982. Plant productivity and environment. Science 218 (4571), 443–448.

Bulgarelli, D., Rott, M., Schlaeppi, K., van Themaat, E.V.L., Ahmadinejad, N., Assenza, F., Rauf, P., Huettel, B., Reinhardt, R., Schmelzer, E., Peplies, J., Gloeckner, F.O., Amann, R., Eickhorst, T., Schulze-Lefert, P., 2012. Revealing structure and assembly cues for Arabidopsis root-inhabiting bacterial microbiota. Nature 488 (7409), 91–95.

Ceballos, I., Ruiz, M., Fernández, C., Peña, R., Rodríguez, A., Sanders, I.R., 2013. The in vitro mass-produced model mycorrhizal fungus, *Rhizophagus irregularis*, significantly increases yields of the globally important food security crop cassava. PLoS One 8 (8), e70633.

Chang, P., Gerhardt, K.E., Huang, X.D., Yu, X.M., Glick, B.R., Gerwing, P.D., Greenberg, B.M., 2014. Plant growth-promoting bacteria facilitate the growth of barley and oats in salt-impacted soil: implications for phytoremediation of saline soils. Int. J. Phytoremediation 16 (11), 1133–1147.

Chaparro, J.M., Sheflin, A.M., Manter, D.K., Vivanco, J.M., 2012. Manipulating the soil microbiome to increase soil health and plant fertility. Biol. Fertil. Soils 48 (5), 489–499.

Chibuike, G.U., Obiora, S.C., 2014. Heavy metal polluted soils: effect on plants and bioremediation methods. Appl. Environ. Soil Sci. 2014.

Cho, S.M., Kang, B.R., Han, S.H., Anderson, A.J., Park, J.Y., Lee, Y.H., Cho, B.H., Yang, K.Y., Ryu, C.M., Kim, Y.C., 2008. 2R, 3R-butanediol, a bacterial volatile produced by Pseudomonas chlororaphis O6, is involved in induction of systemic tolerance to drought in Arabidopsis thaliana. Mol. Plant Microbe Interact. 21 (8), 1067–1075.

Cramer, G.R., Urano, K., Delrot, S., Pezzotti, M., Shinozaki, K., 2011. Effects of abiotic stress on plants: a systems biology perspective. BMC Plant Biol. 11 (1), 1–14.

Creed, D., 1984. The photophysics and photochemistry of the near-UV absorbing amino acids–i. Tryptophan and its simple derivatives. Photochem. Photobiol. 39 (4), 537–562.

Cuartero, J., Yeo, A.R., Flowers, T.J., 1992. Selection of donors for salt-tolerance in tomato using physiological traits. New Phytol. 121 (1), 63–69.

Deguchi, S., Shimazaki, Y., Uozumi, S., Tawaraya, K., Kawamoto, H., Tanaka, O., 2007. White clover living mulch increases the yield of silage corn via arbuscular mycorrhizal fungus colonization. Plant and Soil 291 (1–2), 291–299.

Dessaux, Y., Grandclement, C., Faure, D., 2016. Engineering the rhizosphere. Trends Plant Sci. 21, 266–278.

Dweck, H.K., Ebrahim, S.A., Thoma, M., Mohamed, A.A., Keesey, I.W., Trona, F., Lavista-Llanos, S., Svatoš, A., Sachse, S., Knaden, M., Hansson, B.S., 2015. Pheromones mediating copulation and attraction in *Drosophila*. Proc. Natl. Acad. Sci. U. S. A. 112 (21), E2829–E2835.

Farwell, A.J., Vesely, S., Nero, V., Rodriguez, H., McCormack, K., Shah, S., Dixon, D.G., Glick, B.R., 2007. Tolerance of transgenic canola plants (*Brassica napus*) amended with plant growth-promoting bacteria to flooding stress at a metal-contaminated field site. Environ. Pollut. 147 (3), 540–545.

Flowers, T.J., 2004. Improving crop salt tolerance. J. Exp. Bot. 55 (396), 307–319.

Gou, W., Tian, L., Ruan, Z., Zheng, P.E.N.G., Chen, F.U.C.A.I., Zhang, L., Cui, Z., Zheng, P., Li, Z., Gao, M., Shi, W., Hu, J., 2015. Accumulation of choline and glycine betaine and drought stress tolerance induced in maize (*Zea mays*) by three plant growth promoting rhizobacteria (PGPR) strains. Pak. J. Bot. 47 (2), 581–586.

Guerena, D.T., Lehmann, J., Thies, J.E., Enders, A., Karanja, N., Neufeldt, H., 2015. Partitioning the contributions of biochar properties to enhanced biological nitrogen fixation in common bean (*Phaseolus vulgaris*). Biol. Fertil. Soils 51 (4), 479–491.

Gupta, S., Pandey, S., 2019. ACC deaminase producing bacteria with multifarious plant growth promoting traits alleviates salinity stress in French bean (*Phaseolus vulgaris*) plants. Front. Microbiol. 10, 1506.

Hao, L.Y., Willis, D.K., Andrews-Polymenis, H., McClelland, M., Barak, J.D., 2012. Requirement of siderophore biosynthesis for plant colonization by *Salmonella enterica*. Appl. Environ. Microbiol. 78 (13), 4561–4570.

Hasanuzzaman, M., Bhuyan, M.H.M., Zulfiqar, F., Raza, A., Mohsin, S.M., Mahmud, J.A., Fujita, M., Fotopoulos, V., 2020. Reactive oxygen species and antioxidant defense in plants under abiotic stress: Revisiting the crucial role of a universal defense regulator. Antioxidants 9 (8), 681.

Hepper, C.M., 1975. Extracellular polysaccharides of soil bacteria. In: Walker, N. (Ed.), Soil microbiology, a critical review. Wiley, New York, pp. 93–111.

Hollósy, F., 2002. Effects of ultraviolet radiation on plant cells. Micron 33 (2), 179–197.

Inbaraj, P., 2021. Plant-microbe interactions in alleviating abiotic stress—a Mini Review. Front. Agron. 3, 28.

Kang, C.H., So, J.S., 2016. Heavy metal and antibiotic resistance of ureolytic bacteria and their immobilization of heavy metals. Ecol. Eng. 97, 304–312.

Kavroulakis, N., Doupis, G., Papadakis, I.E., Ehaliotis, C., Papadopoulou, K.K., 2018. Tolerance of tomato plants to water stress is improved by the root endophyte *Fusarium solani* FsK. Rhizosphere 6, 77–85.

Khajeh-Hosseini, M., Powell, A.A., Bingham, I.J., 2003. The interaction between salinity stress and seed vigour during germination of soyabean seeds. Seed Sci. Technol. 31 (3), 715–725.

Khalid, M.F., Hussain, S., Ahmad, S., Ejaz, S., Zakir, I., Ali, M.A., Ahmed, N., Anjum, M.A., 2019. Impacts of abiotic stresses on growth and development of plants. In: Plant tolerance to environmental stress. CRC Press, pp. 1–8.

Khan, A., Singh, A.V., 2021. Multifarious effect of ACC deaminase and EPS producing *Pseudomonas* sp. and *Serratia marcescens* to augment drought stress tolerance and nutrient status of wheat. World J. Microbiol. Biotechnol. 37 (12), 1–17.

Khan, N., Bano, A., Rahman, M.A., Guo, J., Kang, Z., Babar, M.A., 2019. Comparative physiological and metabolic analysis reveals a complex mechanism involved in drought tolerance in chickpea (*Cicer arietinum* L.) induced by PGPR and PGRs. Sci. Rep. 9 (1), 1–19.

Khan, A., Upadhayay, V.K., Panwar, M., Singh, A.V., 2020. Soil microbiota: a key bioagent for revitalization of soil health in hilly regions. In: Goel, R., Soni, R., Suyal, D.C. (Eds.), Microbiological Advancements for Higher Altitude Agro-Ecosystems & Sustainability. Springer, Singapore, pp. 183–200.

Kim, S.B., Timmusk, S., 2013. A simplified method for gene knockout and direct screening of recombinant clones for application in *Paenibacillus polymyxa*. PLoS One 8 (6), e68092.

Kohler, J., Caravaca, F., Carrasco, L., Roldan, A., 2007. Interactions between a plant growth-promoting rhizobacterium, an AM fungus and a phosphate-solubilising fungus in the rhizosphere of *Lactuca sativa*. Appl. Soil Ecol. 35 (3), 480–487.

Kramer, P.J., Boyer, J.S., 1995. Water Relations of Plants and Soils. Academic press.

Kumar, S., 2020. Abiotic stresses and their effects on plant growth, yield and nutritional quality of agricultural produce. Int. J. Agric. Food Sci. 4, 367–378.

Lange, M., Eisenhauer, N., Sierra, C.A., Bessler, H., Engels, C., Griffiths, R.I., Mellado-Vázquez, P.G., Malik, A.A., Roy, J., Scheu, S., Steinbeiss, S., Thomson, B.C., Trumbore, S.E., Gleixner, G., 2015. Plant diversity increases soil microbial activity and soil carbon storage. Nat. Commun. 6 (1).

Levitt, J., 1980. Responses of plants to environmental stress, volume 1: chilling, freezing, and high temperature stresses. Academic Press.

Li, L., Li, S.M., Sun, J.H., Zhou, L.L., Bao, X.G., Zhang, H.G., Zhang, F.S., 2007. Diversity enhances agricultural productivity via rhizosphere phosphorus facilitation on phosphorus-deficient soils. Proc. Natl. Acad. Sci. U. S. A. 104 (27), 11192–11196.

Lim, J.H., Ahn, C.H., Jeong, H.Y., Kim, Y.H., Kim, S.D., 2011. Genetic monitoring of multi-functional plant growth promoting rhizobacteria *Bacillus subtilis* AH18 and *Bacillus licheniformis* K11 by multiplex and real-time polymerase chain reaction in a pepper farming field. J. Korean Soc. Appl. Biol. Chem. 54 (2), 221–228.

Ma, W., Guinel, F.C., Glick, B.R., 2003a. *Rhizobium leguminosarum* biovar viciae 1-aminocyclopropane-1-carboxylate deaminase promotes nodulation of pea plants. Appl. Environ. Microbiol. 69 (8), 4396–4402.

Ma, W., Sebestianova, S., Sebestian, J., Burd, G.I., Guinel, F., Glick, B.R., 2003b. Prevalence of 1-aminocyclopropaqne-1-carboxylate in deaminase in *Rhizobia* spp. Antonie Van Leeuwenhoek 83, 285–291.

Mader, P., Fliebbach, A., Dubois, D., Gunst, L., Fried, P., Niggli, U., 2002. Soil fertility and biodiversity in organic farming. Science 296, 1694–1697.

Martínez-Alvarez, V., González-Ortega, M.J., Martin-Gorriz, B., Soto-García, M., Maestre-Valero, J.F., 2018. Seawater desalination for crop irrigation—current status and perspectives. In: Emerging Technologies for Sustainable Desalination Handbook, pp. 461–492.

McNear Jr., D.H., 2013. The rhizosphere-roots, soil and everything in between. Nat. Educ. Knowl. 4 (3), 1.

Mendes, R., Kruijt, M., De Bruijn, I., Dekkers, E., van der Voort, M., Schneider, J.H., Piceno, Y.M., DeSantis, T.Z., Andersen, G.L., Bakker, P.A., Raaijmakers, J.M., 2011. Deciphering the rhizosphere microbiome for disease-suppressive bacteria. Science 332 (6033), 1097–1100.

Panke-Buisse, K., Poole, A.C., Goodrich, J.K., Ley, R.E., Kao-Kniffin, J., 2015. Selection on soil microbiomes reveals reproducible impacts on plant function. ISME J. 9, 980–989.

Parihar, M., Rakshit, A., Rana, K., Tiwari, G., Jatav, S.S., 2020. The effect of arbuscular mycorrhizal fungi inoculation in mitigating salt stress of pea (*Pisum sativum* L.). Commun. Soil Sci. Plant Anal. 51 (11), 1545–1559.

Pasbani, B., Salimi, A., Aliasgharzad, N., Hajiboland, R., 2020. Colonization with arbuscular mycorrhizal fungi mitigates cold stress through improvement of antioxidant defense and accumulation of protecting molecules in eggplants. Sci. Hortic. 272, 109575.

Pieterse, C.M., Zamioudis, C., Berendsen, R.L., Weller, D.M., Van Wees, S.C., Bakker, P.A., 2014. Induced systemic resistance by beneficial microbes. Annu. Rev. Phytopathol. 52, 347–375.

Pittelkow, C.M., Liang, X., Linquist, B.A., van Groenigen, K.J., Lee, J., Lundy, M.E., van Kessel, C., 2014. Productivity limits and potentials of the principles of conservation agriculture. Nature 517 (7534), 365–368.

Rastogi, R.P., Kumar, A., Tyagi, M.B., Sinha, R.P., 2010. Molecular mechanisms of ultraviolet radiation-induced DNA damage and repair. J. Nucleic Acids 2010.

Ruíz-Lozano, J.M., 2003. Arbuscular mycorrhizal symbiosis and alleviation of osmotic stress, new perspectives for molecular studies. Mycorrhiza 13, 309–317.

Ryan, P.R., Dessaux, Y., Thomashow, L.S., Weller, D.M., 2009. Rhizosphere engineering and management for sustainable agriculture. Plant and Soil 321 (1), 363–383.

Saleem, M., Arshad, M., Hussain, S., Bhatti, A.S., 2007. Perspective of plant growth promoting rhizobacteria (PGPR) containing ACC deaminase in stress agriculture. J. Ind. Microbiol. Biotechnol. 34 (10), 635–648.

Sati, D., Pandey, S.C., Pande, V., Upreti, S., Gouri, V., Joshi, T., Gangola, S., Debbarma, P., Pandey, A., Samant, M., 2020. Toward an enhanced understanding of plant growth promoting microbes for sustainable agriculture. In: Recent Advancements in Microbial Diversity. Academic Press, pp. 87–112.

Sati, D., Pande, V., Pandey, S.C., Samant, M., 2021. Recent advances in PGPR and molecular mechanisms involved in drought stress tolerance. Preprints, 2021050331, https://doi.org/10.20944/preprints202105.0331.v1.

Selvakumar, G., Panneerselvam, P., Ganeshamurthy, A.N., 2012. Bacterial mediated alleviation of abiotic stress in crops. In: Maheshwari, D.K. (Ed.), Bacteria in Agrobiology: Stress Management. Springer, Berlin, Heidelberg, pp. 205–224.

Sen, S., Chandrasekhar, C.N., 2014. Effect of PGPR on growth promotion of rice (*Oryza sativa* L.) under salt stress. Asian J. Plant Sci. Res. 4 (5), 62–67.

Siipola, S.M., Kotilainen, T., Sipari, N., Morales, L.O., Lindfors, A.V., Robson, T.M., Aphalo, P.J., 2015. Epidermal UV-A absorbance and whole-leaf flavonoid composition in pea respond more to solar blue light than to solar UV radiation. Plant Cell Environ. 38 (5), 941–952.

Singh, A.V., Khan, A., Joshi, M., 2021. Plant–microbe interaction: a sustainable strategy to elevate salinity tolerance in plants. In: Microbes and Signaling Biomolecules Against Plant Stress. Springer, Singapore, pp. 37–54.

Song, G., Wang, M., Zeng, B., Zhang, J., Jiang, C., Hu, Q., Geng, G., Tang, C., 2015. Anther response to high-temperature stress during development and pollen thermotolerance heterosis as revealed by pollen tube growth and in vitro pollen vigor analysis in upland cotton. Planta 241 (5), 1271–1285.

Staxén, I., Bergounioux, C., Bornman, J.F., 1993. Effect of ultraviolet radiation on cell division and microtubule organization in *Petunia hybrida* protoplasts. Protoplasma 173 (1), 70–76.

Subramanian, P., Mageswari, A., Kim, K., Lee, Y., Sa, T., 2015. Psychrotolerant endophytic *Pseudomonas* sp. strains OB155 and OS261 induced chilling resistance in tomato plants (*Solanum lycopersicum* Mill.) by activation of their antioxidant capacity. Mol. Plant Microbe Interact. 28 (10), 1073–1081.

Supanekar, S., Sorty, A., Raut, A., 2013. Study of catechol siderophore from a newly isolated *Azotobacter* sp. SUP-III for its antimicrobial activity. J. Microbiol. Biotechnol. Food Sci. 3, 270–273.

Tanveer, M., Yousaf, U., 2020. Plant single-cell biology and abiotic stress tolerance. In: Plant Life under Changing Environment. Academic Press, pp. 611–626.

Vandermeer, J., van Noordwijk, M., Anderson, J., Ong, C., Perfecto, I., 1998. Global change and multi-species agroecosystems: concepts and issues. Agric. Ecosyst. Environ. 67 (1), 1–22.

Vargas, M.A., Mendes, I.C., Hungria, M., 2000. Response of field-grown bean (Phaseolus vulgaris L.) to Rhizobium inoculation and nitrogen fertilization in two Cerrados soils. Biol. Fertil. Soils 32 (3), 228–233.

Venkateswarlu, B., Shanker, A.K., 2009. Climate change and agriculture: adaptation and mitigation strategies. Indian J. Agron. 54, 226–230.

Vinocur, B., Altman, A., 2005. Recent advances in engineering plant tolerance to abiotic stress: achievements and limitations. Curr. Opin. Biotechnol. 16 (2), 123–132.

Walder, F., Van Der Heijden, M.G., 2015. Regulation of resource exchange in the arbuscular mycorrhizal symbiosis. Nat. Plants 1 (11), 1–7.

Wan, Y., Luo, S., Chen, J., Xiao, X., Chen, L., Zeng, G., Liu, C., He, Y., 2012. Effect of endophyte-infection on growth parameters and Cd-induced phytotoxicity of Cd-hyperaccumulator *Solanum nigrum* L. Chemosphere 89 (6), 743–750.

Waqas, M.A., Kaya, C., Riaz, A., Farooq, M., Nawaz, I., Wilkes, A., Li, Y., 2019. Potential mechanisms of abiotic stress tolerance in crop plants induced by thiourea. Front. Plant Sci. 10, 1336.

Yadav, S.K., 2010. Heavy metals toxicity in plants: an overview on the role of glutathione and phytochelatins in heavy metal stress tolerance of plants. S. Afr. J. Bot. 76 (2), 167–179.

Yang, J., Kloepper, J.W., Ryu, C.M., 2009. Rhizosphere bacteria help plants tolerate abiotic stress. Trends Plant Sci. 14, 1–4.

Yang, Y., Han, X., Liang, Y., Ghosh, A., Chen, J., Tang, M., 2015. The combined effects of Arbuscular Mycorrhizal Fungi (AMF) and lead (Pb) stress on Pb accumulation, plant growth parameters, photosynthesis, and antioxidant enzymes in *Robinia pseudoacacia* L. PLoS One 10, e0145726.

Zaremba, T.G., LeBon, T.R., Millar, D.B., Smejkal, R.M., Hawley, R.J., 1984. Effects of ultraviolet light on the in vitro assembly of microtubules. Biochemistry 23 (6), 1073–1080.

Zimmerman, A.R., Gao, B., Ahn, M.Y., 2011. Positive and negative carbon mineralization priming effects among a variety of biochar-amended soils. Soil Biol. Biochem. 43 (6), 1169–1179.

CHAPTER 14

Role of microorganisms in alleviation of arsenic toxicity in plants

Amir Khan[a], Bharti Kukreti[a], Govind Makarana[b], Deep Chandra Suyal[c], Ajay Veer Singh[a], and Saurabh Kumar[b]

[a]Department of Microbiology, College of Basic Sciences and Humanities, Govind Ballabh Pant University of Agriculture and Technology, Pantnagar, US Nagar, Uttarakhand, India, [b]ICAR-Research Complex for Eastern Region, Patna, Bihar, India, [c]Akal College of Basic Sciences, Eternal University, Rajgarh, Himachal Pradesh, India

Chapter outline

Introduction

Arsenic is a toxic metalloid present on the earth's crust. More than 300 million people worldwide are exposed to arsenic (Kumar et al., 2021). It is found in both inorganic and organic forms in the environment. The inorganic forms of arsenic are more prevalent and

Unravelling Plant-Microbe Synergy. https://doi.org/10.1016/B978-0-323-99896-3.00007-2

toxic than are the organic forms. These inorganic forms include trivalent arsenite [As(III)] and pentavalent arsenate [As(V)]. On the other hand, the major organic forms of arsenic include monomethylarsine (MMA), dimethylarsine (DMA), and trimethylarsine (TMA) (Turpeinen et al., 2002). As(III) has been reported to be more toxic than As(V) (Kumar et al., 2022). As(V) is majorly found in the oxic environment, whereas As(III) is present in anaerobic conditions like water-flooded paddy soils (Mawia et al., 2021). As(V) mimics the structural similarity of phosphate and thus interferes in cellular energy production (Finnegan and Chen, 2012). However, As(III) interacts with the thiol groups present in proteins, thus inactivating enzyme functions (Finnegan and Chen, 2012). The chemical and biological properties of soil determine the distribution and abundance of different arsenic species present in the soil. Soil pH, organic matter concentration, and soil phosphate and iron affect the availability of soluble arsenic species in soils (Mawia et al., 2021). Acidic pH leads to high available concentrations of As in soils due to the solubilization of arsenic-binding species. High soil organic matter is linked to the dissolution of Fe oxyhydroxides that increase As mobility (Ma et al., 2014). High phosphate in soil reduces the transport of As(V) in plants through competitive inhibition. Microorganisms present in soil lead to the biotransformation of arsenic in soil through oxidation–reduction and methylation–demethylation processes (Di et al., 2019).

High levels of As are present in the groundwater of several countries. Both anthropogenic and geogenic sources are responsible for As contamination in soil and water systems. In arsenic-affected regions, people are exposed to arsenic by drinking arsenic-contaminated water and consuming arsenic-contaminated food. Extreme levels of As in soil (agriculture) and water (irrigation) lead to the accumulation of arsenic in plant and food grains, which, when consumed, is detrimental to human health (Gillispie et al., 2015). Uncontrolled consumption of As leads to several types of cancers (Kumar et al., 2021). Thus, arsenic accumulation in food grains is a major threat to food safety (Kumar et al., 2022). Considering the detrimental effects of arsenic on plant and human health, there is an immediate need to understand the role of microorganisms in the alleviation of arsenic stress in plants. Arsenic-resistant microorganisms could be utilized for promoting plant growth and reducing arsenic toxicity in plants. Therefore, understanding the role of microorganisms in lowering arsenic toxicity in plants is essential for sustainable and safe agriculture.

The status of arsenic contamination in food crops

In cereals, rice accumulates severalfold higher arsenic than do other cereal crops (Upadhyay et al., 2020). The amount of arsenic in rice depends on plant genotypes, geographic location, and processing of rice. Inorganic arsenic and dimethylarsine (DMA) are the major arsenic species found in rice grains (Zhao et al., 2013). Brown rice contains more arsenic than does white rice. As per the World Health Organization (WHO), the permissible limit for inorganic As in rice is 0.2 $mg\,kg^{-1}$ for white rice and 0.3 $mg\,kg^{-1}$ for brown rice (Atiaga-Franco et al., 2019). However, for rice-based food items intended for infants, this limit is

0.1 mg kg^{-1}. The average concentration of inorganic arsenic in rice varies and is found to be less than 0.1 mg kg^{-1} in India, 0.13 mg kg^{-1} in Bangladesh, 0.14 mg kg^{-1} in Taiwan and China, and 0.19 mg kg^{-1} in Japan (Upadhyay et al., 2019). In Spain, the total arsenic content in rice varies from 0.05 to 0.82 mg kg^{-1}. In USA, Italy, and Spain the highest mean arsenic levels vary between 0.156 and 0.19 mg kg^{-1}. In Pakistan, inorganic As is found to be 92.5 ± 41.88 μg kg^{-1} and 79.21 ± 76.42 μg kg^{-1} in raw and cooked rice, respectively (Rasheed et al., 2018), whereas the mean total arsenic content was 47.47 ± 30.72 38 μg kg^{-1} and 71.65 ± 74.7 μg kg^{-1}, respectively.

Wheat is the second largest staple crop in the world with a consumption of 730.9 million tons (Rasheed et al., 2018). However, little is known about arsenic mobilization in wheat plants and grains. However, it is assumed that wheat is less prone to arsenic accumulation as it is grown in aerobic conditions and has less affinity to silica (Shi et al., 2015). The total As concentration in wheat varies from 0.010 to 0.500 mg kg^{-1} (Adomako et al., 2011). Mondal et al. (2021) reported an increase in arsenic content in wheat in the arsenic-contaminated regions of India and raised concerns about arsenic consumption based on arsenic-contaminated wheat-based diets. In Pakistan, the concentration of inorganic arsenic content in wheat was 116.38 ± 51.38 μg kg^{-1} and the mean total arsenic content was 105 ± 61.47 μg kg^{-1} (Rasheed et al., 2018). Contrary to this, in pulses, contamination of only inorganic arsenic has been reported (Bhattacharya et al., 2010). In West Bengal, India, 0.076–0.168 mg kg^{-1} arsenic was reported in pulses and oilseeds (Rahaman et al., 2013). Human consumption of arsenic-contaminated beans has been found in Brazil where 11%–23% of arsenic comes from beans (Ciminelli et al., 2017). In pulses studied in West Bengal, India, peas had an arsenic content of 1.30 mg kg^{-1}, whereas mung bean was reported to have 0.314 mg kg^{-1} (Santra et al., 2013).

Arsenic-resistant microorganisms

Microbes have adopted a variety of strategies to tolerate the prevalence of As stresses such as efflux, compartmentalization, complexation, or metal–ion reduction or to utilize them as terminal electron acceptors under anaerobic respiration. For sustaining the life of microbes, it is necessary to maintain a balanced homeostasis of the crucial nutrients and nonessential toxic ions (Jackson et al., 2005). Like other organisms, arsenic at elevated levels is fatal to microbes. To date, many bacteria, archaea, cyanobacteria, and fungi have been reported to sustain and overcome arsenic toxicity by transforming the arsenic species (Oremland and Stolz, 2005). Microorganisms transform arsenic via redox reaction (in inorganic forms) and interconversion of inorganic form into organic form (methylation and demethylation). Microbe-based redox reactions transform the highly toxic arsenite [As (III)] into the lesser harmful arsenate form [As(V)]. Green (1918) first identified the microbe-mediated metabolism of arsenic (an arsenate-reducing strain *Bacterium arsenoreducens* and an arsenite-oxidizing strain *Bacillus arsenoxydans*). Interestingly, a *Thermus* strain demonstrated

Table 14.1: Arsenic-resistant microorganisms.

Organism	Reported arsenic tolerance level	References
	Bacteria	
Methylobacterium oryzae	580 µM As(V)	Alcántara-Martínez et al. (2018)
Serratia marcescens	15 mM As(III), 500 mM As(V)	Botes et al. (2007)
Pseudomonas putida strains RS-5	66.7 mM As(V)	Chang et al. (2008)
Bacillus flexus	280 mM As(V), 32 mM As(III)	Das et al. (2016)
Exiguobacterium, *Aeromonas* spp.	5 mM As(V)	Anderson and Cook (2004)
Acidithiobacillus caldus	100 mM As(V), 5 mM As(III)	Dopson et al. (2001), Kotze et al. (2006)
Kocuria flava and *Bacillus vietnamensis*	*K. flava* can tolerate up to 35 mM, whereas *B. vietnamensis* can tolerate 20 mM of As(III)	Mallick et al. (2018)
Leptospirillum ferriphilum	60 mM As(V) as well as As(III)	Tuffin et al. (2006)
Acidithiobacillus ferrooxidans	14.7 g L^{-1}As(III)	Dave et al. (2008)
Brevundimonas aurantiaca PFAB1	310 mM As(V), 90 mM As(III)	Banerjee et al. (2021)
Exiguobacterium sp. S17	150 mM As(V), 10 mM As(III)	Castro-Severyn et al. (2019)
Acinetobacter lwoffii (RJB-2)	125 mM As(V), 50 mM As(III)	Das and Sarkar (2018)
	Fungi	
Trichoderma sp.	650 ppm As(III)	Govarthanan et al. (2018)
Glomus geosporum	70 ppm As(V)	Wu et al. (2015a, b)
Piriformospora indica	100 µM As(V)	Mohd et al. (2017)
Rhizophagus intraradices	60 ppm As(V)	Li et al. (2016)
	Cyanobacteria	
Anabaena doliolum	58 mM As(V), 11 mM As(III)	Srivastava et al. (2009)
Microcystis novacekii	5810PPM As(V), 41 PPM As(III)	Ferreira et al. (2018)
Microcystis aeruginosa	below 10^{-3} mol L^{-1} As(V), below 10^{-5} mol L^{-1} As(III)	Gong et al. (2011)

reduction of As(V) to As(III) using oxidation strategies (Gihring et al., 2001). The important arsenic-resistant microorganisms are listed in Table 14.1.

Bacteria

Due to environmental pressure, selection, and microbial adaptation strategies, a number of bacteria have been found to possess resistance against arsenic stress (Table 14.1). Mallick et al. (2018) isolated two strains of As-resistant bacteria, namely, *Kocuria flava* and *Bacillus vietnamensis*, from the rhizospheric area of mangrove trees in the Sundarbans. Both strains enhanced rice plant growth while decreasing As levels. In a research study by García-González et al. (2016), As-tolerant *Bacillus licheniformis*, *Micrococcus luteus*, and *Pseudomonas fluorescens* were found to produce siderophores and have phosphate solubilization as well as nitrogen fixation characteristics. *M. luteus* inoculation in grapevine offered As resistance with enhanced biomass and antioxidant activity. Zhang et al. (2017)

transferred the As(III) *S*-adenosylmethionine methyltransferase gene from *Chlamydomonas reinhardtii* in the *Rhizobium leguminosarum* strain and observed that the legume symbionts could methylate As(III) to diverse As species. Red clover plants symbiotically cultivated with genetically engineered *R. Leguminosarum* contain 42.4% of methylated arsenic and 0.01–0.02% of overall arsenic volatilization. In research, plants (*Pteris vittata*) have been grown in iron-free media (Fe-free Hoagland solution with $FeAsO_4$ mineral) containing the *Pseudomonas* PG12 strain, which is capable of producing a catecholate-type siderophore. It was observed that Fe dissolution was efficient under treatment of PG12-generated siderophores, resulting in a rise in Fe and As (14.3–78.5 $mg\,kg^{-1}$ in fronds) and an increase in the plant biomass (Liu et al., 2015). Furthermore, the As accumulating abilities of plants (*Pteris vittata*) were significantly higher in the presence of microbial inoculants than in their absence (Singh et al., 2015).

Brevundimonas diminuta NBRI012 has been shown to be an arsenic-resistant, IAA-producing rhizobacterium, which may mitigate the detrimental impacts of arsenic stress on rice and boost rice growth (Singh et al., 2016). Moreover, Singh et al. (2015) investigated As-resistant strains of bacteria from paddy soils, including *Bacillus megaterium*, *B. altitudinis*, and *Lysinibacillus*. The *Lysinibacillus* strains were able to withstand up to 3.256 $g\,L^{-1}$ As(V) and 1.136 gL^{-1}As(III). Das et al. (2014) found 12 potentially arsenic-resistant bacterial cultures from Taiwanese agricultural soil. *Pseudomonas*, *Acinetobacter*, *Comamonas*, and *Klebsiella* As(III) bacteria were reported to have the As(III) oxidizing potential (Das et al., 2014). Moreover, other isolated strains of *Paenibacillus* sp., *Pseudomonas* sp., *Bacillus* sp., *Geobacillus* sp., *Comamonas* sp., and *Enterobacter* sp. also showed other PGP properties.

Earlier, Wang et al. (2012) also isolated As-tolerant strains of bacteria from the rhizospheric region of *P. vittata*, which was capable of oxidizing As(III) (*Acinetobacter*), whereas four strains (*Comamonas*, *Pseudomonas*, *Flavobacterium*, and *Staphylococcus*) showed As(III) oxidation and As(V) reduction. Similarly, *Agrobacterium radiobacter* D14-treated *Populus deltoids* revealed that *P. deltoids* could survive in soil having 300 $mg\,kg^{-1}$ arsenic concentration and exhibited 54% arsenic reduction, which was 43% more than that seen in uninoculated controls. Furthermore, the translocation of As to shoots was found to be enhanced (Wang et al., 2011). The majority of As-tolerant rhizobacteria belongs to *Alphaproteobacteria*, *Gammaproteobacteria*, and *Betaproteobacteria* and are related to genera such as *Pseudomonas*, *Achromobacter*, *Bacillus*, *Brevundimonas*, *Ochrobactrum*, and *Microbacterium* (Purakayastha, 2011). Dhankher et al. (2002) reported the expression level of bacterial enzymes arsenate reductase (ArsC) and GSH synthetase (g-ECS) in *Arabidopsis thaliana*, which resulted in accumulation of As(V) as a GSH–As complex. Furthermore, *E. coli*, transformed with the phytochelatin synthase enzyme (encoded by the *AtPCS* gene) from *A. thaliana*, were able to synthesize phytochelatin and accumulated moderate to high levels of arsenic under arsenic stress conditions (Sauge-Merle et al., 2003; Singh et al., 2008).

Overexpression of ArsR (i.e., the operon responsible for As efflux from cells) in *E. coli* resulted in efficient removal of arsenic (Kostal et al., 2004).

Fungi

Arbuscular mycorrhizal fungi (AFM) are present in around 80% of all plants (Chen et al., 2017). AMF primarily aid in phosphate absorption in plants, but their contribution to conferring stress tolerance has also been reported. Sharma et al. (2017) investigated the ability of AM fungi (*Glomus etunicatum* and *Rhizoglomus intraradices*) to alleviate arsenic stress in wheat plants. As stress affects the root colonization percentage due to fungal inoculants, and mycorrhizae-inoculated wheat plants showed better growth when compared to nonmycorrhizal plants. Mycorrhiza also helped overcome arsenic-induced P deficiency, hence maintaining the P:As ratio and lowering As translocation in low As treatment. In another study by Ceci et al. (2020), who have worked on four fungi (i.e., *Absidia spinosa*, *Metarhizium marquandii*, *Purpureocillium lilacinum*, and *Cephalotrichum nanum*), it was found that all the four fungi were tolerant to arsenic stress and were able to remediate arsenic stress through siderophore production. Spagnoletti et al. (2016) reported the role of AMF inoculation in soybean plants under As-infested soils and revealed that the plant biomass increases after AMF treatment. Molecular research has revealed that *RiPT*, an AMF high-affinity phosphate transporter, and *RiArsA*, a possible As efflux pump, were upregulated, resulting in As elimination. Li et al. (2016) amended *R. intraradices* with six rice cultivars to investigate the responses to AMF colonization, which resulted in a decrease in the inorganic-to-organic-arsenic ratio among all six rice cultivar grains. Wu et al. (2015a, b) used three *Glomus* strains (*Glomus geosporum*, *G. mosseae*, and *G. versiforme*) to alleviate As stress in an upland rice variety and reported that AMF colonization increases rice productivity without raising As concentration in grains. Another feature of AMF plant symbiosis is the role of the 14-3-3 protein that modulates various functions through interactions with important proteins. A recent research study on rice plants treated with an AM fungus (*R. irregularis*) has revealed an increase in the 14–3-3 protein levels during AMF colonization, suggesting a potential role in AMF-assisted arsenic stress reduction in rice plants (Pathare et al., 2016). Garg and Singla (2012) investigated the impact of arsenic stress on pea (*Pisum sativum*) in a symbiotic interaction along with *Glomus mosseae*. There was a concentration-dependent enhancement in the intake, and accretion of As and a reduction in plant biomass (dry) were also reported.

Trichoderma is a filamentous fungus that belongs to the *Ascomycetes* class and is a well-studied major plant growth promoter (Waghunde et al., 2016). It is an efficient plant growth stimulator that improves soil health and can confer stress tolerance (Souza et al., 2015). An array of functional groups present on their surface enable them to interact with metals (Congeevarama et al., 2007). Tripathi et al. (2017) examined the As-sensitive and As-tolerant strain of *Trichoderma* spp. for As(V) toxicity reduction in chickpea plants. Although total As was unaffected by the two strains, biotransformation of inorganic arsenic into organic arsenic was reported under tolerant strains and this outcome was associated with better growth and nutrient content in plants. *T. reesei* NBRI0716 modulated As speciation (66% decrease of inorganic As and organic As) and enhanced grain yield of chickpea plants when cultivated in arsenic-contaminated soils (100 mg kg^{-1}) (Tripathi et al., 2013). This strain might also improve condensed trichome density, nodulation, levels of chlorophyll and proline, and also upregulation of the gene expression via modulating stress-responsive genes. Srivastava et al. (2011) obtained several fungal cultures from arsenic-infested soils (9.45–15.65 mg kg^{-1}) from the West Bengal region, including strains of *Aspergillus*, *Trichoderma*, *Rhizopus*, and *Penicillium*. *Piriformospora indica*, an endophytic fungus, protects rice plants from As toxicity, lowers As levels in the plants' surroundings, and also limits As in colonized roots via immobilization of As into its insoluble form. This fungus also alters plant antioxidant defenses in order to ameliorate As stress (Mohd et al., 2017). Verma et al. (2016) transferred the arsenic methyltransferase (*WaarsM*) gene from *Westerdykella aurantiaca* and inserted it into the *S. cerevisiae* strain (Δ *acr2*) and showed that the *WaarsM*-containing yeast cell (Δ *acr2*) demonstrated elevated As methylation capacity of volatile arsenicals under exposure to 2 ppm of As(III) and 20 ppm of As(V), respectively.

Archaebacteria

As(V) works as an electron acceptor and arsenate reduction occurs simultaneously with the oxidation of inorganic and organic molecules, resulting in the production of energy. The microbes that are involved in these types of reactions are called dissimilatory arsenate-respiring prokaryotes (DARPs), which are commonly observed in several obligative and facultative anaerobic microbes. Several such organisms have been found within the diverse phylogenetic group Archaea. *Pyrobaculum arsenaticum* and *Pyrobaculum aerophilum* are some of the reported archaea, which use hydrogen as an electron donor. Methanogens such as *Methanobacterium formicicum* and *M. bryantiiable* are some other archaea that can participate in As remediation by converting inorganic As into its methylated form (Bentley and Chasteen, 2002).

Cyanobacteria

Photosynthetic prokaryotes carry out oxygenic photosynthesis. They potentially absorb and accumulate As from their surroundings. *Anabaena*, *Synechocystis*, *Phormidium*, and *Oscillatoria* have been recognized as As-resistant dominant taxa in polluted areas. Certain cyanobacterial strains are able to sustain in the presence of elevated levels of arsenate (up to 100 mM), whereas some in low millimolar levels of arsenite (López-Maury et al., 2003). Both *Synechocystis* ArsC and R773 ArsC utilize GSH and Grx as reducing equivalents. The R773 enzyme for catalysis demands only one cysteine, whereas Synechocystis ArsC has three vital cysteines. Compared to the pI258 enzyme, *S. aureus* and *Synechocystis* enzymes appear to be comprised of two independent evolutionary pathways (Li et al., 2003). Shaheen et al. (2007) investigated the survival of various genera/species of cyanobacterial strains in composite cultures amended with different concentrations of arsenic. Their study concluded that at a low level of As in a culture medium, the abundant genera of the cyanobacterial species were *Aphanothece* (unicellular) *Phormidium* (filamentous and nonheterocystous), and *Anabaena, Nostoc, Calothrix Cylindrospermum* (filamentous and heterocystous), whereas at a high concentration, *Aphanothece* and *Cylindrospermum* died; however, taxa such as *Phormidium, Nostoc* spp., *Anabaena*, and *Calothrix* survived without any cell damage. With successive augmentation in concentration, *Anabaena, Nostoc* spp., and *Phormidium* were capable to grow up to the 5000 ppm level; however, genera such as *Nostoc* and *Phormidium* survived even at 10,000 ppm of As in the medium.

Genetics of arsenic resistance in microorganisms

Mechanisms to survive under varying levels of arsenic are present in living organisms. All organisms from humans to prokaryotes have genes involved in the detoxification of arsenic (Ben Fekih et al., 2018). Arsenic-resistant genes are present not only in microorganisms growing under arsenic stress but also in microorganisms present in arsenic-free environments. Prokaryotic arsenic-resistant genes are organized in the *ars* operon. *arsRBC* and *ars*RDABC are the common operons present in bacteria, but variants of these operons are also common. The ArsR protein of this operon is a trans-acting repressor, which binds to the promoter of the *ars* operon. In the presence of arsenic, this repressor is dissociated, which leads to the expression of the *ars* operon. ArsA is an ATPase, which assists in the ArsB-mediated efflux of arsenic (Yang et al., 2012). ArsB is an integral membrane protein, which extrudes the arsenic outside of the cell. ArsC exhibits arsenate reductase activity, which reduces arsenate to arsenite, thus facilitating the export of the latter by ArsB (Yang et al., 2012). ArsD is an arsenite metalloprotein, which transfers arsenite to ArsA ATPase before extrusion by ArsB (Ben Fekih et al., 2018).

Microorganisms-assisted phytoremediation and mechanisms of microorganisms-mediated arsenic bioremediation

The extent of arsenic quantity mixed with soils and its bioavailability determine its toxicity. Generally, As enters the agricultural soil through anthropogenic actions or pesticide and herbicide applications. Furthermore, using contaminated subsurface water for crop irrigation results in As pollution of agricultural land, which is critical because As contamination is a key contributor to food chain contamination (Mandal, 2017). Plants that experience As stress exhibit diverse problems. Because the species of As is structurally similar to that of inorganic phosphate, it can enter a cell by the same route that transports phosphate through the phosphate transporter of the cell membrane, which results in interruption of phosphorylation-dependent metabolic pathways and inhibition of adenosine triphosphate synthesis (Yang et al., 2012). It tightly binds to the vicinal cysteine pairs in compounds like lipoic acids and the enzyme pyruvate dehydrogenase, acting as an effective inhibitor of oxidative metabolism (Lin et al., 2006). By inducing reactive oxygen species (ROS), arsenite binding to protein sulfhydryl groups may cause membrane breakdown and, eventually, cell death. Arsenite-stimulated ROS bonded to reduced glutathione (GSH) depletes the GSH pool and leads to DNA molecule, lipid, and protein damage (Bhattacharjee et al., 2008).

Plants utilize an array of methods to combat As toxicity. Phytoremediation is a plant-based technique that has been successfully exercised to remove metals from contaminated soils (Iqbal and Nasrullah, 2020). Hyperaccumulation, phytoextraction, phytostabilization, and phytotransformation are some of the methods that plants use to eradicate contaminated surroundings. Metal-tolerant plants (such as *P. vittata*, *A. thaliana*, *Corrigiola telephiifolia*, *Pteris cretica*, etc.) are used to reclaim metal-contaminated land and reduce airborne transfer and pollution through a process called phytostabilization. The utilization of metal-accumulating plants for extracting and translocating metals to harvestable sections is known as phytoextraction. Continuous phytoextraction is dependent on a plant's natural ability to store, translocate, and withstand high arsenic concentrations throughout its life cycle. Plants that can take up and uphold >0.1% of metals in their tissues are said to be hyperaccumulating (Brooks, 1998). Hyperaccumulator plants are favorable because they produce concentrated toxicants, making the final removal of contaminant-rich biomass easier. The majority of metal-hyperaccumulating plants grow at a slower pace and produce little biomass. Fast-growing plants, on the other hand, produce a lot of biomass but are usually vulnerable to high metal stress, with the exception being arsenic hyperaccumulator ferns, i.e., *Pityrogramma calomelanos*. Hence, the ability to tolerate and collect high metal mass in vegetative parts as well as the ability to generate large volumes of dry matter are always not mutually exclusive (Ma et al., 2001). Increased pollutant availability in soil solution for plant intake is critical for efficient phytoremediation, especially phytoextraction. Rhizospheric variables such as root exudation, rhizospheric microbial population, and chelating agents are significant in boosting phytoremediation efficacy (Purakayastha, 2011). Microbially enhanced phytoextractions appear to be the potential solution for remediating arsenic-contaminated sites. As documented in a research study, plant-derived exudates boost the proliferation of microorganisms in the rhizosphere, which also enhances As metal remediation (Mishra et al., 2017). In addition, P fertilizers and the rhizospheric bacterial community also boosted As accumulation and biomass in the fern *Pityrogramma calomelanos*. Moreover, the significance of arbuscular mycorrhizal fungi (AMF) application in As phytoremediation has been widely explored (Mitra et al., 2021). Arbuscular mycorrhizal fungi enhance plant biomass while preventing plants from absorbing As, hence hindering the phytoextraction process (Purakayastha, 2011).

Above-ground harvestable vegetative portions in phytoextraction contain a large amount of hazardous metals, increasing the chances of arsenic contamination across the food chain. The use of rhizospheric AM fungi has emerged to be promising in this area as it prevents biomagnifications of As in the food chain.

Microbes have been shown to improve plant development and survival under heavy metal stress by transforming more toxic species into nontoxic by-products. This is possible because microbes have evolved various As-resistant mechanisms to facilitate them and others to survive in the presence of poisonous heavy metals (Ben Fekih et al., 2018). Bacteria also limit metal bioavailability from soil through chelation and precipitation by organic acid production through lowering soil pH, which aids in metal–ion sequestration (Mishra et al., 2017). Besides the rhizosphere, the significance of the phyllosphere as a potential habitat for arsenic-resistant microbes was first described and identified by Rathinasabapathi et al. (2006), who isolated arsenic-resistant epiphytic bacteria, which are genetically correlated with *Variovorax paradoxus*. Therefore, it can be concluded that phyllospheric and rhizospheric microbes could assist plants in As uptake and determine the As bioavailability in soil. A study by Lampis et al. (2015) demonstrated the role of arsenic-resistant rhizobacterial inoculation with plants in successfully eliminating arsenic metal from the soil. In addition to their interactions with plants, microbes have an indispensable role in arsenic biogeochemistry. In soils, different microbes perform arsenic oxidation or reduction, sorption or desorption, and methylation or demethylation, which result in the reduction of the toxicity and bioavailability of arsenic, which benefits not only the species involved but also the surroundings plants and other organisms.

Moreover, microbes' capacity to withstand osmotic and oxidative stresses helps them resist arsenic stress. Microorganisms can lower the toxicity of pollutants by utilizing them as energy sources in their metabolic activities. For heavy metal removal, this technique is the most successful, innovative, and eco-friendly. To combat arsenic stress, both eukaryotic and prokaryotic microorganisms employ similar mechanisms. Active efflux pumps, compartmentalization, redox conversion, extrusion and immobilization, and lowering the sensitivity of the cellular target are only a few examples (Turpeinen, 2002). These strategies are applied individually or in permutations. Arsenic can either co-precipitate or be adsorbed on producing iron when a coagulant is applied. Furthermore, microbe-produced siderophores aid in the mobilization of arsenic by producing an As–siderophore complex and thereby limiting As concentration in soil and protecting plants from As stress (Zawadzka et al., 2007). The fixation of As(III) on iron oxides is a common approach for removing arsenic. The iron oxide is formed only by bacterial action rather than chemical coagulation/precipitation in biological therapy. In both alkaline and acidic soils, iron oxides or hydroxides are primarily engaged in arsenic adsorption (Polemio et al., 1982). Fe arsenate [$Fe_3(AsO_4)^2$] controls arsenic solubility in oxygenated acidic soils, whereas the sulfide of As(III) controls the amount of arsenic in soil solution in anoxic soils, in which microbes deposit iron oxide in the filtering media, providing a vast surface for chemical reactions. During this process, arsenic is accumulated in sludge on a constant basis, which could be removed from these sites (Harvey and Crundwell, 1996). Some of the microbe-assisted As remediation mechanisms are described in following paragraphs.

Oxidation of arsenite

Periplasmic arsenite oxidase (aox) is an enzyme that catalyzes the process of arsenite oxidation in which As (III) is convert into the As (V) form, which is less toxic (Lebrun et al., 2003). Both chemotrophic and heterotrophic bacteria have been reported to produce this enzyme (Dhuldhaj et al., 2013). Chemoautotrophic arsenite-oxidizing (CAO) bacteria utilize CO_2/HCO_3 as their carbon source and oxidize arsenite into arsenate by employing nitrate, oxygen, or Fe^{3+} as a terminal electron acceptor, whereas heterotrophic arsenite-oxidizing bacteria (HAO) utilize organic carbon as their source of carbon and energy (Oremland and Stolz, 2005). A number of microorganisms like *Agrobacterium tumefaciens*, *Thermus thermophilus*, *Pseudomonas arsenitoxidans*, *Thermus aquaticus*, etc. are reported to convert As(III) into As(V) by oxidation (Salmassi et al., 2002). The arsenite oxidase enzyme of *Agrobacterium tumefaciens* is controlled through a sensor kinase (*aoxS*, sensor) and response regulator (*aoxR*, receptor) system (Silver and Phung, 2005). The *aoxB* gene is unique for arsenite oxidizers. Furthermore, a large family of bacteria and archaea contain the arsenite transporter gene (the *ars* operon) on chromosomal or plasmid DNA, which exports the arsenite from the cell and confers As resistance. Bacteria possessing both the arsenite oxidase genes and the arsenite transporter genes demonstrate higher arsenite resistance than do strains with only the arsenite transporter genes (Cai et al., 2009).

Methylation and demethylation

As is methylated upon reduction of As(V) to As(III) and then by reducing and adding methyl groups (Chen and Rosen, 2020). The *Methanobacterium bryantii* (a methanogen)-mediated alteration of As(V) into a modest quantity of volatile methylarsines (MMA) was first reported by McBride and Wolfe (1971). The same process of As(V) conversion was also observed in many bacteria like *Desulfovibrio gigas*, *Desulfovibrio vulgaris*, *Methanobacterium formicicum*, and *Clostridium collagenovorans*. MMAA (monomethylarsonic acid), DMAA (dimethylarsinic acid), and TMAO (trimethylarsine oxide) are all derivatives of MMA (Turpeinen, 2002). Moreover, some microbes are also reported to carry out the demethylation process, which can reduce methylated species to their inorganic state (Chen and Rosen, 2020). Microbe-mediated methylation of inorganic As results in the creation of volatile species that escape into the atmosphere from soil surfaces and water. Methylated forms are a key source of arsine in the environment. Besides reports that methylated As forms are less toxic than are inorganic forms, various reports have documented that methylated As(III) forms are more prone to DNA damage than are the As(V)-methylated species in the order, DMAA(III) > MMAA(III) > As(III), As(V) > MMA (V) > DMAA(V) > TMAO(V) (Rahman et al., 2014). However, arsines have a short 'lifetime' in air, and they quickly reverse to water-soluble species like trimethylarsine oxide (TMAO) and As(V) (Turpeinen, 2002). Hence, the toxicity, mobility, bioavailability, and geocycle of As in soil and groundwater are influenced by demethylation and methylation. Arsenic is removed from the soil milieu by the reduction of As(V) to As(III) and, consequently, As(III) methylation with trimethylarsine (TMA) (Woolson, 1977).

Mobilization and immobilization

The majority of the arsenic found in subterranean aquifers is natural. Microbes aid in arsenic mobilization in aqueous environments (Oremland and Stolz, 2005). Arsenic mobilization in aquifers is favored by various physicochemical circumstances that are changeable, complex, and poorly understood. Microbe-produced siderophores aid in arsenic mobilization by producing an As–siderophore complex (Zawadzka et al., 2007). Arsenic is present mostly as As(III) under strongly reducing circumstances and is freed via desorption from or dissolution of iron oxides. Increased arsenic (V) levels have been reported in locations with high groundwater pH (> 8) under oxic conditions (Smedley and Kinniburgh, 2002). In environments dominated by the As(V) species, arsenic content is positively linked to anion-forming species. Under reducing circumstances, arsenicals can also immobilize. Sulfate-reducing microorganisms that are able to utilize As(V) can lead to the development of immobilized As_2S_3 precipitates. Arsenic is also immobilized during the formation of iron sulfides, thus limiting the bioavailability of As to plants and reducing stress (Zhang et al., 2010).

Other mitigation strategies for reducing arsenic toxicity in plants

Modifications in agronomical practices

Cultivation of arsenic-tolerant and less arsenic-accumulating genotypes of cereals and pulses in arsenic-contaminated areas is a viable approach to reduce arsenic-induced toxicity in food grains. Arsenic species in soils compete with P and Si for their transport inside plant cells. Applications of silicon and phosphorus fertilizers in arsenic-contaminated fields significantly reduce the concentration of total arsenic in roots and shoots in crops (Wu et al., 2015a, b). Therefore, an exogenous application of theses mineral nutrients would decrease the arsenic content in grains. Furthermore, water-saving agricultural technologies such as direct-seeded rice cultivation are sustainable approaches for lowering arsenic content in rice as arsenic uptake in rice has been reported to be lower (0.23–0.26 $mg\,kg^{-1}$) in aerobically managed conditions than in the anaerobic ones (0.60–0.67 $mg\,kg^{-1}$; Talukder et al., 2011). Cultivation of rice varieties with high iron plaque formation could be a feasible approach to decreasing arsenic content in rice (Syu et al., 2013). Furthermore, application of rice straw during paddy cultivation enhanced As release from iron minerals and As(V) reduction to As(III) in soil but decreased As(III) accumulation (Yang et al., 2018).

Applications of nanoparticles

Nanomaterials have extreme properties compared to their bulky counterparts and therefore offer wide applications in agriculture (Shang et al., 2019). The high surface area of nanoparticles makes them more efficient in their functions (Khan et al., 2019). Several nanoparticles in the past have been used for bioremediation of heavy metals (Rizwan et al., 2014; Mallikarjunaiah et al., 2020). Moreover, nanoparticles have also been extensively used for the amelioration of arsenic in plants (Liu et al., 2019). The major nanoparticles used for arsenic amelioration in crops are shown in Table 14.2. Amalgamation of nanoparticles with arsenic-resistant PGPR could be the better approach to arsenic amelioration in agricultural crops in future. Thus, nanoparticles offer potential application in arsenic reduction in food grains.

Genetic modifications in arsenic transporters in plants

The expression of transporters involved in the uptake and transport of arsenic could be altered through genetic engineering to reduce the arsenic accumulation in plants (Mawia et al., 2021). CRISPR/Cas9-based genetic manipulation of As(III) and As(V) transporters in cereals and pulses could be a viable option. The *lsi*2 mutants of rice have been reported to show low arsenic accumulation in the grain (Ma et al., 2008). Similarly, the *AtINT2* and AtINT4 mutants (inositol transporter genes) in *Arabidopsis* exhibit low uptake of arsenic in seeds (Duan et al., 2015). Adoption of genetically modified cultivars of cereals and pulses

Table 14.2: Nanoparticles used for arsenic amelioration in crops.

Nanoparticles	Crop	Effect on arsenic levels	References
SiNp	Maize	Lowered the accumulation of As in plants	Tripathi et al. (2016)
CeO_2	Rice	No significant change in arsenic levels in rice	Wang et al. (2018)
CuO	Rice	Reduced arsenic in shoots and grains	Liu et al. (2018, 2019)
MnO_2	Rice	Total As concentration in rice grains reduced by 65.4%	Zhou et al. (2015)
SiO_2	Rice	Inhibited As uptake in rice	Cui et al. (2020)
TiO_2	Rice	Reduced arsenic bioaccumulation in rice seedlings by 40%–90%	Mawia et al. (2021)
ZnO	Rice	Reduced the accumulation of As(III) in rice roots and shoots. Reduced As(V) in rice roots but not in shoots	Wang et al. (2018)
ZnO	Rice	Decreased As accumulation in shoots	Yan et al. (2021),
ZnO	Soybean	As concentration declined in both roots and shoots of soybean plants	Ahmad et al. (2020)

could reduce arsenic toxicity in plants. However, the transporters for arsenic transport are also involved in the transport of other nutrients, which could possibly impact plant homeostasis.

Conclusions

Utilization of biological agents for alleviation of arsenic toxicity in plants has enormous potential to be applied in agriculture in arsenic-contaminated soils. Arsenic-resistant microorganisms promote plant growth while protecting plants from the detrimental effects of arsenic. These beneficial microorganisms are suitable and sustainable bioinoculants for agriculture. Thus, arsenic-resistant microorganisms are a cost-effective and sustainable resource for reclamation of arsenic-contaminated soils. However, systematic studies have to be conducted to use microorganisms for arsenic abatement in agriculture.

References

Adomako, E.E., Williams, P.N., Deacon, C., Meharg, A.A., 2011. Inorganic arsenic and trace elements in Ghanaian grain staples. Environ. Pollut. 159 (10), 2435–2442.

Ahmad, P., Alyemeni, M.N., Al-Huqail, A.A., Alqahtani, M.A., Wijaya, L., Ashraf, M., Bajguz, A., 2020. Zinc oxide nanoparticles application alleviates arsenic (As) toxicity in soybean plants by restricting the uptake of as and modulating key biochemical attributes, antioxidant enzymes, ascorbate-glutathione cycle and glyoxalase system. Plants 9 (7), 825.

Alcántara-Martínez, N., Figueroa-Martínez, F., Rivera-Cabrera, F., GutiérrezSánchez, G., Volke-Sepúlveda, T., 2018. An endophytic strain of *Methylobacterium* sp. increases arsenate tolerance in *Acacia farnesiana* (L.) Willd: a proteomic approach. Sci. Total Environ. 625, 762–774.

Anderson, C.R., Cook, G.M., 2004. Isolation and characterization of arsenate-reducing bacteria from arsenic contaminated sites in New Zealand. Curr. Microbiol. 48, 341–347.

Atiaga-Franco, O., Otero, X.L., Gallego-Picó, A., Escobar-Castañeda, L., Bravo-Yagüe, J., Carrera-Villacrés, D., 2019. Analysis of total arsenic content in purchased rice from Ecuador. Czech J. Food Sci. 37 (6), 425–431.

Banerjee, A., Sarkar, S., Gorai, S., Kabiraj, A., Bandopadhyay, R., 2021. High arsenic tolerance in *Brevundimonas aurantiaca* PFAB1 from an arsenic-rich Indian hot spring. Electron. J. Biotechnol. 53, 1–7.

Ben Fekih, I., Zhang, C., Li, Y.P., Zhao, Y., Alwathnani, H.A., Saquib, Q., Rensing, C., Cervantes, C., 2018. Distribution of arsenic resistance genes in prokaryotes. Front. Microbiol. 9, 2473.

Bentley, R., Chasteen, T.G., 2002. Microbial methylation of metalloids: arsenic, antimony, and bismuth. Microbiol. Mol. Biol. Rev. 66 (2), 250–271. https://doi.org/10.1128/MMBR.66.2.250-271.2002.

Bhattacharjee, H., Mukhopadhyay, R., Thiyagarajan, S., Rosen, B.P., 2008. Aquaglyceroporins: ancient channels for metalloids. J. Biol. 7 (9), 1–6.

Bhattacharya, P., Samal, A.C., Majumdar, J., Santra, S.C., 2010. Arsenic contamination in rice, wheat, pulses, and vegetables: a study in an arsenic affected area of West Bengal, India. Water Air Soil Pollut. 213 (1), 3–13.

Botes, E., van Heerden, E., Litthauer, D., 2007. Hyper-resistance to arsenic to arsenic in bacteria isolated from an antimony mine in South Africa. S. Afr. J. Sci. 103, 279–281.

Brooks, R.R., 1998. Plants that Hyperaccumulate Heavy Metals. University Press, Cambridge, p. 380.

Cai, L., Liu, G., Rensing, C., Wang, G., 2009. Genes involved in arsenic transformation and resistance associated with different levels of arsenic-contaminated soils. BMC Microbiol. 9, 1–11.

Castro-Severyn, J., Pardo-Esté, C., Sulbaran, Y., Cabezas, C., Gariazzo, V., Briones, A., Morales, N., Séveno, M., Decourcelle, M., Salvetat, N., Remonsellez, F., Saavedra, C.P., 2019. Arsenic response of three altiplanic *Exiguobacterium* strains with different tolerance levels against the metalloid species: a proteomics study. Front. Microbiol. 10, 2161.

Ceci, A., Spinelli, V., Massimi, L., Canepari, S., Persiani, A.M., 2020. Fungi and arsenic: tolerance and bioaccumulation by soil saprotrophic species. Appl. Sci. 10 (9), 3218.

Chang, J.S., Kim, Y.H., Kim, K.W., 2008. The ars genotype characterization of arsenic-resistant bacteria from arsenic-contaminated gold-silver mines in the republic of Korea. Appl. Microbiol. Biotechnol. 80, 155–165.

Chen, J., Rosen, B.P., 2020. The arsenic methylation cycle: how microbial communities adapted methylarsenicals for use as weapons in the continuing war for dominance. Front. Environ. Sci. 8, 43.

Chen, X.W., Wu, F.Y., Li, H., Chan, W.F., Wu, S.C., Wong, M.H., 2017. Mycorrhizal colonization status of lowland rice (*Oryza sativa* L.) in the southeastern region of China. Environ. Sci. Pollut. Res. 24, 5268–5276.

Ciminelli, V.S., Gasparon, M., Ng, J.C., Silva, G.C., Caldeira, C.L., 2017. Dietary arsenic exposure in Brazil: the contribution of rice and beans. Chemosphere 168, 996–1003.

Congeevarama, S., Dhanarani, S., Park, J., Dexilin, M., Thamaraiselui, K., 2007. Biosorption of chromium and nickel by heavy metal resistant fungal and bacterial isolates. J. Hazard. Mater. 146, 270–277.

Cui, J., Li, Y., Jin, Q., Li, F., 2020. Silica nanoparticles inhibit arsenic uptake into rice suspension cells via improving pectin synthesis and the mechanical force of the cell wall. Environ. Sci. Nano 7 (1), 162–171.

Das, J., Sarkar, P., 2018. Remediation of arsenic in mung bean (*Vigna radiata*) with growth enhancement by unique arsenic-resistant bacterium *Acinetobacter lwoffii*. Sci. Total Environ. 624, 1106–1118.

Das, S., Jean, J.S., Kar, S., Chou, M.L., Chen, C.Y., 2014. Screening of plant growth-promoting traits in arsenic-resistant bacteria isolated from agricultural soil and their potential implication for arsenic bioremediation. J. Hazard. Mater. 272, 112–120.

Das, S., Jean, J.S., Chou, M.L., Rathod, J., Liu, C.C., 2016. Arsenite-oxidizing bacteria exhibiting plant growth promoting traits isolated from the rhizosphere of *Oryza sativa* L. implications for mitigation of arsenic contamination in paddies. J. Hazard. Mater. 302, 10–18.

Dave, S.R., Gupta, K.H., Tipre, D.R., 2008. Characterization of arsenic resistant and arsenopyrite oxidizing *Acidithiobacillus ferrooxidans* from Hutti gold leachate and effluents. Bioresour. Technol. 99, 7514–7520.

Dhankher, O.P., Li, Y., Rosen, B.P., Shi, J., Salt, D., Senecoff, J.F., Sashti, N.A., Meagher, R.B., 2002. Engineering tolerance and hyperaccumulation of arsenic in plants by combining arsenate reductase and γ-glutamylcysteine synthetase expression. Nat. Biotechnol. 20 (11), 1140–1145.

Dhuldhaj, U.P., Yadav, I.C., Singh, S., Sharma, N.K., 2013. Microbial interactions in the arsenic cycle: adoptive strategies and applications in environmental management. Rev. Environ. Contam. Toxicol. 224, 1–38.

Di, X., Beesley, L., Zhang, Z., Zhi, S., Jia, Y., Ding, Y., 2019. Microbial arsenic methylation in soil and uptake and metabolism of methylated arsenic in plants: a review. Int. J. Environ. Res. Public Health 16 (24), 5012.

Dopson, M., Lindström, E.B., Hallberg, K.B., 2001. Chromosomally encoded arsenical resistance of the moderately thermophilic acidophile *Acidithiobacillus caldus*. Extremophiles 5 (4), 247–255. https://doi.org/10.1007/s007920100196.

Duan, G.L., Hu, Y., Schneider, S., McDermott, J., Chen, J., Sauer, N., Rosen, B.P., Daus, B., Liu, Z., Zhu, Y.G., 2015. Inositol transporters AtINT2 and AtINT4 regulate arsenic accumulation in *Arabidopsis* seeds. Nat. Plants 2 (1), 1–6.

Ferreira, F.A.G., Franco, M.W., Oliveira, D.M.D., Magalhães, S.M.S., Barbosa, F.A.R., 2018. Arsenic tolerance of *Microcystis novacekii* (Komárek-Compère, 1974) and its arsenic decontamination potential. Braz. Arch. Biol. Technol. 61.

Finnegan, P., Chen, W., 2012. Arsenic toxicity: the effects on plant metabolism. Front. Physiol. 3, 182.

García-González, I., Quemada, M., Gabriel, J.L., Hontoria, C., 2016. Arbuscular mycorrhizal fungal activity responses to winter cover crops in a sunflower and maize cropping system. Appl. Soil Ecol. 102, 10–18.

Garg, N., Singla, P., 2012. The role of *Glomus mosseae* on key physiological and biochemical parameters of pea plants grown in arsenic contaminated soil. Sci. Hortic. 143, 92–101.

Gihring, T.M., Druschel, G.K., McCleskey, R.B., Hamers, R.J., Banfield, J.F., 2001. Rapid arsenite oxidation by *Thermus aquaticus* and *Thermus thermophilus*: field and laboratory investigations. Environ. Sci. Technol. 35, 3857–3862.

Gillispie, E.C., Sowers, T.D., Duckworth, O.W., Polizzotto, M.L., 2015. Soil pollution due to irrigation with arsenic-contaminated groundwater: current state of science. Curr. Pollut. Rep. 1 (1), 1–12.

Gong, Y., Ao, H., Liu, B., Wen, S., Wang, Z., Hu, D., Zhang, X., Song, L., Liu, J., 2011. Effects of inorganic arsenic on growth and microcystin production of a *Microcystis* strain isolated from an algal bloom in Dianchi Lake, China. Chin. Sci. Bull. 56 (22), 2337–2342.

Govarthanan, M., Mythili, R., Selvankumar, T., Kamala-Kannan, S., Kim, H., 2018. Myco-phytoremediation of arsenic- and lead-contaminated soils by *Helianthus annuus* and wood rot fungi, *Trichoderma* sp. isolated from decayed wood. Ecotoxicol. Environ. Saf. 151, 279–284.

Green, H.H., 1918. Description of a bacterium which oxidizes arsenite to arsenate, and of one which reduces arsenate to arsenite, isolated from a cattle-dipping Etank. S. Afr. J. Sci. 14, 465–467.

Harvey, P.I., Crundwell, F.K., 1996. The effect of As(III) on the growth of *Thiobacillus ferrooxidans* in an electrolytic cell under controlled redox potential. Miner. Eng. 9, 1059–1068.

Iqbal, R.K., Nasrullah, 2020. Phytoremediation of arsenic in soil and water. Research and Reviews. J. Bot. Sci. 8 (3), 9–16.

Jackson, C.R., Horrison, K.G., Dugas, S.L., 2005. Enumeration and characterization of culturable arsenate resistant bacteria in large estuary. Syst. Appl. Microbiol. 28, 727–734.

Khan, I., Saeed, K., Khan, I., 2019. Nanoparticles: properties, applications and toxicities. Arab. J. Chem. 12 (7), 908–931.

Kostal, J., Yang, R., Wu, C.H., Mulchandani, A., Chen, W., 2004. Enhanced arsenic accumulation in engineered bacterial cells expressing ArsR. Appl. Environ. Microbiol. 70 (8), 4582–4587.

Kotze, A.A., Tuffin, M.I., Deane, S.M., Rawlings, D.Y., 2006. Cloning and characterization of the chromosomal arsenic resistance genes from *Acidithiobacillus caldus* and enhanced arsenic resistance on conjugal transfer of ars genes located on transposon Tn *AtcArs*. Microbiology 152, 3551–3560.

Kumar, A., Ali, M., Kumar, R., Kumar, M., Sagar, P., Pandey, R.K., Ghosh, A.K., 2021. Arsenic exposure in Indo Gangetic plains of Bihar causing increased cancer risk. Sci. Rep. 11 (1), 1–16.

Kumar, S., Choudhary, A.K., Suyal, D.C., Makarana, G., Goel, R., 2022. Leveraging arsenic resistant plant growth-promoting rhizobacteria for arsenic abatement in crops. J. Hazard. Mater. 425, 127965.

Lampis, S., Santi, C., Ciurli, A., Andreolli, M., Vallini, G., 2015. Promotion of arsenic phytoextraction efficiency in the fern *Pteris vittata* by the inoculation of As-resistant bacteria: a soil bioremediation perspective. Front. Plant Sci. 6, 80.

Lebrun, E., Brugna, M., Baymann, F., Muller, D., Lievremont, D., Lett, M.C., Nitschke, W., 2003. Arsenite oxidase, an ancient bioenergetic enzyme. Mol. Biol. Evol. 20 (5), 686–693.

Li, R., Haile, J.D., Kennelly, P.J., 2003. An arsenate reductase from Synechocystis sp. strain PCC 6803 exhibits a novel combination of catalytic characteristics. J. Bacteriol. 185 (23), 6780–6789.

Li, H., Chen, X.W., Wong, M.H., 2016. Arbuscular mycorrhizal fungi reduced the ratios of inorganic/organic arsenic in rice grains. Chemosphere 145, 224–230.

Lin, Y.F., Walmsley, A.R., Rosen, B.P., 2006. An arsenic metallochaperone for an arsenic detoxification pump. Proc. Natl. Acad. Sci. 103 (42), 15617–15622.

Liu, X., Yang, G.M., Guan, D.X., Ghosh, P., Ma, L.Q., 2015. Catecholate-siderophore produced by As-resistant bacterium effectively dissolved $FeAsO_4$ and promoted *Pteris vittata* growth. Environ. Pollut. 206, 376–381.

Liu, J., Simms, M., Song, S., King, R.S., Cobb, G.P., 2018. Physiological effects of copper oxide nanoparticles and arsenic on the growth and life cycle of rice (*Oryza sativa* japonica 'Koshihikari'). Environ. Sci. Technol. 52 (23), 13728–13737.

Liu, J., Wolfe, K., Potter, P.M., Cobb, G.P., 2019. Distribution and speciation of copper and arsenic in rice plants (Oryza sativa japonica 'Koshihikari') treated with copper oxide nanoparticles and arsenic during a life cycle. Environ. Sci. Technol. 53 (9), 4988–4996.

López-Maury, L., Florencio, F.J., Reyes, J.C., 2003. Arsenic sensing and resistance system in the cyanobacterium *Synechocystis* sp. strain PCC 6803. J. Bacteriol. 185 (18), 5363–5371.

Ma, L.Q., Komar, K.M., Tu, C., Zhang, W., Cai, Y., Kennelley, E.D., 2001. A fern that hyperaccumulates arsenic. Nature, 409–579.

Ma, J.F., Yamaji, N., Mitani, N., Xu, X.Y., Su, Y.H., McGrath, S.P., Zhao, F.J., 2008. Transporters of arsenite in rice and their role in arsenic accumulation in rice grain. Proc. Natl. Acad. Sci. 105 (29), 9931–9935.

Ma, R., Shen, J., Wu, J., Tang, Z., Shen, Q., Zhao, F.J., 2014. Impact of agronomic practices on arsenic accumulation and speciation in rice grain. Environ. Pollut. 194, 217–223.

Mallick, I., Bhattacharyya, C., Mukherji, S., Dey, D., Sarkar, S.C., Mukhopadhyay, U.K., Ghosh, A., 2018. Effective rhizoinoculation and biofilm formation by arsenic immobilizing halophilic plant growth promoting bacteria (PGPB) isolated from mangrove rhizosphere: a step towards arsenic rhizoremediation. Sci. Total Environ. 610, 1239–1250.

Mallikarjunaiah, S., Pattabhiramaiah, M., Metikurki, B., 2020. Application of nanotechnology in the bioremediation of heavy metals and wastewater management. In: Nanotechnology for Food, Agriculture, and Environment. Springer, Cham, pp. 297–321.

Mandal, P., 2017. An insight of environmental contamination of arsenic on animal health. Emerg. Contam. 3 (1), 17–22.

Mawia, A.M., Hui, S., Zhou, L., Li, H., Tabassum, J., Lai, C., Wang, J., Shao, G., Wei, X., Tang, S., Luo, J., Hu, P., 2021. Inorganic arsenic toxicity and alleviation strategies in rice. J. Hazard. Mater. 408, 124751.

McBride, B.C., Wolfe, R.S., 1971. Biosynthesis of dimethylasrine by a methanobacterium. Biochemistry 10, 4312–4317.

Mishra, J., Singh, R., Arora, N.K., 2017. Alleviation of heavy metal stress in plants and remediation of soil by rhizosphere microorganisms. Front. Microbiol. 8, 1706.

Mitra, D., Saritha, B., Janeeshma, E., Gusain, P., Khoshru, B., Nouh, F.A.A., Rani, A., Olatunbosun, A.N., Ruparelia, J., Rabari, A., Mosquera-Sánchez, L.P., Mondal, R., Verma, D., Panneerselvam, P., Mohapatra, P.K.D., Sierra, G.B.E., 2021. Arbuscular mycorrhizal fungal association boosted the arsenic resistance in crops with special responsiveness to rice plant. Environ. Exp. Bot. 193, 104681.

Mohd, S., Shukla, J., Kushwaha, A.S., Mandrah, K., Shankar, J., Arjaria, N., Saxena, P.N., Narayan, R., Roy, S.K., 2017. Endophytic fungi *Piriformospora indica* mediated protection of host from arsenic toxicity. Front. Microbiol. 8, 754.

Mondal, D., Rahman, M.M., Suman, S., Sharma, P., Siddique, A.B., Rahman, M.A., Bari, A.F., Kumar, R., Bose, N., Singh, S.K., Ghosh, A., Polya, D.A., 2021. Arsenic exposure from food exceeds that from drinking water in endemic area of Bihar, India. Sci. Total Environ. 754, 142082.

Oremland, R.S., Stolz, J.F., 2005. Arsenic, microbes and contaminated aquifers. Trends Microbiol. 13, 45–48.

Pathare, V., Srivastava, S., Sonawane, B.V., Suprasanna, P., 2016. Arsenic stress affects the expression profile of genes of 14-3-3 proteins in the shoot of mycorrhiza colonized rice. Physiol. Mol. Biol. Plants 22 (4), 515–522.

Polemio, M., Senesi, N., Buto, S.A., 1982. Soil contamination by metals. A survey in industrial and rural areas of Southern Italy. Sci. Total Environ. 25, 71–79.

Purakayastha, T.J., 2011. Microbial remediation of arsenic contaminated soil. In: Sherameti, I., Varma, A. (Eds.), Detoxification of Heavy Metals. Springer, Berlin, Heidelberg, pp. 221–260.

Rahaman, S., Sinha, A.C., Pati, R., Mukhopadhyay, D., 2013. Arsenic contamination: a potential hazard to the affected areas of West Bengal, India. Environ. Geochem. Health 35 (1), 119–132.

Rahman, M.A., Hogan, B., Duncan, E., Doyle, C., Krassoi, R., Rahman, M.M., Naidu, R., Lim, R.P., Maher, W., Hassler, C., 2014. Toxicity of arsenic species to three freshwater organisms and biotransformation of inorganic arsenic by freshwater phytoplankton (*Chlorella* sp. CE-35). Ecotoxicol. Environ. Saf. 106, 126–135.

Rasheed, H., Kay, P., Slack, R., Gong, Y.Y., 2018. Arsenic species in wheat, raw and cooked rice: exposure and associated health implications. Sci. Total Environ. 634, 366–373.

Rathinasabapathi, B., Ma, L.Q., Srivastava, M., 2006. Arsenic hyperaccumulating ferns and their application to phytoremediation of arsenic contaminated sites. Floricult. Ornam. Plant Biotechnol. 3 (32), 304–311.

Rizwan, M., Singh, M., Mitra, C.K., Morve, R.K., 2014. Ecofriendly application of nanomaterials: nanobioremediation. J. Nanopart. 2014.

Salmassi, T.M., Venkateswaren, K., Satomi, M., Newman, D.K., Hering, J.G., 2002. Oxidation of arsenite by *Agrobacterium albertimagni*, AOL15, sp. nov., isolated from Hot Creek, California. Geomicrobiol J. 19 (1), 53–66.

Santra, S.C., Samal, A.C., Bhattacharya, P., Banerjee, S., Biswas, A., Majumdar, J., 2013. Arsenic in foodchain and community health risk: a study in Gangetic West Bengal. Procedia Environ. Sci. 18, 2–13.

Sauge-Merle, S., Cuiné, S., Carrier, P., Lecomte-Pradines, C., Luu, D.T., Peltier, G., 2003. Enhanced toxic metal accumulation in engineered bacterial cells expressing *Arabidopsis thaliana* phytochelatin synthase. Appl. Environ. Microbiol. 69 (1), 490–494.

Shaheen, R., Mahmud, R., Sen, J., 2007. A study on arsenic decontaminating cyanobacteria of an arsenic affected soil. J. Soil Nat. 1 (2), 23–29.

Shang, Y., Hasan, M., Ahammed, G.J., Li, M., Yin, H., Zhou, J., 2019. Applications of nanotechnology in plant growth and crop protection: a review. Molecules 24 (14), 2558.

Sharma, S., Anand, G., Singh, N., Kapoor, R., 2017. Arbuscular mycorrhiza augments arsenic tolerance in wheat (*Triticum aestivum* L.) by strengthening antioxidant defense system and thiol metabolism. Frontiers in. Plant Sci. 8, 906.

Shi, G.L., Zhu, S., Meng, J.R., Qian, M., Yang, N., Lou, L.Q., Cai, Q.S., 2015. Variation in arsenic accumulation and translocation among wheat cultivars: the relationship between arsenic accumulation, efflux by wheat roots and arsenate tolerance of wheat seedlings. J. Hazard. Mater. 289, 190–196.

Silver, S., Phung, L.T., 2005. Genes and enzymes involved in bacterial oxidation and reduction of inorganic arsenic. Appl. Environ. Microbiol. 71, 599–608.

Singh, S., Mulchandani, A., Chen, W., 2008. Highly selective and rapid arsenic removal by metabolically engineered Escherichia coli cells expressing *Fucus vesiculosus* metallothionein. Appl. Environ. Microbiol. 74 (9), 2924–2927.

Singh, S., Shrivastava, A., Barla, A., Bose, S., 2015. Isolation of arsenic resistant bacteria from Bengal delta sediments and their efficacy in arsenic removal from soil in association with *Pteris vittata*. Geomicrobiol J. 32, 712–723.

Singh, N., Marwa, N., Mishra, J., Verma, P.C., Rathaur, S., Singh, N., 2016. *Brevundimonas diminuta* mediated alleviation of arsenic toxicity and plant growth promotion in *Oryza sativa* L. Ecotoxicol. Environ. Saf. 125, 25–34.

Smedley, P.L., Kinniburgh, D.G., 2002. A review of the source, behavior and distribution of arsenic in natural waters. Appl. Geochem. 17, 517–568.

Souza, R.D., Ambrosini, A., Passaglia, L.M., 2015. Plant growth-promoting bacteria as inoculants in agricultural soils. Genet. Mol. Biol. 38, 401–419.

Spagnoletti, F.N., Balestrasse, K., Lavado, R.S., Giacometti, R., 2016. Arbuscular mycorrhiza detoxifying response against arsenic and pathogenic fungus in soybean. Ecotoxicol. Environ. Saf. 133, 47–56.

Srivastava, A.K., Bhargava, P., Thapar, R., Rai, L.C., 2009. Differential response of antioxidative defense system of *Anabaena Doliolum* under arsenite and arsenate stress. J. Basic Microbiol. 49, 63–72.

Srivastava, P.K., Vaish, A., Dwivedi, S., Chakrabarty, D., Singh, N., Tripathi, R.D., 2011. Biological removal of arsenic pollution by soil fungi. Sci. Total Environ. 409 (12), 2430–2442.

Syu, C.H., Jiang, P.Y., Huang, H.H., Chen, W.T., Lin, T.H., Lee, D.Y., 2013. Arsenic sequestration in iron plaque and its effect on as uptake by rice plants grown in paddy soils with high contents of As, iron oxides, and organic matter. Soil Sci. Plant Nutr. 59 (3), 463–471.

Talukder, A.S.M.H.M., Meisner, C.A., Sarkar, M.A.R., Islam, M.S., 2011. Effect of water management, tillage options and phosphorus status on arsenic uptake in rice. Ecotoxicol. Environ. Saf. 74 (4), 834–839.

Tripathi, P., Singh, P.C., Mishra, A., Chauhan, P.S., Dwivedi, S., Bais, R.T., Tripathi, R.D., 2013. Trichoderma: a potential bioremediator for environmental clean up. Clean Techn. Environ. Policy 15 (4), 541–550.

Tripathi, D.K., Singh, S., Singh, V.P., Prasad, S.M., Chauhan, D.K., Dubey, N.K., 2016. Silicon nanoparticles more efficiently alleviate arsenate toxicity than silicon in maize cultiver and hybrid differing in arsenate tolerance. Front. Environ. Sci. 4, 46.

Tripathi, P., Singh, P.C., Mishra, A., Srivastava, S., Chauhan, R., Awasthi, S., Nautiyal, C.S., 2017. Arsenic tolerant *Trichoderma* sp. reduces arsenic induced stress in chickpea (*Cicer arietinum*). Environ. Pollut. 223, 137–145.

Tuffin, I.M., Hector, S.B., Deane, S.M., Rawling, D.E., 2006. Resistance determinants of highly arsenic resistant strain of *Leptospirillum ferriphilum* isolated from a commercial biooxidation tank. Appl. Environ. Microbiol. 72, 2247–2253.

Turpeinen, R., 2002. Interactions between Metals, Microbes and Plants: Bioremediation of Arsenic and Lead Contaminated Soils. A dissertation in environmental ecology, Faculty of Science, University of Helsinki, Neopoli, Lahti.

Turpeinen, R., Pantsar-Kallio, M., Kairesalo, T., 2002. Role of microbes in controlling the speciation of arsenic and production of arsines in contaminated soils. Sci. Total Environ. 285 (1–3), 133–145.

Upadhyay, M.K., Shukla, A., Yadav, P., Srivastava, S., 2019. A review of arsenic in crops, vegetables, animals and food products. Food Chem. 276, 608–618.

Upadhyay, M.K., Majumdar, A., Suresh Kumar, J., Srivastava, S., 2020. Arsenic in rice agro-ecosystem: solutions for safe and sustainable rice production. Front. Sustain. Food Syst. 4, 53.

Verma, S., Verma, P.K., Meher, A.K., Dwivedi, S., Bansiwal, A.K., Pande, V., Srivastava, P.K., Verma, P.C., Tripathi, R.D., Chakrabarty, D., 2016. A novel arsenic methyltransferase gene of *Westerdykella aurantiaca* isolated from arsenic contaminated soil: phylogenetic, physiological, and biochemical studies and its role in arsenic bioremediation. Metallomics 8 (3), 344–353.

Waghunde, R.R., Shelake, R.M., Sabalpara, A.N., 2016. *Trichoderma*: a significant fungus for agriculture and environment. Afr. J. Agric. Res. 11 (22), 1952–1965.

Wang, Q., Xiong, D., Zhao, P., Yu, X., Tu, B., Wang, G., 2011. Effect of applying an arsenic-resistant and plant growth-promoting rhizobacterium to enhance soil arsenic phytoremediation by *Populus deltoides* LH05-17. J. Appl. Microbiol. 111, 1065–1074.

Wang, X., Rathinasabapathi, B., Oliveira, L.M.D., Guilherme, L.R., Ma, L.Q., 2012. Bacteria-mediated arsenic oxidation and reduction in the growth media of arsenic hyperaccumulator *Pteris vittata*. Environ. Sci. Technol. 46, 11259–11266.

Wang, X., Sun, W., Zhang, S., Sharifan, H., Ma, X., 2018. Elucidating the effects of cerium oxide nanoparticles and zinc oxide nanoparticles on arsenic uptake and speciation in rice (*Oryza sativa*) in a hydroponic system. Environ. Sci. Technol. 52 (17), 10040–10047.

Woolson, E.A., 1977. Generation of alkylarsines from soil. Weed Sci. 25, 412–416.

Wu, F., Hu, J., Wu, S., Wong, M.H., 2015a. Grain yield and arsenic uptake of upland rice inoculated with arbuscular mycorrhizal fungi in As-spiked soils. Environ. Sci. Pollut. Res. 22 (12), 8919–8926.

Wu, C., Zou, Q., Xue, S., Mo, J., Pan, W., Lou, L., Wong, M.H., 2015b. Effects of silicon (Si) on arsenic (As) accumulation and speciation in rice (*Oryza sativa* L.) genotypes with different radial oxygen loss (ROL). Chemosphere 138, 447–453.

Yan, S., Wu, F., Zhou, S., Yang, J., Tang, X., Ye, W., 2021. Zinc oxide nanoparticles alleviate the arsenic toxicity and decrease the accumulation of arsenic in rice (*Oryza sativa* L.). BMC Plant Biol. 21 (1), 1–11.

Yang, H.C., Fu, H.L., Lin, Y.F., Rosen, B.P., 2012. Pathways of arsenic uptake and efflux. Curr. Top. Membr. 69, 325–358.

Yang, Y.P., Zhang, H.M., Yuan, H.Y., Duan, G.L., Jin, D.C., Zhao, F.J., Zhu, Y.G., 2018. Microbe mediated arsenic release from iron minerals and arsenic methylation in rhizosphere controls arsenic fate in soil-rice system after straw incorporation. Environ. Pollut. 236, 598–608.

Zawadzka, A.M., Crawford, R.L., Paszczynski, A.J., 2007. Pyridine-2,6-bis(thiocarboxylic acid) produced by *Pseudomonas stutzeri* KC reduces chromium(VI) and precipitates mercury, cadmium, lead and arsenic. Biometals 20, 145–158.

Zhang, M., Wang, Y., Zhao, D., Pan, G., 2010. Immobilization of arsenic in soils by stabilized nanoscale zero-valent iron, iron sulfide (FeS), and magnetite (Fe_3O_4) particles. Chin. Sci. Bull. 55 (4), 365–372.

Zhang, J., Xu, Y., Cao, T., Chen, J., Rosen, B.P., Zhao, F.J., 2017. Arsenic methylation by a genetically engineered Rhizobium-legume symbiont. Plant Soil 416, 259–269.

Zhao, F.J., Zhu, Y.G., Meharg, A.A., 2013. Methylated arsenic species in rice: geographical variation, origin, and uptake mechanisms. Environ. Sci. Technol. 47 (9), 3957–3966.

Zhou, S., Peng, L., Lei, M., Pan, Y., Lan, D., 2015. Control of As soil to rice transfer (*Oryza sativa* L.) with nano-manganese dioxide. Acta Sci. Circumstantiae 35, 855–861.

CHAPTER 15

Chemistry of plant microbe synergy in the rhizosphere

Aparna B. Gunjal

Department of Microbiology, Dr. D. Y. Patil, Arts, Commerce and Science College, Pimpri, Pune, Maharashtra, India

Chapter outline

Introduction

Microorganisms include bacteria, *Actinobacteria*, fungi, yeasts, algae, and protozoa. There are beneficial microorganisms found in the rhizosphere of plants. Many different types of interactions exist between these microorganisms and plants. The rhizosphere is the soil region rich in beneficial microorganisms close to the roots of various crops (Mendes et al., 2013). Microorganisms near the rhizosphere include plant growth-promoting rhizobacteria (PGPR), which help in nitrogen fixation and mycorrhizae, exhibit antimicrobial activity against various pathogens, impart tolerance to abiotic and biotic stresses, etc. (Mendes et al., 2013; Chandra et al., 2018a, 2018b, 2020; Chandra and Sharma, 2020). These microorganisms near the region of the roots help in the growth of various plants and other crops by assisting in the uptake of macro- and micronutrients, producing

Unravelling Plant-Microbe Synergy. https://doi.org/10.1016/B978-0-323-99896-3.00009-6

iron-chelating compounds (siderophores), solubilizing phosphorus and potassium, fixing nitrogen, producing enzymes, viz., chitinase, cellulase, pectinase, xylanase, etc., and also by exhibiting biocontrol activity (Hassan et al., 2019). The microorganisms near the rhizosphere are shown in Fig. 15.1, and their benefits are displayed in Fig. 15.2.

Endophytic bacteria are those that are found in plant tissues without causing any harm to the hosts. These endophytic bacteria exist in close association with plant tissues and can enter the

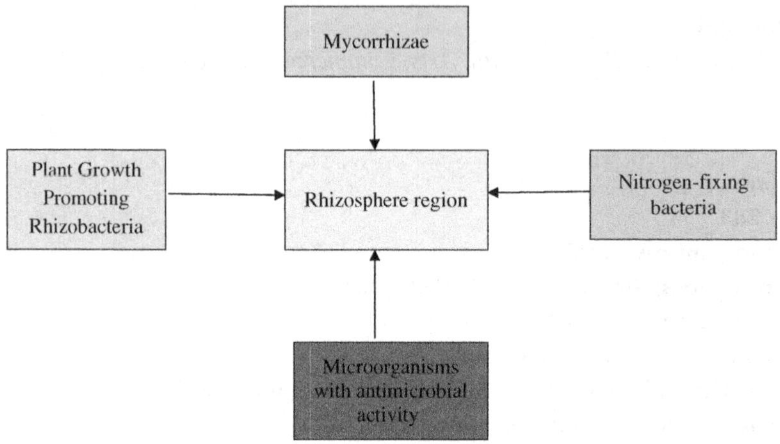

Fig. 15.1
Microorganisms near the rhizosphere.

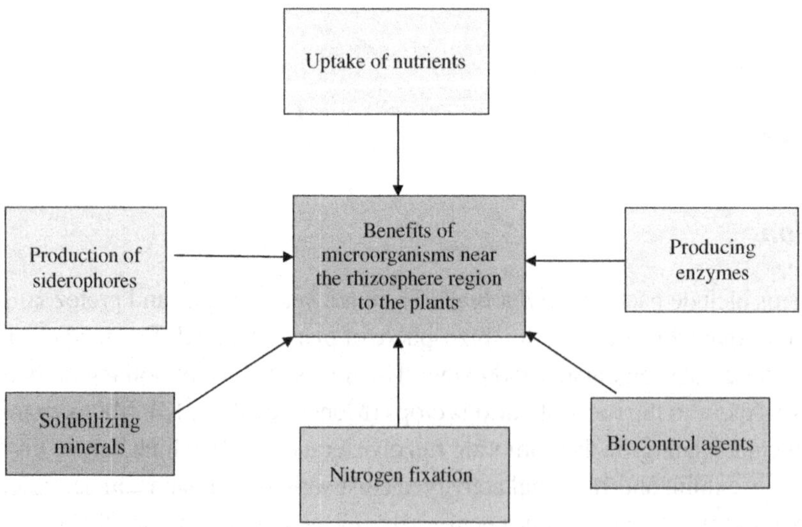

Fig. 15.2
Benefits of microorganisms near the rhizosphere to plants.

host plants through the roots in an active or passive manner (White et al., 2019). Endophytes can enter the host plants in a passive manner through the cracks present on the surface of the root tips. They can enter the host plants in an active manner through the attachment and proliferation caused by lipopolysaccharides, the presence of flagella, and pili. These help in endophytic colonization in various host plants (Afzal et al., 2019). There are interactions between endophytes and host plants. Endophytes establish themselves in the roots of hosts plants in three steps:

(i) First, they enter the roots of the host plants.
(ii) Next, they anchor themselves to the roots.
(iii) Finally, they firmly attach to the surface of the roots.

Rhizospheric bacteria can harbor around 10^{12} cells in one gram of soil. Microbial populations are found in different regions of the root zones and change over time. Population and microbial diversity are closely associated with exudates as well as various interactions that take place in the rhizosphere (Somers et al., 2004). The interactions between microorganisms and consumers and predators help understand the rhizospheric processes.

Beneficial microorganisms

A group of useful microorganisms are presented in Table 15.1. Microorganisms and their interactions with plants in the rhizosphere are shown in Fig. 15.3.

Plant growth-promoting rhizobacteria (PGPR)

Roots are complex organs and consist of root tips, meristems, and lateral roots. These are important in plant–microbe interactions. Root tips help in *Rhizobial* colonization, (Desbrosses and Stougaard, 2011). Root hairs and lateral roots confer plant growth properties (Combes-Meynet et al., 2011). PGPR help in plant growth promotion by producing compounds that help in the uptake of iron called 'siderophores', converting insoluble phosphorus and potassium into soluble forms, and aiding in nitrogen fixation and production of various plant growth hormones, viz., indole acetic acid (IAA), gibberellins, cytokinins, and enzymes, viz., 1-aminocyclopropane-1-carboxylic acid (ACC) deaminase, cellulase, chitinase, pectinase,

Table 15.1: List of beneficial microorganisms.

Microorganisms	Examples
Plant growth-promoting rhizobacteria	*Pseudomonas, Bacillus, Serratia, Burkholderia, Herbaspirillum, Acetobacter, Arthrobacter, Flavobacterium*
Nitrogen-fixing bacteria	*Azotobacter, Azospirillum, Rhizobium, Frankia, Beijerinckia*
Mycorrhizae	Arbuscular mycorrhiza
Microorganisms with biocontrol activity	*Trichoderma* sp.

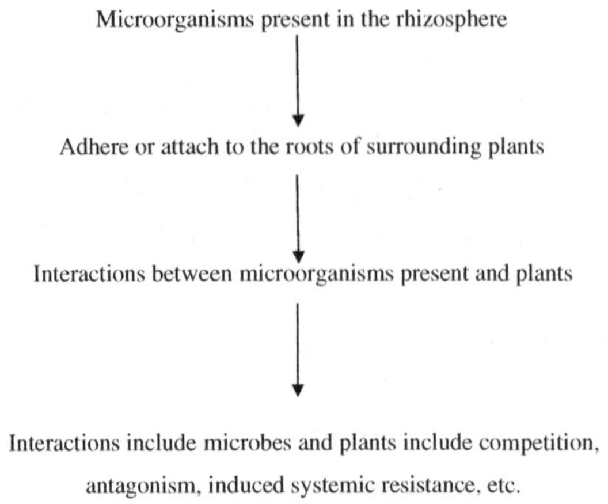

Fig. 15.3
Microorganisms and their interactions with plants in the rhizosphere.

protease, etc. (Jha and Saraf, 2015; Backer et al., 2018; Basu et al., 2021; Chandra et al., 2019a,b). Some PGPR also exhibit antimicrobial activity against plant pathogens.

Arbuscular mycorrhiza fungi

Arbuscular mycorrhiza fungi (AMF) are soil-borne fungi, which help plants take up water, nutrients, and also overcome abiotic stresses (Sun et al., 2018). There are two types of mycorrhizae: endomycorrhizae and ectomycorrhizae (Fig. 15.4). In endomycorrhizae, fungi create intraradical structures (i.e., in the roots of plants). In ectomycorrhizae, fungi create a structure known as the 'mantle' (or sheath) outside the plant roots (Begum et al., 2019). There are interactions between arbuscular mycorrhiza and microorganisms in the rhizosphere. Examples of endomycorrhizal fungi include *Glomus mosseae*, *Glomus aggregatum*, *Glomus etunicatum*, *Glomus deserticola*, *G. clarum*, *G. monosporum*, *Paraglomus brasilianum*, and *Gigaspora margarita*. Examples of ectomycorrhizal fungi include *Rhizopogon villosulus*, *Rhizopogon luteolus*, *R. amylopogon*, *R. fulvigleba*, *Pisolithus tinctorius*, *Suillus granulatus*, *Scleroderma cepa*, *Laccaria bicolor*, and *L. laccata*.

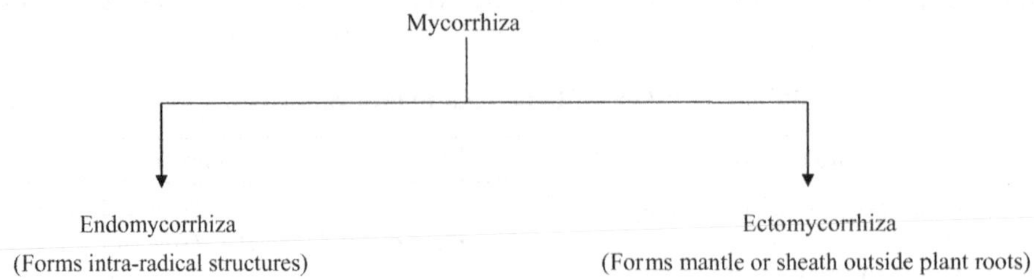

Fig. 15.4
Types of mycorrhiza.

Trichoderma: A biocontrol agent

Trichoderma belongs to the family *Hypocreaceae* whose members are common in the rhizosphere. There are 10,000 species of *Trichoderma* (Waghunde et al., 2016). Initially, *Trichoderma* strains look white and cottonish and then turn into yellow-green to dark-green tufts with the formation of concentric ring-like zones on the medium. Biocontrol is a mechanism in which beneficial microorganisms can be used to control insects, pests, etc., and the *Trichoderma* sp. can be used as a strong biocontrol agent (Woo et al., 2014). Some of the species of *Trichoderma* are *T. harzianum*, *T. afarasin*, *T. afroharzianum*, *T. atrobrunneum*, *T. camerunense*, *T. endophyticum*, *T. neotropicale*, *T. pyramidale*, *T. rifaii*, *T. simmonsii*, etc. (Zin and Badaluddin, 2020).

The rhizosphere: The main hotspot for microbial communities

The rhizosphere is a hotspot for a diversity of beneficial microorganisms (Kramer et al., 2016). The main characteristic feature of the rhizosphere is rhizodeposition, in which the roots of various plants and crops release various primary and secondary metabolites into the surrounding soil (Mohanram and Kumar, 2019). These exudates are responsible for the rhizosphere characteristics of changing the soil chemistry near the area of the roots of plants and also acting as substrates that help in the growth of beneficial microorganisms present in the soil (Yang and Crowley, 2000). The composition of crop root exudates can change depending on the accessibility of location, macro- and micronutrients available to plants, growth of plants, etc. (Malusa et al., 2016). This further creates good selective pressure in the rhizosphere with specific microorganisms in the rhizosphere. Microbes in the rhizosphere can enter the roots, shoots, and other tissues of plants (Thapa and Prasanna, 2018). Therefore, there exist mutual symbioses between plants and microbes. Microorganisms present near the region of the roots of plants and crops play an important role. The microbial processes that occur in the rhizosphere are pathogenesis, competition, enhancement of the growth of plants, colonization, antagonism, induced systemic resistance, etc. (Pii et al., 2015). Therefore, plant–microbe interactions can be beneficial, neutral, or harmful, depending on the types of microorganisms and environmental conditions.

Microbial signaling molecules and quorum sensing

Plants produce many signals, which help in microbial colonization, and respond to microorganisms in the rhizosphere (Haichar et al., 2014). Plants are modified by genetic engineering, and this allows them to release different signal molecules, thereby enhancing microbial gene expression in the rhizosphere. Many of the microbial functions are controlled via quorum-sensing signaling molecules (Abisado et al., 2018). These quorum-sensing signaling molecules by the microbes also help in the interactions between plants and

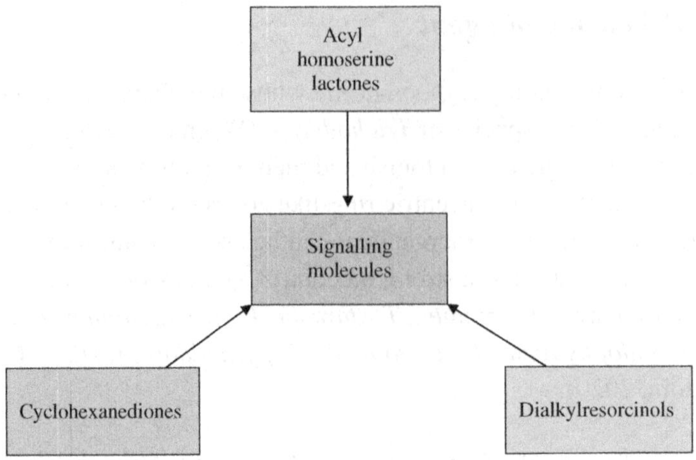

Fig. 15.5
Signaling molecules.

microorganisms. Quorum-signaling molecules, viz., *N*-acyl-L-homoserine lactones (AHLs) and volatile organic compounds from *Pseudomonas* sp., help in plant morphogenetic processes (Mohanram and Kumar, 2019). The AHLs produced through the plasma membranes disperse in the rhizosphere, where they regulate many microbial processes. The signaling molecules are displayed in Fig. 15.5.

Cell-to-cell communication among microorganisms is called 'quorum sensing'. The common quorum-sensing signals by rhizobacteria are acyl homoserine lactones (AHLs). The behavior of quorum-sensing signals can vary, but they have many advantages as they help develop symbiotic associations (e.g., the symbiotic association between legumes and *Rhizobia*) (Kumar and Dubey, 2020).

Root exudates as plant-to-microbe signals

There are various signaling molecules and interactions between plants and microbes. Root exudates play an important role in this regard and shape microbial communities (Kristin and Miranda, 2013). These root exudates include sugars, amino acids, organic acids, and phytochemical compounds present in the cytoplasm of plants (Musilova et al., 2016). Even signal molecules, i.e., jasmonic acid and salicylic acid, in the rhizosphere are found to play a role in the chemical synergy between plants and microorganisms (Doornbos et al., 2011).

Root exudates depend on the species of plants (Chaparro et al., 2013, 2014) and the microbial activity in the rhizosphere (Oburger et al., 2013). Many microorganisms secrete signaling compounds in the rhizosphere, viz., IAA, gibberellins, cytokinins, enzymes, organic acids, lipopolysaccharides (Dangl et al., 2013), antibiotics, and volatile signals. PGPR produce various quorum-sensing (QS) molecules (e.g., *N*-acyl-homoserine lactones, AHLs) in order to monitor the gene expression. The AHLs synthesized by PGPR have been shown to enhance root development and also elicit induced systemic resistance (ISR). This helps plants overcome attacks by pathogenic microorganisms. Thus, these AHLs are important signaling factors for plants to protect against pathogens. This is a systemic reaction, in which inoculation with PGPR makes plants resistant to pathogens. Many microorganisms are able to relieve signals by breaking molecules synthesized by plants as well as microorganisms in the rhizosphere through these quorum signals. Induced systemic resistance (ISR) and systemic acquired resistance (SAR) play a role in plant–microbe interactions. Root exudates are specific to plant species and are involved in microbe–plant and root–root interactions (Mommer et al., 2016; Sasse et al., 2018). PGPR establish symbiosis with plants by producing plant growth-promoting compounds, viz., IAA, gibberellins, cytokinins, volatile organic compounds, (VOCs), enzymes, etc., which modify the plant transcriptome (Fahad et al., 2015).

Various mechanisms in plant–microbe interactions

The various processes in plant–microbe interactions are shown in Fig. 15.6. They include antagonism for biological control of plant pathogens, colonization, competition, and induced systemic resistance.

Antagonistic mechanisms for biological control of plant pathogens

Microorganisms exhibit antagonistic activities toward many plant pathogens. Such microorganisms include bacteria, viz., *Agrobacterium*, *Bacillus* sp., *Streptomyces*, *Burkholderia*, and *Pseudomonas* sp., and fungi, viz., *Trichoderma*, *Pythium*, and *Fusarium*

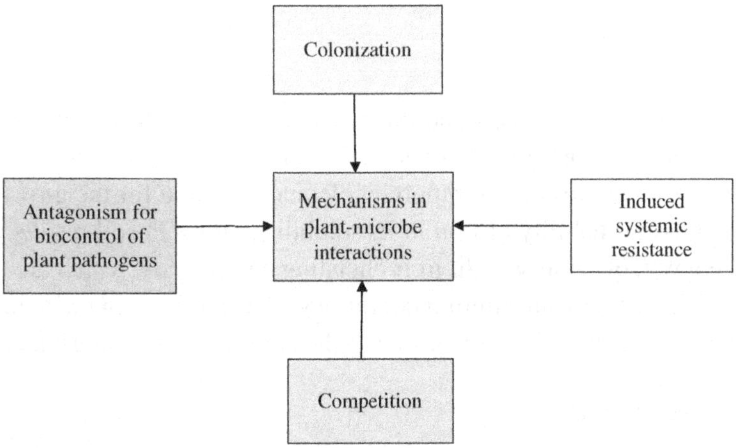

Fig. 15.6
Mechanisms in plant-microbe interactions.

Table 15.2: List of beneficial microorganisms exhibiting antagonistic activities against plant pathogens.

Bacteria	Fungi
Agrobacterium *Bacillus* sp. *Streptomyces* *Burkholderia* *Pseudomonas* sp.	*Trichoderma* *Pythium* *Fusarium* sp.

sp. The beneficial microorganisms exhibiting antagonistic activities against plant pathogens are shown in Table 15.2. The direct mechanisms of antagonistic activities by these microorganisms toward plant pathogens can include competition for nutrients, inhibition of plant pathogens by secretion of secondary metabolites by antagonistic microorganisms, breakdown of factors responsible for pathogenicity, etc. (Vishwakarma et al., 2020). The indirect mechanisms include change in the microbial community in the rhizospheric region and root anatomy.

Colonization

Colonization is also important in plant–microbe interactions. Microbial cells attach to the surface, divide, and form biofilms. The steps involved in colonization are recognition, adherence, invasion, colonization, and establishment of various interactions. The plant roots start communicating with microorganisms in the soil by producing chemical signals. These chemical signals are recognized by microorganisms, which leads to the initiation of colonization. PGPR, after reaching the roots, follow similar mechanisms for colonization (Nihorimbere et al., 2011).

Competition

There exists competition for nutrients and trace elements among the diverse microorganisms present in soils. There is competition between microorganisms for suitable sites near the root surfaces. For example, iron is an important element required for the growth of all microorganisms. Less availability of iron in its solubilized form in soil results in competition. Siderophores are low-molecular-weight iron-chelating compounds, which are synthesized by some microorganisms under iron-limiting conditions. This makes iron unavailable to other soil microorganisms, which will not grow due to the absence of it (Nihorimbere et al., 2011).

Induced systemic resistance

Plant growth-promoting rhizobacteria can prevent the growth of pathogens not only by antagonism but also by helping plants defend themselves by a phenomenon called induced

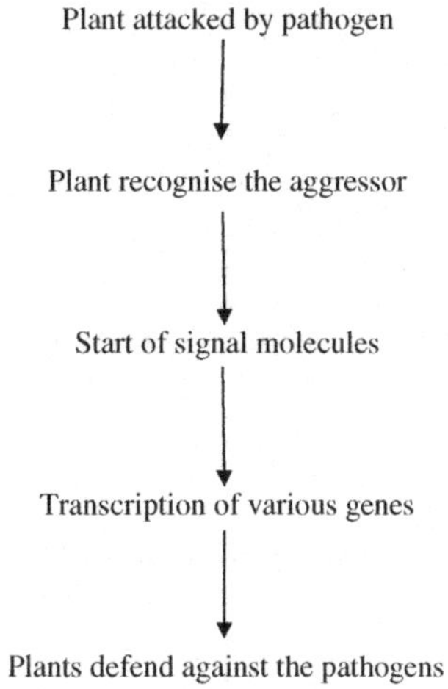

Fig. 15.7
Steps in induced systemic resistance by plants to defend themselves against pathogens.

systemic resistance. Plants show defense reactions against many biotic and abiotic stresses by ISR. Plants will first recognize the aggressor, followed by start of signal molecules and transcription of various genes, which help plants defend themselves against pathogens. The defense molecules include phytoalexins, pathogenesis-related (PR) proteins (chitinases, β-1, 3-glucanases, proteinase inhibitors, etc.), and lignins for reinforcement of cell walls (Nihorimbere et al., 2011). The steps in ISR by plants to defend against pathogens are shown in Fig. 15.7.

Antioxidants in plant–microbe interactions

Antioxidants also play an extremely essential role in plant–microbe interactions. These antioxidants have a role to play in signaling pathways as they produce important biological compounds. These biological molecules or compounds are termed 'secondary metabolites', which are associated with growth and development. Antioxidants also perform important functions at the time of oxidative stress produced by abiotic stimulants or attack by pathogenic microbes. The signaling and synthesis of antioxidants also help understand how plants can withstand stress environmental conditions (Singh et al., 2021). The antioxidants in plant–microbe interactions are shown in Fig. 15.8.

Conclusions

Plant–microbe interactions are extremely important in nature. Various mechanisms and signaling factors help in plant–microbe interactions. These interactions are usually beneficial. Antioxidants and quorum sensing also play an essential role in these plant and microorganism

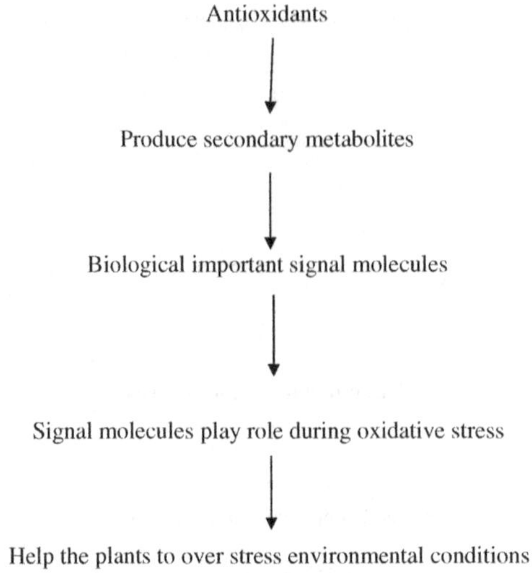

Fig. 15.8
Antioxidants in plant-microbe interactions.

interactions. Further research on genes responsible for plant–microbe interactions and signaling pathways needs to be carried out. Moreover, the molecular aspects of plant–microbe interactions are highly important, which need to be focused on a large scale.

References

Abisado, R.G., Benomar, S., Klaus, J.R., Dandekar, A.A., Chandler, J.R., 2018. Bacterial quorum sensing and microbial community interactions. mBio 9 (3), e02331-e02317.

Afzal, I., Shinwari, Z., Sikandar, S., Shahzad, S., 2019. Plant beneficial endophytic bacteria: mechanisms, diversity, host range and genetic determinants. Microbiol. Res. 221, 36–49.

Backer, R., Rokem, J., Ilangumaran, G., Lamont, J., Praslickova, D., Ricci, E., Subramanian, S., Smith, D., 2018. Plant growth-promoting rhizobacteria: context, mechanisms of action, and roadmap to commercialization of biostimulants for sustainable agriculture. Front. Plant Sci. 9, 1–17.

Basu, A., Prasad, P., Das, S., Kalam, S., Sayyed, R., Reddy, M., Enshasy, H., 2021. Plant growth promoting rhizobacteria (PGPR) as green bioinoculants: recent developments, constraints, and prospects. Sustainability 13, 1140.

Begum, N., Qin, C., Ahanger, M., Raza, S., Khan, M., Ashraf, M., Ahmed, N., Zhang, L., 2019. Role of arbuscular mycorrhizal fungi in plant growth regulation: implications in abiotic stress tolerance. Front. Plant Sci. 10, 1068.

Chandra, D., Sharma, A.K., 2020. Field evaluation of consortium of bacterial inoculants producing ACC deaminase on growth, nutrients and yield components of rice and wheat. J. Crop. Sci. Biotechnol. 24 (3), 293–305.

Chandra, D., Srivastava, R., Glick, B.R., Sharma, A.K., 2018a. Drought-tolerant *Pseudomonas* spp. improve the growth performance of finger millet (*Eleusine coracana* (L.) Gaertn.) under non-stressed and drought-stressed conditions. Pedosphere 28, 227–240.

Chandra, D., Srivastava, R., Sharma, A., 2018b. Influence of IAA and ACC deaminase producing fluorescent pseudomonads in alleviating drought stress in wheat (*Triticum aestivum*). Agric. Res. 7, 290–299.

Chandra, D., Srivastava, R., Gupta, V.V., Franco, C.M., Paasricha, N., Saifi, S.K., Tuteja, N., Sharma, A.K., 2019a. Field performance of bacterial inoculants to alleviate water stress effects in wheat (*Triticum aestivum* L.). Plant Soil 441 (1), 261–281.

Chandra, D., Srivastava, R., Gupta, V.V., Franco, C.M., Sharma, A.K., 2019b. Evaluation of ACC-deaminase-producing rhizobacteria to alleviate water-stress impacts in wheat (*Triticum aestivum* L.) plants. Can. J. Microbiol. 65 (5), 387–403.

Chandra, D., Srivastava, R., Glick, B.R., Sharma, A.K., 2020. Rhizobacteria producing ACC deaminase mitigate water-stress response in finger millet (*Eleusine coracana* (L.) Gaertn.). 3 Biotech 10, 65.

Chaparro, J.M., Badri, D.V., Bakker, M.G., Sugiyama, A., Manter, D.K., Vivanco, J.M., 2013. Root exudation of phytochemicals in *Arabidopsis* follows specific patterns that are developmentally programmed and correlate with soil microbial functions. PLoS One 8, e55731.

Chaparro, J.M., Badri, D.V., Vivanco, J.M., 2014. Rhizosphere microbiome assemblage is affected by plant development. ISME J. 8, 790–803.

Combes-Meynet, E., Pothier, J.F., Moenne-Loccoz, Y., Prigent-Combaret, C., 2011. The *Pseudomonas* secondary metabolite 2, 4-diacetylphloroglucinol is a signal inducing rhizoplane expression of *Azospirillum* genes involved in plant-growth promotion. Mol. Plant Microbe Interact. 24, 271–284.

Dangl, J.L., Horvath, D.M., Staskawicz, B.J., 2013. Pivoting the plant immune system from dissection to deployment. Science 341, 746–751.

Desbrosses, G.J., Stougaard, J., 2011. Root nodulation: a paradigm for how plant-microbe symbiosis influences host developmental pathways. Cell Host Microbe 10 (4), 348–358.

Doornbos, R.F., Geraats, B.P., Kuramae, E.E., Van Loon, L.C., Bakker, P.A., 2011. Effects of jasmonic acid, ethylene, and salicylic acid signalling on the rhizosphere bacterial community of *Arabidopsis thaliana*. Mol. Plant Microbe Interact. 24, 395–407.

Fahad, S., Hussain, S., Bano, A., Saud, S., Hassan, S., Shan, D., 2015. Potential role of phytohormones and plant growth-promoting rhizobacteria in abiotic stresses: consequences for changing environment. Environ. Sci. Pollut. Res. 22, 4907–4921.

Haichar, F., Santaella, C., Heulin, T., Achouak, W., 2014. Root exudates mediated interactions below ground. Soil Biol. Biochem. 77, 69–80.

Hassan, M., McInroy, J.A., Kloepper, J.W., 2019. The interactions of rhizodeposits with plant growth-promoting rhizobacteria in the rhizosphere: a review. Agriculture 9 (7), 142.

Jha, C., Saraf, M., 2015. Plant growth promoting rhizobacteria (PGPR): a review. J. Agric. Res. Dev. 5 (2), 108–119.

Kramer, S., Dibbern, D., Moll, J., Huenninghaus, M., Koller, R., Krueger, D., Marhan, S., Urich, T., Wubet, T., Bonkowski, M., Buscot, F., Lueders, T., Kandeler, E., 2016. Resource partitioning between bacteria, fungi, and protists in the detritusphere of an agricultural soil. Front. Microbiol. 7, 1524.

Kristin, A., Miranda, H., 2013. The root microbiota—a fingerprint in the soil? Plant Soil 370, 671–686.

Kumar, A., Dubey, A., 2020. Rhizosphere microbiome: engineering bacterial competitiveness for enhancing crop production. J. Adv. Res. 24, 337–352.

Malusa, E., Pinzari, F., Canfora, L., 2016. Efficacy of biofertilizers: challenges to improve crop production. In: Singh, D.P., Singh, H.P., Prabha, R. (Eds.), Microbial Inoculants in Sustainable Agricultural Productivity. Springer, Mumbai, India, pp. 17–40.

Mendes, R., Garbeva, P., Raaijmakers, J., 2013. The rhizosphere microbiome: significance of plant beneficial, plant pathogenic, and human pathogenic microorganisms. FEMS Microbiol. Rev. 37 (5), 634–663.

Mohanram, S., Kumar, P., 2019. Rhizosphere microbiome: revisiting the synergy of plant-microbe interactions. Ann. Microbiol. 69, 307–320.

Mommer, L., Hinsinger, P., Prigent-Combaret, C., Visser, E.W., 2016. Advances in the rhizosphere: stretching the interface of life. Plant Soil 407, 1–8.

Musilova, L., Ridl, J., Polivkova, M., Macek, T., Uhlik, O., 2016. Effects of secondary plant metabolites on microbial populations: changes in community structure and metabolic activity in contaminated environments. Int. J. Mol. Sci. 17, 1205.

Nihorimbere, V., Ongena, M., Smargiassi, M., Thonart, P., 2011. Beneficial effect of the rhizosphere microbial community for plant growth and health. Biotechnol. Agron. Soc. Environ. 15 (2), 327–337.

Oburger, E., Dell'Mour, M., Hann, S., Wieshammer, G., Puschenreiter, M., Wenzel, W.W., 2013. Evaluation of a novel tool for sampling root exudates from soil-grown plants compared to conventional techniques. Environ. Exp. Bot. 87, 235–247.

Pii, Y., Mimmo, T., Tomasi, N., Terzano, R., Cesco, S., Crecchio, C., 2015. Microbial interactions in the rhizosphere: beneficial influences of plant growth-promoting rhizobacteria on nutrient acquisition process. A review. Biol. Fertil. Soils 51, 403–415.

Sasse, J., Martinoia, E., Northen, T., 2018. Feed your friends: do plant exudates shape the root microbiome? Trends Plant Sci. 23, 25–41.

Singh, B., Vaishnav, A., Sayyed, R.Z., 2021. Antioxidants in Plant-Microbe Interaction. Publisher Springer, Singapore.

Somers, E., Vanderleyden, J., Srinivasan, M., 2004. Rhizosphere bacterial signalling: a love parade beneath our feet. Crit. Rev. Microbiol. 30, 205–235.

Sun, Z., Song, J., Xin, X., Xie, X., Zhao, B., 2018. Arbuscular mycorrhizal fungal proteins 14-3-3- are involved in Arbuscule formation and responses to abiotic stresses during AM symbiosis. Front. Microbiol. 5, 9–19.

Thapa, S., Prasanna, R., 2018. Prospecting the characteristics and significance of the phyllosphere microbiome. Ann. Microbiol. 68, 229–245.

Vishwakarma, K., Kumar, N., Shandilya, C., Mohapatra, S., Bhayana, S., Varma, A., 2020. Revisiting plant-microbe interactions and microbial consortia application for enhancing sustainable agriculture: a review. Front. Microbiol. 11, 3195.

Waghunde, R.R., Shelake, R.M., Sabalpara, A.N., 2016. *Trichoderma*: a significant fungus for agriculture and environment. Afr. J. Agric. Res. 11 (22), 1952–1965.

White, J., Kingsley, K., Zhang, Q., Verma, R., Obi, N., Dvinskikh, S., Elmore, M., Verma, S., Gond, S., Kowalski, K., 2019. Review: endophytic microbes and their potential applications in crop management. Pest Manag. Sci. 75 (10), 2558–2565.

Woo, S.L., Ruocco, M., Vinale, F., Nigro, M., Marra, R., Lombardi, N., Lorito, M., 2014. *Trichoderma*-based products and their widespread use in agriculture. Open Mycol. J. 8 (1), 71–126.

Yang, C.H., Crowley, D.E., 2000. Rhizosphere microbial community structure in relation to root location and plant iron nutritional status. Appl. Environ. Microbiol. 66 (1), 345–351.

Zin, N., Badaluddin, N., 2020. Biological functions of *Trichoderma* spp. for agriculture applications. Ann. Agric. Sci. 65 (2), 168–178.

Index

Note: Page numbers followed by *f* indicate figures and *t* indicate tables.

Q

R

S

Made in the USA
Las Vegas, NV
16 October 2025

32380170R00175